New Frontiers in Statistical Distributions

New Frontiers in Statistical Distributions

Edited by
Calanthia Wright

MURPHY & MOORE
www.murphy-moorepublishing.com

Published by Murphy & Moore Publishing,
1 Rockefeller Plaza,
New York City, NY 10020, USA

ISBN: 978-1-63987-388-3

Cataloging-in-Publication Data

New frontiers in statistical distributions / edited by Calanthia Wright.
p. cm.
Includes bibliographical references and index.
ISBN 978-1-63987-388-3
1. Distribution (Probability theory). 2. Statistics. 3. Mathematical statistics. I. Wright, Calanthia.
QA273.6 .H36 2022
519.24--dc23

For information on all Murphy & Moore Publications
visit our website at www.murphy-moorepublishing.com

MURPHY & MOORE

Contents

Preface

This book aims to highlight the current researches and provides a platform to further the scope of innovations in this area. This book is a product of the combined efforts of many researchers and scientists, after going through thorough studies and analysis from different parts of the world. The objective of this book is to provide the readers with the latest information of the field.

A parameterized mathematical function that provides the probabilities of different outcomes for a random variable is referred to as a statistical distribution. These distributions are applied to continuous, discrete, and bivariate random variables. Some of the most prominent statistical distributions are exponential, Erlang, gamma, beta, Weibull, normal, lognormal, left-truncated normal, right-truncated normal, triangular, discrete uniform, binomial, geometric and Pascal. The knowledge of statistical distributions is fundamental for researchers in almost all disciplines. Understanding when and how to apply each of the distributions in research studies, with a goal to identify the distribution that best applies to the study is also crucial. This book traces the progress of this field and highlights some of its key concepts and applications. Some of the diverse topics covered herein address the varied branches that fall under statistical distributions. This book will serve as a reference to a broad spectrum of readers.

I would like to express my sincere thanks to the authors for their dedicated efforts in the completion of this book. I acknowledge the efforts of the publisher for providing constant support. Lastly, I would like to thank my family for their support in all academic endeavors.

Editor

Preface

1

Simulation of polyhedral convex contoured distributions

Wolf-Dieter Richter* and Kay Schicker

*Correspondence:
wolf-dieter.richter@uni-rostock.de
Institute of Mathematics, University of Rostock, Ulmenstr. 69, Haus 3, Rostock, 18057, Germany

Abstract

In low dimensions, the relatively easily implementable acceptance-rejection method for generating polyhedral convex contoured uniform distributions is compared to more sophisticated particular methods from the literature, and applied to drug combination studies. Based upon a stochastic representation, the method is extended to the general class of polyhedral convex contoured distributions of known dimension. Based upon a geometric measure representation, an algorithm for simulating corresponding probabilities of rather arbitrary random events is derived.

Keywords: Acceptance-rejection method, Polyhedral convex contoured uniform distribution, Minkowski functional, Geometric and stochastic representation, The class of polyhedral convex contoured distributions, Simulation of distributions and probabilities

Mathematics Subject Classification (2000): MSC 65C05, MSC 62E15, MSC 68U20

1 Introduction

The need of uniform distributions on convex polyhedra appears for various purposes of statistical modeling. In the field of computational visualization of unstructured data, uniform designs on convex polyhedra are used as space filling designs to create three dimensional pictures on the computer, see e.g. (Cutler et al. 2004; Rocchini and Cignoni 2000). The simulation of uniform distributions on convex polyhedra applies in the field of drug combination studies as described in (Fang and Yang 2000; Tan et al. 2003; Tian et al. 2009) which is of special interest in this paper. In (Richter and Schicker 2014) the simulation of a uniform distribution on a tetrahedron is used to detect optimal production schedules.

(Tian et al. 2009) consider a combination study of two drugs with linear dose-response curves. These curves represent the low dose for agents like ionizing radiation, enzyme inhibitions or mutagens. If x_i denotes the dose of drug A_i, $i = 1, 2$ then the single dose-response curves of A_1 and A_2 are assumed to be represented by

$$f_1(x_1) = \alpha_1 + \beta_1 x_1 \text{ and } f_2(x_2) = \alpha_2 + \beta_2 x_2 = \alpha_1 + \beta_2 \left(x_2 - \frac{\alpha_1 - \alpha_2}{\beta_2} \right),$$

where $\alpha_i \in \mathbb{R}$ and $\beta_i \in \mathbb{R}$, $i = 1, 2$ denote regression coefficients. If $x_2^* = x_2 - (\alpha_1 - \alpha_2)/\beta_2$ then the domain for dose ranges of interest is assumed to be

$$\mathbb{P}(a_0, b_0) = \{(x_1, x_2^*)^T \in \mathbb{R} : a_0 \leq \alpha_1 + \beta_1 x_1 + \beta_2 x_2^* \leq b_0, x_1 > 0, x_2^* > 0\} \tag{1}$$

which is an irregular tetragon in $\mathbb{R}^2$. The constants a_0 and b_0 are pre-specified constants, chosen by pharmacological experts. For further details, we refer to (Tian et al. 2009). Since in (Tan et al. 2003) it is proved that a uniform design on $\mathbb{P}(a_0, b_0)$ maximizes the minimum power of the F-test to detect significant deviations from additive actions of the drugs, it is of importance to be able to effectively simulate the uniform distribution on general tetragons $\mathbb{P}(a_0, b_0)$.

In (Tian et al. 2009) several algorithms for the simulation of uniformly distributed random points on convex polyhedra are presented for particular two, three, and n-dimensional cases. The presented algorithms require a pool of analytical as well as probabilistic methods, in dimensions $n > 2$ also numerical methods especially for integration and root computation. They give challenging and individual solutions for the simulation of uniformly distributed random points on special subclasses of convex polyhedra.

In this paper, we present a rather simple but general simulation algorithm using basic geometric-analytical properties of the class of convex polyhedra as well as basic probabilistic simulation methods. This algorithm reduces the mathematical complexity and, since numerical methods for integration and root computation are not necessary, the computational complexity, too. Furthermore, the structure of this algorithm will be the same for every convex polyhedron in $\mathbb{R}^n$ which makes the algorithm flexible to use. Its basic idea is to effectively construct the Minkowski functional of a convex polyhedron and then to apply a fast and simple acceptance-rejection method.

It turns out that this idea simply extends to simulate both distributions from much more general classes of probability distributions and probabilities of rather arbitrary random events measured with respect to any of these distributions. The construction and simulation of multivariate probability distributions and their event probabilities is one of the current challenges in probability theory and statistics. It is possible to characterize a multivariate density by the geometry of its density level sets. For an overview about this broad field of research, we refer to (Arnold et al. 2008; Balkema et al. 2010; Fang et al. 1990; Fernandez and Osiewalski 1995; Gupta and Song 1997; Kamiya et al. 2008; Richter 2009; 2013; 2014; 2015a; 2015b; Richter and Schicker 2017; Sarabia and Gomez-Deniz 2008). Convex polyhedral distributions are characterized by contours being the topological boundaries of convex polyhedra and can be considered being a subclass of polyhedral star-shaped distributions that are studied in (Richter and Schicker 2017). To simulate them needs just the additional independent simulation of a positive random variable playing the role of a certain generalized radius variable of a generalized ball. This way, we widen the flexibility of our approach by describing how arbitrary continuous distributions, having convex polyhedral density level sets, can be constructed and simulated with the new method.

The paper is structured as follows. We describe the main ways of describing convex polyhedra and their Minkowski functionals in Section 2.1 and 2.2, respectively. In Section 3 we present the acceptance-rejection algorithm for generating the uniform distribution on an arbitrary convex polyhedron in $\mathbb{R}^n$. Comparisons of the general acceptance-rejection algorithm with particular algorithms from the literature and applications to drug combination studies are given in Section 4. A general technique from

(Richter 2014, 2015a; Richter and Schicker 2017) is summarized in Section 5.1 to construct the general class of polyhedral convex contoured distributions from the particular polyhedral convex contoured uniform distribution. Sections 5.2 and 5.3 deal with applications of the presented geometric and stochastic representations to simulating probabilities of random events and to linear transformations, respectively. A final discussion including some remarks on the closely related recent paper (Nolan 2016) and several examples are presented in Section 6 and the Appendix, respectively.

2 Representations of convex polyhedra

2.1 $\mathcal{V}$- and $\mathcal{H}$-representations of convex polyhedra

Representations of convex polyhedra are considered in the broad literature of convex geometry, see e.g. (Böhm and Hertel 1980; Ziegler 1995). Here, we recall two standard representations of convex polyhedra that are very fruitful for the purposes in this paper. It is said that a convex polyhedron $\mathbb{P}$, $\mathbb{P} \subset \mathbb{R}^n$, is defined by its vertices $p_1, \ldots, p_l \in \mathbb{R}^n$ if

$$\mathbb{P} = conv(\{p_1, \ldots, p_l\}), \tag{2}$$

where $conv(M)$ denotes the convex hull of the point set M. Alternatively, it is possible to represent $\mathbb{P}$ as the intersection of suitably chosen closed half-spaces. In this case there exist a matrix $A \in \mathbb{R}^{k \times n}$ and a vector $b \in \mathbb{R}^k$ such that

$$\mathbb{P} = \{x \in \mathbb{R}^n : Ax \leq b\}, \tag{3}$$

where "$\leq$" is interpreted componentwise. Note that (2) and (3) are equivalent and there exists software to convert a given representation (2) into a representation (3), and vice versa. In this regard, we refer to (Avis 2000; Fukuda and Prodon 1996). For an illustration of the construction of (2) and (3), see Example 3 in the Appendix.

2.2 Minkowski functionals of convex polyhedra

In case one wants to decide according to an acceptance-rejection algorithm whether a point x from a sample space $\mathbb{R}^n$ belongs to the Euclidean ball of radius r, or not, one can equivalently observe whether the Euclidean norm of x is less or equal to r, $||x|| \leq r$. If B denotes the Euclidean unit ball in $\mathbb{R}^n$, and h_B the Minkowski functional of B defined by $h_B(z) = \inf\{r > 0 : z \in rB\}$, $z \in \mathbb{R}^n$, then $||x|| \leq r$ iff $h_B(x) \leq r$. Imagine B is replaced by a convex polyhedron $\mathbb{P} \in \mathbb{R}^n$ containing the origin as an inner point, $0_n \in int(\mathbb{P})$, where we define $int(\mathbb{P})$ to be the interior of $\mathbb{P}$. It is then possible to specify vector b in representation (3) to be a vector with exclusively positive real entries, i.e. $b \in \mathbb{R}^k_+$. From this, we can conclude that

$$\mathbb{P} = \{x \in \mathbb{R}^n : A'x \leq \mathbb{1}_n\},$$

where $A = (a_{ij})_{i \in \{1,\ldots,k\}, j \in \{1,\ldots,n\}}$ is chosen according to (3), $A' = (A'_1, \ldots, A'_k)^T = (a_{ij}/b_i)_{i \in \{1,\ldots,k\}, j \in \{1,\ldots,n\}}$ and $\mathbb{1}_n$ denotes the n-dimensional vector of ones. Thus

$$\begin{aligned}\mathbb{P} &= \{x \in \mathbb{R}^n : A'_1 x \leq 1, A'_2 x \leq 1, \ldots, A'_k x \leq 1\} \\ &= \{x \in \mathbb{R}^n : \max\{A'_1 x, A'_2 x, \ldots, A'_k x\} \leq 1\}.\end{aligned}$$

The last equations show that it is possible to represent a convex polyhedron $\mathbb{P} \subset \mathbb{R}^n$ with the help of the functional $\max\{A_1'x, A_2'x, \dots, A_k'x\}$ and since

$$\begin{aligned}\inf\{r > 0 : x/r \in \mathbb{P}\} &= \inf\{r > 0 : \max\{A_1'x/r, \dots, A_k'x/r\} \leq 1\} \\ &= \inf\{r > 0 : \max\{A_1'x, \dots, A_k'x\} \leq r\} = \max\{A_1'x, \dots, A_k'x\}, x \in \mathbb{R}^n\end{aligned}$$

we observe that the function

$$h_{\mathbb{P}}(x) = \max\{A_1'x, \dots, A_k'x\}, x \in \mathbb{R}^n \tag{4}$$

is the Minkowski functional of $\mathbb{P}$. One can represent thus the polyhedron as $\mathbb{P} = \{x \in \mathbb{R}^n : h_{\mathbb{P}}(x) \leq 1\}$. A particular case is dealt with in Example 4 in the Appendix.

3 A general acceptance-rejection simulation algorithm (ARSA)

In this section we introduce an algorithm to simulate uniformly distributed random points on an arbitrarily given multivariate convex polyhedron. To this end, we combine the representations presented in the latter section with the method of acceptance-rejection sampling that was introduced first in (von Neumann 1951). To start with, we assume $\mathbb{P} \subset \mathbb{R}^n$ to be a convex polyhedron, given by representation (3), and choose a cuboid Q containing $\mathbb{P}$, $Q \supseteq \mathbb{P}$. We simulate uniformly distributed random points $x \in Q$ on Q and accept them as uniformly distributed on $\mathbb{P}$, if $x \in \mathbb{P}$, i.e. $Ax \leq b$ according to (3). In Algorithm 1 we summarize the algorithm generating polyhedral convex contoured uniform distributions.

Algorithm 1: Sampling algorithm for uniform distributions on convex polyhedra.

Assumption *$\mathbb{P} \subset \mathbb{R}^n$ is a convex polyhedron, represented by $\mathbb{P} = \{x \in \mathbb{R}^n : Ax \leq b\}$;*
$Q \supseteq \mathbb{P}$, where $Q = [q_{min,1}, q_{max,1}] \times \dots \times [q_{min,n}, q_{max,n}]$

Input: $A, b, q_{min,1}, \dots, q_{min,n}, q_{max,1}, \dots, q_{max,n}$

Output: $X = (X_1, \dots, X_n)^T$ uniformly distributed on $\mathbb{P}$

Algorithm

1. Sample uniformly and independent X_1 from $[q_{min,1}, q_{max,1}], \dots, X_n$ from $[q_{min,n}, q_{max,n}]$ until $A(X_1, \dots, X_n)^T \leq b$.

Return X

The proofs of that this method stops within a finite time and that the resulting sample points are uniformly distributed on $\mathbb{P}$ can be done analogously to those in Appendix 1 of (Kalke and Richter 2013). A suitable cuboid Q can be found, applying representation (2) of $\mathbb{P}$. One can always find $q_{min,i}$ and $q_{max,i}$, $i = 1, \dots, n$ by

$$q_{min,i} = \min\{p_j[i], j = 1, \dots, l\} \text{ and } q_{max,i} = \max\{p_j[i], j = 1, \dots, l\}, i = 1, \dots, n,$$

where $p_j[i]$ denotes the ith component of vertex p_j. Applying (3) one can numerically solve for every $i = 1, \dots, n$ the optimization problems: minimize $q_{min,i} = x_i$ (maximize $q_{max,i} = x_i$), subject to $A(x_1, \dots, x_n)^T \leq b$.

4 Simulations in particular classes of convex polyhedra

4.1 Uniform distributions in tetragons in $\mathbb{R}^2$

4.1.1 *Method of Tian, Fang, Tan, Qin and Tang*

In (Tian et al. 2009) a particular algorithm is presented to simulate uniformly distributed random points on an arbitrary two dimensional tetragon that is declared by its vertices a_1, a_2, a_3 and a_4, $\mathbb{P}(a_1, a_2, a_3, a_4) \subset \mathbb{R}^2$. It is shown there that a uniformly on $\mathbb{P}(a_1, a_2, a_3, a_4)$ distributed random vector X allows the stochastic representation

$$X \stackrel{d}{=} (a_1, a_2, a_3, a_4) \begin{pmatrix} (1-Y_1)(1-Y_2) \\ Y-1(1-Y_2) \\ Y_1 Y_2 \\ (1-Y_1)Y_2 \end{pmatrix},$$

where the random vector $Y = (Y_1, Y_2)^T$ follows the density function $f_Y(y_1, y_2) = \frac{d_0 + d_1 y_1 + d_2 y_2}{d_0 + 0.5(d_1 + d_2)}$ with $d_0 = |a_2 - a_1, a_4 - a_1|$, $d_1 = |a_2 - a_1, a_3 - a_4|$ and $d_2 = |a_3 - a_2, a_4 - a_1|$ where $|.,.|$ denotes a determinant. The simulation of Y is realized, applying the inverse transform sampling of the marginal distributions of Y that can be found in (Tian et al. 2009).

4.1.2 *The alternative method:* **ARSA**

Our general acceptance-rejection simulation Algorithm 1 presented in Section 3 offers an alternative method to that of the latter section and applies immediately. One has only to calculate the matrix A and the vector b from the given vertices $a_1, \ldots, a_4$ such that (3) holds. To this, it is possible to apply the algorithms described in (Fukuda and Prodon 1996; Avis 2000). In the following, however, we present an alternative method that uses only elementary theory from linear algebra. To this end, let the points $a_i = (x_i, y_i)^T$, $i = 1, \ldots, 4$, be ordered anticlockwise and let $(x, g_{a_i, a_j}(x))^T$ denote a point from the line through the points a_i and a_j, $j = i + 1$ if $i = 1, 2, 3$ and $j = 1$ if $i = 4$. Then,

$$\frac{g_{a_i a_j}(x) - y_i}{y_j - y_i} = \frac{x - x_i}{x_j - x_i}$$

is equivalent to

$$(y_j - y_i)x - (x_j - x_i)g_{a_i a_j}(x) = x_i y_j - x_j y_i,$$

where $(y_j - y_i, -(x_j - x_i))^T$ denotes the outer normal vector of the line through the points a_i and a_j. With $m_{i,j} = x_i y_j - x_j y_i$, $i, j \in \{1, \ldots, 4\}$, $i \neq j$,

$$A = \begin{pmatrix} y_2 - y_1 & -(x_2 - x_1) \\ y_3 - y_2 & -(x_3 - x_2) \\ y_4 - y_3 & -(x_4 - x_3) \\ y_1 - y_4 & -(x_1 - x_4) \end{pmatrix} \text{ and } b = \begin{pmatrix} m_{1,2} \\ m_{2,3} \\ m_{3,4} \\ m_{4,1} \end{pmatrix},$$

one gets the representation $\mathbb{P}(a_1, a_2, a_3, a_4) = \{(x, y)^T \in \mathbb{R}^2 : A(x, y)^T \leq b\}$. A particular numerical situation and a more general application to the drug combination study are considered in Examples 5 and 6, respectively, in the Appendix.

4.2 Uniform distributions in triangular prisms in $\mathbb{R}^3$

4.2.1 *Method of Tian, Fang, Tan, Qin and Tang*

Apart from the simulation of uniform distributions on tetragons in (Tian et al. 2009), the presented method is transferred there to the class of triangular prisms in $\mathbb{R}^3$. A triangular prism $\mathbb{P}(a_1, a_2, a_3, a_4, a_5, a_6)$ is defined by its six vertices a_1, a_2, a_3, a_4, a_5 and $a_6 \in \mathbb{R}^3$, where $conv(\{a_1, a_2, a_3\})$ and $conv(\{a_4, a_5, a_6\})$ each forms a triangular base of $\mathbb{P}(a_1, \ldots, a_6)$ and $\overline{a_1a_4}$, $\overline{a_2a_5}$ and $\overline{a_3a_6}$ each forms an edge of $\mathbb{P}(a_1, \ldots, a_6)$. It is shown in (Tian et al. 2009) that a uniformly on $\mathbb{P}(a_1, \ldots, a_6)$ distributed random vector X allows the stochastic representation

$$X \stackrel{d}{=} (a_1, a_2, a_3, a_4, a_5, a_6) \begin{pmatrix} (1 - Y_1 - Y_2)(1 - Y_3) \\ Y_1(1 - Y_3) \\ Y_2(1 - Y_3) \\ (1 - Y_1 - Y_2)Y_3 \\ Y_1 Y_2 \\ Y_2 Y_3 \end{pmatrix},$$

where $Y = (Y_1, Y_2, Y_3)^T$ follows the density $f_Y(y_1, y_2, y_3)$ given by (3.2)-(3.6) in (Tian et al. 2009). By deriving marginal and conditional densities of f_Y a conditional sampling for Y is presented there using numerical methods.

4.2.2 ARSA *for triangular prisms in* $\mathbb{R}^3$

As in the previous section, it is again possible to apply our general Algorithm 1 by calculating the matrix A, the vector b and a cuboid Q for triangular prisms. This can be done computationally as described in Section 2.1 or by adopting the **ARSA** similarly as we did already for the simulation of uniform distributions on tetragons in Section 4.1.2.

Given the vertices $a_j = (a_{j1}, a_{j2}, a_{j3})$, $j = 1, \ldots, 6$ of the triangular prisma $\mathbb{S}(a_1, \ldots, a_6)$, one can represent the planes of which $conv(\{a_1, a_2, a_3\})$, $conv(\{a_1, a_2, a_4, a_5\})$, $conv(\{a_2, a_3, a_5, a_6\})$, $conv(\{a_1, a_3, a_4, a_6\})$ and $conv(\{a_4, a_5, a_6\})$ are subsets as $\{(x, y, z)^T \in \mathbb{R}^3 : n_i \cdot (x, y, z)^T = \lambda_i\}$, $i = 1, \ldots, 5$, respectively, where

$$n_1 = (a_2 - a_1) \times (a_3 - a_1),\ \lambda_1 = n_1 a_1,\ n_2 = (a_2 - a_1) \times (a_4 - a_1),\ \lambda_2 = n_2 a_1,$$
$$n_3 = (a_5 - a_2) \times (a_3 - a_2),\ \lambda_3 = n_3 a_2,\ n_4 = (a_1 - a_3) \times (a_6 - a_3),\ \lambda_4 = n_4 a_3,$$
$$n_5 = (a_5 - a_4) \times (a_6 - a_4),\ \lambda_5 = n_5 a_4,$$

the symbol "$\times$" denotes the vector cross product, n_i is a normal vector of the considered plane and λ_i is the Euclidean distance of the plane from the origin, $i = 1, \ldots, 5$. With $A' = (n_1, n_2, n_3, n_4, n_5)^T$ and $b' = (\lambda_1, \lambda_2, \lambda_3, \lambda_4, \lambda_5)^T$ one gets the representation $\mathbb{P}(a_1, \ldots, a_6) = \{(x, y, z)^T \in \mathbb{R}^3 : A(x, y, z)^T \leq b\}$ by considering $A'a_i$ for every $i = 1, \ldots, 6$ and comparing $A'a_i$ with b' componentwise. If a component of $A'a_i$ is greater than the component of b', one multiplies the considered row of A' and the component of b' by -1 and continues with the next edge and the transformed A' and b'. This method results in matrix A and vector b and thus the desired representation of $\mathbb{P}(a_1, \ldots, a_6)$, and is summarized by Algorithm 2. A suitable cuboid Q can be chosen as

$$\begin{aligned} Q = & [\min\{a_{11}, \ldots, a_{61}\}, \max\{a_{11}, \ldots, a_{61}\}] \times \\ & [\min\{a_{12}, \ldots, a_{62}\}, \max\{a_{12}, \ldots, a_{62}\}] \times [\min\{a_{13}, \ldots, a_{63}\}, \max\{a_{13}, \ldots, a_{63}\}]. \end{aligned}$$

For a particular numerical illustration and a more general application to the drug combination study, see Examples 7 and 8, respectively, in the Appendix.

Algorithm 2: Calculate representation (3) of $\mathbb{P}(a_1,\dots,a_6)$ from $\mathbb{S}(a_1,\dots,a_6)$.

Input: $A', b', a_1, a_2, a_3, a_4, a_5, a_6$
Output: A, b

Algorithm
  $A_0 := A'; b_0 := b'$;
  for $i := 1$ to 6 step 1 **do**
    $z_i := A_{i-1}a_i$;
    for $j := 1$ to 5 step 1 **do**
      if $z_i[j] > b_{i-1}[j]$ **then**
        $A_{i-1}[j,] := (-1) \cdot A_{i-1}[j,]$;
        $b_{i-1}[j] := (-1) \cdot b_{i-1}[j]$;
      end if
    end for
    $A_i := A_{i-1}; b_i := b_{i-1}$;
  end for
Return $A := A_6, b := b_6$

5 Polyhedral convex contoured distributions

In this section we are going beyond uniformity and show a flexible method to construct arbitrary continuous distributions having convex polyhedra as density level sets, which we call polyhedral convex contoured distributions. Our starting point is the consideration of the uniform distribution on a convex polyhedron in the latter sections. In Section 5.1 we summarize certain stochastic and geometric representations and linear transformation methods from (Richter 2014; 2015a) and (Richter and Schicker 2017) for the particular classes of convex polyhedra and polyhedral convex contoured distributions, respectively. Specific applications of representations (6) and (4) below are presented in Sections 5.2 and 5.3, respectively.

5.1 Geometric and stochastic representations

Since the class of convex polyhedra is both a subclass of the class of star bodies and a subclass of convex bodies, we can apply the stochastic and geometric representations from these cases, to construct polyhedral convex contoured distributed random vectors. To this end, let $\mathbb{P} \subset \mathbb{R}^n$ be a convex polyhedron having the origin 0_n in its interior, i.e. $0_n \in int(\mathbb{P})$. Let $G = (G_1,\dots,G_n)^T$ be a random vector uniformly distributed on $\mathbb{P}$, then the random vector $U = G/h_{\mathbb{P}}(G)$ is generalized uniformly distributed on the topological boundary $\mathbb{S} = \{x \in \mathbb{R}^n : h_{\mathbb{P}}(x) = 1\}$ of $\mathbb{P}$, $U \sim \omega_{\mathbb{S}}$. The probability measure $\omega_{\mathbb{S}}$ is defined for every Borel measurable set $A \in (\mathfrak{B}^n \cap \mathbb{S})$ by $\omega_{\mathbb{S}}(A) = \mathfrak{O}_{\mathbb{S}}(A)/\mathfrak{O}_{\mathbb{S}}(\mathbb{S})$, where $\mathfrak{O}_{\mathbb{S}}$ is called the star-generalized surface measure according to (Richter 2014). It is defined considering the central projection cone $CPC(A) = \{x \in \mathbb{R}^n : x/h_{\mathbb{P}}(x) \in A\}$ and the star sector of radius $\rho > 0$, $sector(A,\rho) = CPC(A) \cap [\rho\mathbb{P}]$ for every $A \in (\mathfrak{B}^n \cap \mathbb{S})$, by

$$\mathfrak{O}_{\mathbb{S}}(A) = f'(1), \text{where } f(\rho) = \mu(sector(A,\rho)), A \in (\mathfrak{B}^n \cap \mathbb{S}).$$

While this definition deals with taking derivatives of volumes of suitably chosen sectors of convex polyhedra, it is also possible to introduce an equivalent way of defining $\mathfrak{O}_{\mathbb{S}}$ using integration and replacing the Euclidean norm of the vector normal to the sphere in the defining integral of the Euclidean surface content by a suitably chosen non-Euclidean norm. If $h_{\mathbb{P}}$ denotes a norm and $N(\vartheta) = (\nabla\eta(\vartheta), -1)^T$ is the outer normal vector to the norm sphere $\mathbb{S}$ at the point $(\vartheta^T, \eta(\vartheta))^T$, then, with notations as in (4), $\mathfrak{O}_{\mathbb{S}}$ satisfies the dual norm representation

$$\mathfrak{O}_{\mathbb{S}}(A) = \int\limits_{G(A)} \max\{p_1^T N(\vartheta), \dots, p_l^T N(\vartheta)\} d\vartheta,\ A \in (\mathfrak{B}^n \cap \mathbb{S})^+, \tag{5}$$

where $G(A) = \{\vartheta \in \mathbb{R}^{n-1} : \exists\eta = \eta(\vartheta) \text{with } (\vartheta^T, \eta)^T \in A\}$ and $(\mathfrak{B}^n \cap \mathbb{S})^+$ denotes the Borel σ-field on the upper half-sphere of $\mathbb{S}$. For the proof of (5) and further integral representations of the star-generalized surface measure of star-shaped polyhedra, we refer to (Richter and Schicker 2017).

Let us recall that since $0_n \in int(\mathbb{P})$ it is possible to calculate the Minkowski functional $h_{\mathbb{P}}$ of $\mathbb{P}$ as described in Section 2.2.

Let $g : \mathbb{R}_+ \to \mathbb{R}_+$ satisfy the assumptions $0 < \mathcal{I}(g) < \infty$ where $\mathcal{I}(g) = \int\limits_0^\infty r^{n-1}g(r)dr$. We call g a density generating function (*dgf*) and a random vector X taking values in $\mathbb{R}^n$ polyhedral convex contoured distributed, if X allows the stochastic representation

$$X \stackrel{d}{=} R_g \cdot U, \tag{6}$$

where R_g and U are stochastically independent, $U \sim \omega_{\mathbb{S}}$ and R_g follows the density $f(r) = \frac{1}{\mathcal{I}(g)} r^{n-1} g(r),\ r > 0$. Note that the density of X can be represented as

$$\varphi_{g,\mathbb{P}}(x) = C(g, \mathbb{P}) g(h_{\mathbb{P}}(x)), x \in \mathbb{R}^n, \tag{7}$$

where $C(g, \mathbb{P}) = 1/(n\mu(\mathbb{P})\mathcal{I}(g))$ is a normalizing constant and μ denotes the Lebesgue measure in $\mathbb{R}^n$. The probability measure having the density $\varphi_{g,\mathbb{P}}$ will be denoted by $\Phi_{g,\mathbb{P}}$ and $\mathbb{P}$ will be called the contour defining polyhedral convex body or the contour defining convex polyhedron. We recall the geometric measure representation formula of $\Phi_{g,K}(B)$ for every $B \in \mathfrak{B}^n$,

$$\Phi_{g,\mathbb{P}}(B) = C(g, \mathbb{P}) \int\limits_0^\infty r^{n-1} g(r) \mathfrak{O}_{\mathbb{S}} \left(\left[\frac{1}{r} B \right] \cap \mathbb{S} \right) dr = \frac{1}{\mathcal{I}(g)} \int\limits_0^\infty r^{n-1} g(r) \mathfrak{F}_{\mathbb{S}}(B, r) dr \tag{8}$$

where $r \to \mathfrak{F}_{\mathbb{S}}(B, r) = \omega_{\mathbb{S}} \left(\left[\frac{1}{r} B \right] \cap \mathbb{S} \right)$ denotes the star intersection-proportion function (ipf) of the set B. Note that if $\mathfrak{O}_{\mathbb{S}}$ satisfies the dual norm representation (5) it holds

$$\Phi_{g,\mathbb{P}}(B) = C(g, \mathbb{P}) \int\limits_0^\infty \int\limits_{G([\frac{1}{r}B] \cap \mathbb{S})} r^{n-1} g(r) \max \left\{ p_1^T N(\vartheta), \dots, p_l^T N(\vartheta) \right\} d\vartheta\, dr.$$

For further specific geometric representation formulae of polyhedral star-shaped distributions, having star-shaped polyhedra as contour defining polyhedron, see again (Richter and Schicker 2017).

For the choice of a *dgf* g there are various possibilities. For some basic types of density generating functions, we refer to (Fang et al. 1990; Richter and Schicker 2017). Note that if we choose $g(r) = I_{(0,1]}(r)$, where $I_{(0,1]}$ is the indicator function on the interval $(0, 1]$, we get the uniform distribution on $\mathbb{P}$.

Finally, we want to show that in certain situations it is possible to find exact representations of $\mathfrak{F}_{\mathbb{S}}(B, r)$ in Eq. (8). If we consider $\Phi_{g,\mathbb{P}}(B(\rho))$, where $B(\rho) = \rho\mathbb{P},\ \rho > 0$, then $\mathfrak{F}_{\mathbb{S}}(B(\rho), r) = I_{[0,\rho)}(r),\ r > 0$, and

$$\Phi_{g,\mathbb{P}}(B(\rho)) = \frac{1}{\mathcal{I}(g)} \int_0^\rho r^{n-1} g(r) dr.$$

Let us consider R_g^2, then

$$\frac{d}{dy} P(R_g^2 < y) = \frac{d}{dy} \Phi_{g,\mathbb{P}}(B(\sqrt{y})) = \frac{y^{n/2-1} g(\sqrt{y})}{2\mathcal{I}(g)}, y > 0.$$

In slightly modified notation, this function is known from (Richter 1991) as the density of the g-generalized χ^2-distribution with n degrees of freedom.

5.2 Simulating probabilities $\Phi_{g,\mathbb{P}}(B)$

In this section we use the geometric measure representation formula (8) of multivariate polyhedral convex contoured probability distributions for simulating single values of this distribution by simulating the corresponding *dgf* g and the generalized uniform distribution $\omega_{\mathbb{S}}$. Given a *dgf* g, a contour defining convex polyhedron $\mathbb{P} \subset \mathbb{R}^n$ and a set $B \subseteq \mathfrak{B}^n$ we can approximate $\Phi_{g,\mathbb{P}}(B)$ with the following method. Choose an integer $k \in \mathbb{N}$, simulate k uniformly on $\mathbb{P}$ distributed points $X_{(j)}, j = 1, \dots, k$ using Algorithm 1. For every $j = 1, \dots, k$ generate generalized uniformly on $\mathbb{S}$ distributed random points $Y_{(j)}$ by $Y_{(j)} = X_{(j)}/h_{\mathbb{P}}(X_{(j)})$. Now, choose $u \in \mathbb{R}^+$ as upper boundary of the integration interval of $r \in (0, u]$ and an increment $1/v$, $v \in \mathbb{N}^+$. For every t, $t = 0, 1, \dots, uv$ consider $t/v \cdot Y_{(j)}$, $j = 1, \dots, k$, count how often $(t/v \cdot Y_{(j)}) \in B$ and store the number in $p(t/v)$,

$$p(t/v) = \# \left\{ j \in \{1, \dots, k\} : \frac{t}{v} Y_{(j)} \in B \right\},$$

where # means "number of cases". Now approximate $\mathfrak{F}_{\mathbb{S}}(B, t/v)$ by $p(t/v)/k$ and calculate

$$integrand(t/v) = (t/v)^{n-1} \cdot g(t/v) \cdot p(t/v)/k.$$

$\Phi_{g,\mathbb{P}}(B)$ is now approximated by numerical integration of $integrand(t/v)$ from 0 to u using the mesh points t/v, $t = 0, 1, \dots, uv$. For the numerical integration use common rules like e.g. Simpson's integration rule. This method is summarized in Algorithm 3.

Example 1

We consider the symmetric octagon having the vertices $p_1 = (1, 0)^T$, $p_2 = (3/4, 3/4)^T$, $p_3 = (0, 1)^T$, $p_4 = (-3/4, 3/4)^T$, $p_5 = (-1, 0)^T$, $p_6 = (-3/4, -3/4)^T$, $p_7 = (0, -1)^T$ and $p_8 = (3/4, -3/4)^T$ as contour defining convex polyhedron $\mathbb{P}$. In (Richter and Schicker 2017) it is shown that

$$h_{\mathbb{P}}(x, y) = \max\{|x|, |y|\} + \frac{1}{3} \min\{|x|, |y|\}, (x, y)^T \in \mathbb{R}^2.$$

We choose the multinormal-type density generator $g(r) = e^{-r^2/2}$ and consider the polyhedral convex contoured density

$$\varphi_{g,\mathbb{P}}(x, y) = \frac{1}{6} e^{-\frac{1}{2}(\max\{|x|,|y|\} + \frac{1}{3}\min\{|x|,|y|\})^2}, (x, y)^T \in \mathbb{R}^2.$$

For a particular illustration of $\varphi_{g,\mathbb{P}}$ and a simulation of a sample point cloud of the corresponding distribution $\Phi_{g,\mathbb{P}}$, we refer to Fig. 1.

For a concrete application of Algorithm 3 we consider now the set $B = \{(x, y)^T \in \mathbb{R}^2 : x, y \geq 0\}$ and calculate $\Phi_{g,\mathbb{P}}(B)$. Because of the symmetry of $\mathbb{P}$, we can conclude in this case that $\Phi_{g,\mathbb{P}}(B) = 0.25$. Applying Algorithm 3, we choose $v = 100$ and approximate $\Phi_{g,\mathbb{P}}(B)$ with different values of k, $k \in \{10^3, 10^4, 10^5\}$. Since R_g from (6) follows the density $f(r) =$

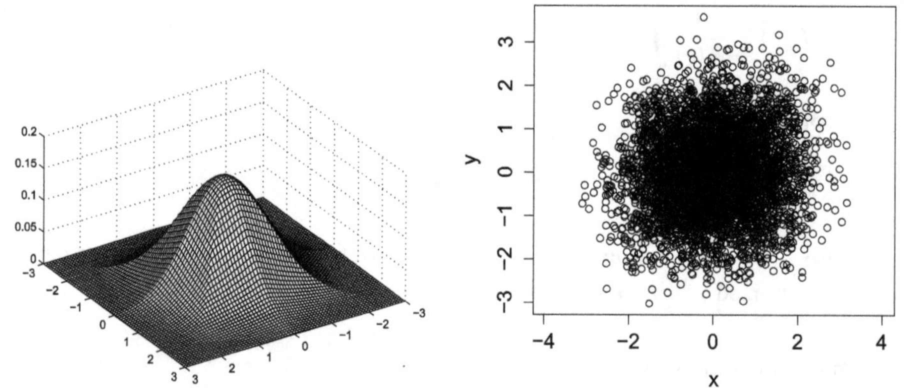

Fig. 1 Density $\varphi_{g,\mathbb{P}}$ (*left*) and a simulated cloud of 5000 points of the distribution $\Phi_{g,\mathbb{P}}$ (*right*)

$re^{-r^2/2}$ and the 99.999 percent quantile of the distribution of R_g is 4.7985, we choose $u = 5$. As numerical integration method we use the Simpson rule in Algorithm 3. Note that it is also possible to apply *integrand* from Algorithm 3 for adaptive numerical integration rules that calculate the mesh points more flexibly than static numerical integration rules. For a comparison we used the adaptive Simpson quadrature, the Gauss-Kronrod quadrature

---

Algorithm 3: Simulation of $\Phi_{g,\mathbb{P}}(B)$.

Input: g, $\mathbb{P}$, $h_{\mathbb{P}}$ B, k, u, v
Output: $\Phi_{g,\mathbb{P}}(B)$

Algorithm
Initializing *r and integrand are both vectors of length* $u*v+1$
    for $j := 1$ to k step 1 **do**
        Simulate $X_{(j)}$ with Algorithm 1;
    end for
    for $j := 1$ to k step 1 **do**
        $Y_{(j)} = X_{(j)}/h_{\mathbb{P}}(X_{(j)})$;
    end for
    $count := 1$;
    for $i := 0$ to u step $1/v$ **do**
        $p := 0$;
        for $j := 1$ to k step 1 **do**
            if $i \cdot Y_{(j)} \in B$ **then**
                $p := p + 1$;
            end if;
        end for
        $r[count] := i$;
        $integrand\,[count] := i^{n-1} \cdot g(i) \cdot p/k$;
        $count := count + 1$;
    end for
    Calculate $\Phi_{g,\mathbb{P}}(B)$ using numerical integration of *integrand* with vector of mesh points r;
Return $\Phi_{g,\mathbb{P}}(B)$

---

and the adaptive Lobatto quadrature. These standard routines are available for example in *MATLAB* or in the package *pracma* of the statistic package *R*. The numerical results are shown in Table 1.

5.3 Linear transformations

For another application of the construction method from Section 5.1 it is assumed that the contour defining convex polyhedron $\mathbb{P}$ contains 0_n in its interior. In this section, we will show how to apply this method for arbitrarily chosen convex polyhedra. Let $\mathbb{P} \subset \mathbb{R}^n$ be an arbitrary convex polyhedron in the n-dimensional space and let $c \in int(\mathbb{P})$ be an arbitrary element of the interior of $\mathbb{P}$. The shifted set $\mathbb{P} - c$ will be denoted by $\mathbb{P}_c$, thus $0_n \in int(\mathbb{P}_c)$. Applying now Section 2.2, we can determine $h_{\mathbb{P}_c}$ and can stochastically represent a random vector X_c by applying (6). It follows

$$X_c \stackrel{d}{=} R_g \cdot U_c, \text{where } U_c \sim \omega_{\mathbb{S}_c}$$

and $\mathbb{S}_c$ is the boundary of $\mathbb{P}_c$. Furthermore, the density of X_c is

$$\varphi_{g,\mathbb{P}_c}(x) = C(g, \mathbb{P}_c) g(h_{\mathbb{P}_c}(x)), x \in \mathbb{R}^n.$$

The class of convex polyhedral distributions $\Phi_{g,\mathbb{P}_c}$ is a subclass of the class of polyhedral star-shaped distributions that is considered in (Richter and Schicker 2017). Specific properties, simulation methods as well as examples of application of convex polyhedral distributions are studied in the current article. According to (Richter and Schicker 2017) we can transform X_c linearly to construct a convex polyhedral distributed random vector Y that follows a density with density level sets being located where we chose the contour defining convex polyhedron $\mathbb{P}$. Doing this, it is proven in Theorem 7.1 of (Richter and Schicker 2017) that $Y \stackrel{d}{=} X_c + c$ satisfies the desired properties and follows the density

$$\varphi_{g,\mathbb{P}_c+c}(y) = C(g, \mathbb{P}_c) g(h_{\mathbb{P}_c}(y - c)), y \in \mathbb{R}^n.$$

Example 2 We consider the trapezoid $\mathbb{P}$ having vertices $p_1 = (4,4)^T$, $p_2 = (7,4)^T$, $p_3 = (6,6)^T$ and $p_4 = (5,6)^T$ as contour defining convex polygon, and choose $c = (5,5)^T$ to shift $\mathbb{P}$ by c, see Fig. 2.

Now, $\mathbb{P}_c$ can be represented by (3) as

$$\mathbb{P}_c = \left\{ \begin{pmatrix} x \\ y \end{pmatrix} \in \mathbb{R}^2 : \begin{pmatrix} 0 & -3 \\ 2 & 1 \\ 0 & 1 \\ -2 & 1 \end{pmatrix} \begin{pmatrix} x \\ y \end{pmatrix} \leq \begin{pmatrix} 3 \\ 3 \\ 1 \\ 1 \end{pmatrix} \right\}$$

and thus, according to Section 2.2,

$$h_{\mathbb{P}_c}(x, y) = \max\{-y, 2/3x + y/3, y, -2x + y\}, (x, y)^T \in \mathbb{R}^2.$$

Table 1 Numerical evaluations of $\Phi_{g,\mathbb{P}}(B)$

Method	k:	10^3	10^4	10^5
Simpson rule		0.2359971	0.247997	0.249377
Adaptive Simpson rule		0.2359991	0.2479991	0.2493791
Gauss-Kronrod quadrature		0.2359991	0.2479991	0.2493791
Adaptive Lobatto quadrature		0.2359993	0.2479992	0.2493792

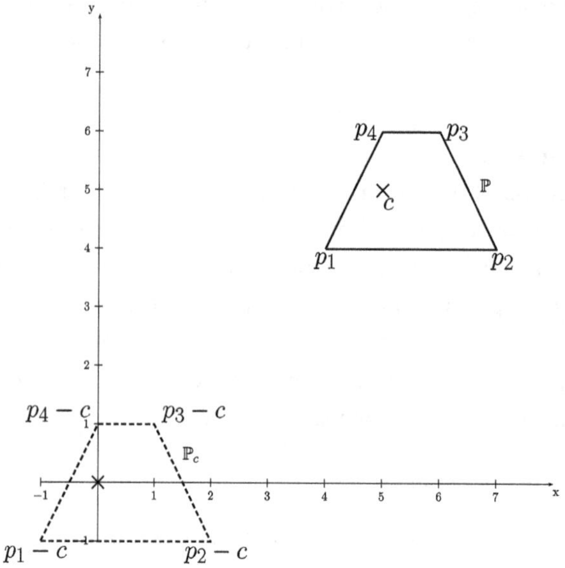

Fig. 2 Contour defining trapezoid $\mathbb{P}$ and the shifted set $\mathbb{P}_c$ from Example 2

Hence,

$$\varphi_{g,\mathbb{P}_c}(x, y) = \frac{1}{8\mathcal{I}(g)} g(\max\{-y, 2/3x + y/3, y, -2x + y\}), (x, y)^T \in \mathbb{R}^2$$

and

$$\varphi_{g,\mathbb{P}_c+c}(x, y) = \frac{1}{8\mathcal{I}(g)} g(\max\{-(y-c), 2/3(x-c) + (y-c)/3, y-c, \\ -2(x-c) + (y-c)\}), (x, y)^T \in \mathbb{R}^2.$$

If we choose, e.g., the density generating function of the uniform distribution, $g_1(r) = I_{[0,1]}(r)$, then $\mathcal{I}(g_1) = 1/2$,

$$\varphi_{g_1,\mathbb{P}_c}(x, y) = \frac{1}{4} I_{[0,1]}(\max\{-y, 2/3x + y/3, y, -2x + y\}), (x, y)^T \in \mathbb{R}^2$$

and

$$\varphi_{g_1,\mathbb{P}_c+c}(x, y) = \frac{1}{4} I_{[0,1]}(\max\{-(y-c), 2/3(x-c) + (y-c)/3, y-c, \\ -2(x-c) + (y-c)\}), (x, y)^T \in \mathbb{R}^2.$$

If we choose alternatively the Kotz-type *dgf* $g_2(r) = r^{18}e^{-20r^2}$, then $\mathcal{I}(g_2) = \Gamma(10)/(2 \cdot 20^{10})$,

$$\varphi_{g_2,\mathbb{P}_c}(x, y) = \frac{20^{10}}{4\Gamma(10)} (h_{\mathbb{P}_c}(x, y))^{18} e^{-20(h_{\mathbb{P}_c}(x,y))^2}, (x, y)^T \in \mathbb{R}^2$$

and

$$\varphi_{g_2,\mathbb{P}_c+c}(x, y) = \frac{20^{10}}{4\Gamma(10)} (h_{\mathbb{P}_c}(x-c, y-c))^{18} e^{-20(h_{\mathbb{P}_c}(x-c,y-c))^2}, (x, y)^T \in \mathbb{R}^2.$$

For an illustration of $\varphi_{g_1,\mathbb{P}_c}$, $\varphi_{g_1,\mathbb{P}_c+c}$, $\varphi_{g_2,\mathbb{P}_c}$ and $\varphi_{g_2,\mathbb{P}_c+c}$, see Figs. 3, 4 and 5.

6 Discussion

(Tian et al. 2009) presented a method for simulating uniform distributions on polyhedral regions which always needs challenging case sensitive representations when it is applied

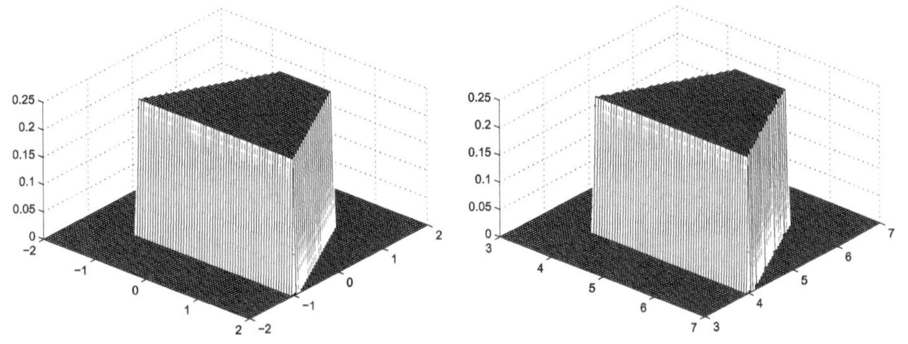

Fig. 3 $\varphi_{g_1,\mathbb{P}_c}$ (*left*) and $\varphi_{g_1,\mathbb{P}_c+c}$ (*right*) from Example 2

to drug combination study. These preparations include the derivation of sophisticated stochastic representations of the uniformly distributed random vectors as presented in Sections 4.1.1 and 4.2.1 as well as non-trivial distributional considerations for the representing random variables sketched also there. Here we establish an alternative approach based upon an acceptance-rejection algorithm which may relatively easily and flexibly be adopted to rather different situations. It will also allow to consider in the future dose-response curves and domains for dose ranges different from that introduced according to (Tian et al. 2009) in Section 1. Moreover, our method allows simulating general polyhedral convex contoured distributions $\Phi_{g,\mathbb{P}}$ and any particular probability $\Phi_{g,\mathbb{P}}(B)$ where the density generating function g allows modeling both heavy and light distribution tails and centers and $\mathbb{P}$ is an arbitrary convex polyhedron. Note, however, that if the probability $\Phi_{g,\mathbb{P}}(B)$ is small which is, e.g., the case if B is a "thin" set or has large distance from the center of the distribution $\Phi_{g,\mathbb{P}}$ then relatively large sample sizes are needed for good approximations via simulation. In such and related cases it may be a useful alternative to try a direct numerical approximation of $\Phi_{g,\mathbb{P}}(B)$ using the geometric measure representation (8) as indicated in Example 1. Fields of successful applications of this type are surveyed, e.g., in (Richter 2014; 2015a; 2015b). The reader is encouraged to find other applications.

Now that the reader has become familiar with the role the stochastic and geometric representations (6) and (8) may play in simulation it is about time to shortly compare

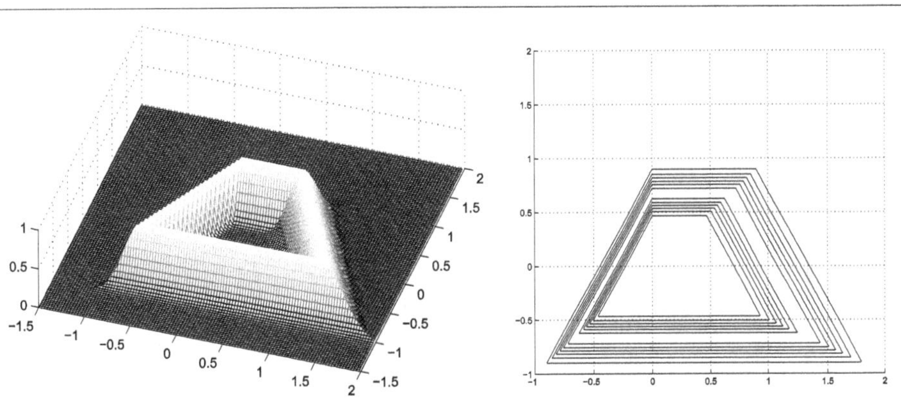

Fig. 4 $\varphi_{g_2,\mathbb{P}_c}$ (*left*), and its density level sets (*right*) from Example 2

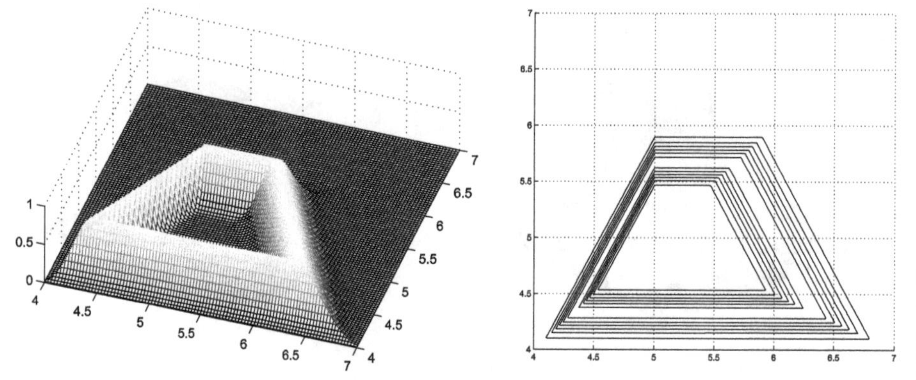

Fig. 5 $\varphi_{g_2,\mathbb{P}_c+c}$ (*left*), and its density level sets (*right*) from Example 2

the approaches presented here and in (Nolan 2016) with each other. The latter one also refers to a stochastic representation like (6) and exploits it in three particular cases for exact simulation. Note that probability models were studied for these particular cases to a certain extent already in (Henschel and Richter 2002) and Richter (2009). In other cases, the stochastic representation is taken in (Nolan 2016) to motivate an approximative simulation method. It would be an interesting aspect of future work to study the resulting approximation accuracy in this approach.

From a purely descriptive point of view, i.e. if we are just given data, visualized by clouds of points, the methods presented here and in (Nolan 2016) may be considered to be attractive alternatives for modeling polyhedral contoured data clouds. But notice that the approach in (Nolan 2016) allows for approximative consideration of much more generalized spherical and related distributions, assuming however an explicit representation of their gauge function is given. It is another aspect of future work to develop exact simulation methods for broader distribution classes.

Appendix

Example 3 Consider triangle $\mathbb{P}$ in Fig. 6. Since $\mathbb{P}$ has vertices $p_1 = (1, -2), p_2 = (2, 1)$ and $p_3 = (-2, 2)$, it can be represented according to (2) as $\mathbb{P} = conv(\{p_1, p_2, p_3\})$. Choosing the closed half-spaces $H_1 = \{(x, y) \in \mathbb{R}^2 : 3x - y \leq 5\}, H_2 = \{(x, y) \in \mathbb{R}^2 : x + 4y \leq 6\}$ and $H_3 = \{(x, y) \in \mathbb{R}^2 : -4x - 3y \leq 2\}$, triangle $\mathbb{P}$ can be equivalently represented according to (3) as intersection of these half-spaces,

$$\mathbb{P} = \left\{ \begin{pmatrix} x \\ y \end{pmatrix} \in \mathbb{R}^2 : \underbrace{\begin{pmatrix} 3 & -1 \\ 1 & 4 \\ -4 & -3 \end{pmatrix}}_{=A} \begin{pmatrix} x \\ y \end{pmatrix} \leq \underbrace{\begin{pmatrix} 5 \\ 6 \\ 2 \end{pmatrix}}_{=b} \right\}.$$

Example 4 This example is a continuation of the previous one. We consider again triangle $\mathbb{P}$ in Fig. 6. Using representation (3) of $\mathbb{P}$ from Example 3, it follows

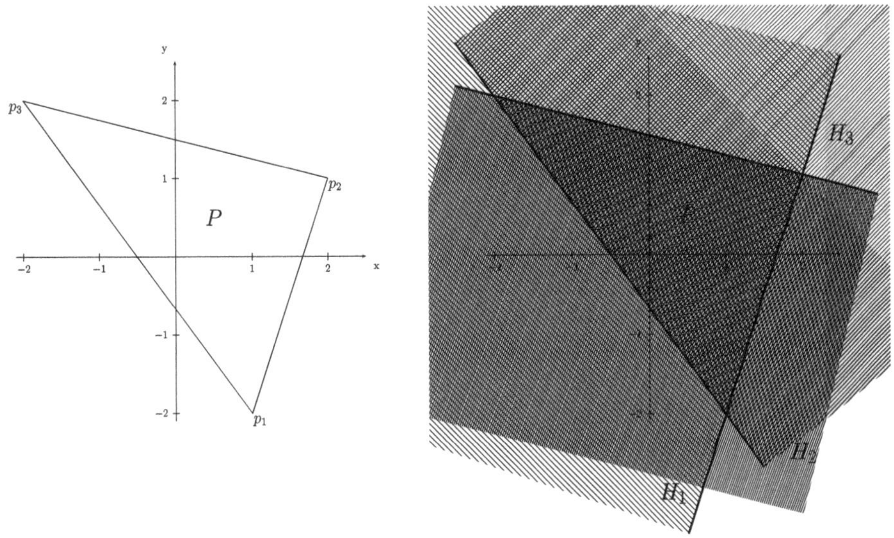

Fig. 6 Triangle $\mathbb{P}$ defined by its vertices (*left*) and by suitably chosen closed half-spaces (*right*)

$A_1' = (3/5, -1/5), A_2' = (1/6, 2/3), A_3' = (-2, -3/2)$, and according to (4) the Minkowski functional $h_{\mathbb{P}}$ can be represented as

$$h_{\mathbb{P}}(x, y) = \max\{3/5x - 1/5y, 1/6x + 2/3y, -2x - 3/2y\}, (x, y)^T \in \mathbb{R}^2.$$

Example 5 Here we illustrate our consideration in Section 4.1.2. To this end, let $\mathbb{P}(a_1, a_2, a_3, a_4)$ be the square with vertices $a_1 = (1, 1)^T, a_2 = (2, 1)^T, a_3 = (2, 2)^T$ and $a_4 = (1, 2)^T$. Then

$$A = \begin{pmatrix} 0 & -1 \\ 1 & 0 \\ 0 & 1 \\ -1 & 0 \end{pmatrix} \text{ and } b = \begin{pmatrix} -1 \\ 2 \\ 2 \\ -1 \end{pmatrix}.$$

Example 6 Here we apply **ARSA** to the drug combination study in (Tian et al. 2009) using the method of Section 4.1.2. In the Introduction we mentioned that the simulation of uniformly distributed random points on the tetragon $\mathbb{P}(a_1, a_2, a_3, a_4)$ is of particular interest for the consideration of the combination study of two drugs with linear dose-response curves. Consider again the tetragonal domain $\mathbb{P}(a_0, b_0)$ in (1), the matrix A and vector b from (3) can be determined from the representation

$$\begin{aligned} \mathbb{P}(a_0, b_0) &= \{(x_1, x_2^*)^T \in \mathbb{R}^2 : -\beta_1 x_1 - \beta_2 x_2^* \leq \alpha_1 - a_0, \beta_1 x_1 + \beta_2 x_2^* \leq b_0 - \alpha_1, \\ &\quad - x_1 \leq 0, -x_2^* \leq 0\} \\ &= \left\{ \begin{pmatrix} x_1 \\ x_2^* \end{pmatrix} \in \mathbb{R}^2 : \underbrace{\begin{pmatrix} -\beta_1 & -\beta_2 \\ \beta_1 & \beta_2 \\ -1 & 0 \\ 0 & -1 \end{pmatrix}}_{=A} \begin{pmatrix} x_1 \\ x_2^* \end{pmatrix} \leq \underbrace{\begin{pmatrix} \alpha_1 - a_0 \\ b_0 - \alpha_1 \\ 0 \\ 0 \end{pmatrix}}_{=b} \right\}. \end{aligned} \tag{9}$$

Now, our Algorithm 1 applies.

For a numerical illustration we consider $LY - 168$ from (Tian et al. 2009) which is said to be a potential anti melanoma agent and has the single dose-response curve $f_1(x_1) = 101.91 - 31.17x_1$. $LY-168$ will be considered with Sorafenib which is marketed as Nevaxar by Bayer for the treatment of advanced renal cell cancer and also of advanced thyroid cancer in the USA. Its dose-response curve is given by $f(x_2) = 111.85 - 9.56x_2 = 101.91 - 9.56(x_2 - 1.04)$. If the chosen dose ranges from 20% to 80% and $x_2^* = x_2 - 1.04$, the experimental domain is given by

$$\mathbb{P}(20,80) = \left\{ \begin{pmatrix} x_1 \\ x_2^* \end{pmatrix} \in \mathbb{R}^2 : 20 \le 101.91 - 31.17x_1 - 9.56x_2^* \le 80, x_1 \ge 0, x_2^* \ge 0 \right\}$$

$$\underset{(9)}{=} \left\{ \begin{pmatrix} x_1 \\ x_2^* \end{pmatrix} \in \mathbb{R}^2 : \underbrace{\begin{pmatrix} 31.17 & 9.56 \\ -31.17 & -9.56 \\ -1 & 0 \\ 0 & -1 \end{pmatrix}}_{=A} \begin{pmatrix} x_1 \\ x_2^* \end{pmatrix} \le \underbrace{\begin{pmatrix} 81.91 \\ -21.91 \\ 0 \\ 0 \end{pmatrix}}_{=b} \right\}.$$

Since $\mathbb{P}(20,80)$ has vertices $a_1 = (0.7029195, 0)^T$, $a_2 = (2.627847, 0)^T$, $a_3 = (0, 8.567992)^T$ and $a_4 = (0, 2.291841)^T$, we can choose $Q = [0, 2.627847] \times [0, 8.567992]$ and apply Algorithm 1 with input A, b, 0, 0, 2.627847, 8.567992. Figure 7 shows the result of an acceptance-rejection simulation of uniformly distributed random points on $\mathbb{P}(20,80)$ with acceptance rates of about 46 percent.

Example 7 Let $\mathbb{P}(a_1, a_2, a_3, a_4, a_5, a_6)$ be the triangular prism with vertices $a_1 = (1,1,0)^T, a_2 = (4,1,0)^T, a_3 = (5,3,0)^T, a_4 = (3,1,3)^T, a_5 = (4,1,1)^T$ and $a_6 = (13/3, 5/3, 3)^T$. For an illustration of the triangular prism, see Fig. 8.

Here,

$$A' = \begin{pmatrix} 0 & 0 & 6 \\ 0 & -9 & 0 \\ -2 & 1 & 0 \\ -6 & 12 & 4 \\ 4/3 & -8/3 & 2/3 \end{pmatrix} \text{ and } b' = \begin{pmatrix} 0 \\ -9 \\ -7 \\ 6 \\ 10/3 \end{pmatrix}.$$

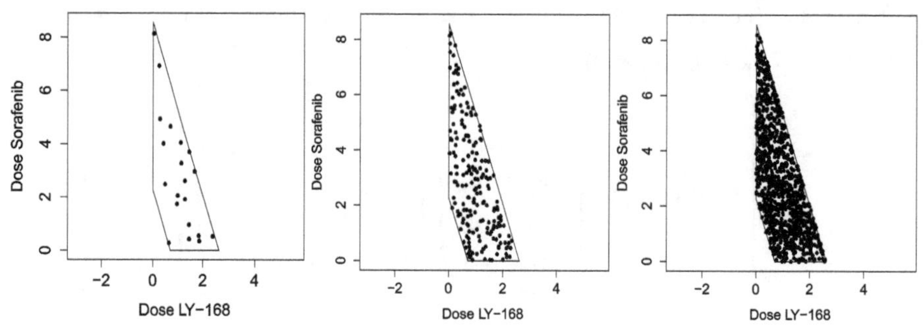

Fig. 7 20 (*left*), 200 (*middle*) and 1000 (*right*) uniformly distributed points on $\mathbb{P}(20,80)$

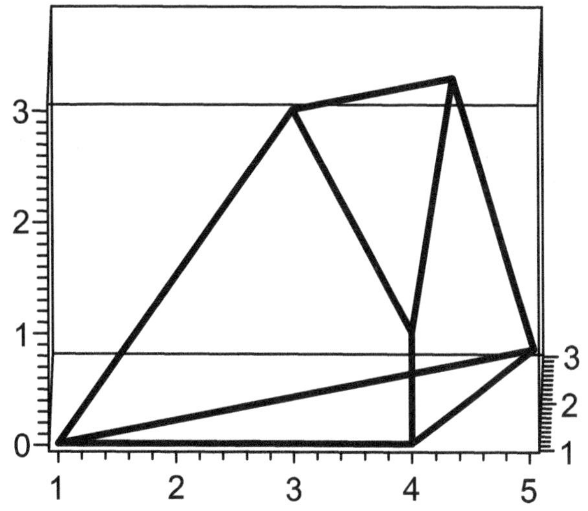

Fig. 8 Triangular prism $\mathbb{P}(a_1, \ldots, a_6)$ in Example 7

Applying Algorithm 2, it follows that

$$z_1 = \begin{pmatrix} 0 \\ -9 \\ -1 \\ 6 \\ -4/3 \end{pmatrix}, \text{thus } A_1 = \begin{pmatrix} 0 & 0 & 6 \\ 0 & -9 & 0 \\ 2 & -1 & 0 \\ -6 & 12 & 4 \\ 4/3 & -8/3 & 2/3 \end{pmatrix} \text{ and } b_1 = \begin{pmatrix} 0 \\ -9 \\ 7 \\ 6 \\ 10/3 \end{pmatrix}.$$

Moreover, $z_2 = (0, -9, 7, -12, 8/3)^T, A_2 = A_1$ and $b_2 = b_1$, $z_3 = (0, -27, 7, 6, -4/3)^T, A_3 = A_2$ and $b_3 = b_2$,

$$z_4 = \begin{pmatrix} 18 \\ -9 \\ 5 \\ 6 \\ 10/3 \end{pmatrix}, A_4 = \begin{pmatrix} 0 & 0 & -6 \\ 0 & -9 & 0 \\ 2 & -1 & 0 \\ -6 & 12 & 4 \\ 4/3 & -8/3 & 2/3 \end{pmatrix} \text{ and } b_4 = b_3,$$

$z_5 = (-6, -9, 7, -8, 10/3)^T, A_5 = A_4$ and $b_5 = b_4$, $z_6 = (-18, -15, 7, 6, 10/3)^T, A_6 = A_5$ and $b_6 = b_5$. Finally it follows, with $A = A_6$ and $b = b_6$, that $\mathbb{P}(a_1, \ldots, a_6) = \{(x, y, z)^T \in \mathbb{R}^3 : A(x, y, z)^T \leq b\}$.

Example 8 Here we apply **ARSA** to the drug combination study in (Tian et al. 2009) using the method of Section 4.2.2. The simulation of uniform distributions in triangular prisms in $\mathbb{R}^3$ is required in experimental designs for the combination study of three drugs with log-linear-dose-response curves that are, according to (Berenbaum 1989) and (Tian et al. 2009), found in a wide variety of systems such as antibiotics, narcotics, cromoglycate and others. The single dose-response curves for drugs A_i, $i = 1, 2, 3$ are assumed to be $f_i(x_i) = \alpha_i + \beta \log(x_i)$, $i = 1, 2, 3$. According to (Berenbaum 1989; Tan et al. 2003; Tian et al. 2009) the additive model at the combination dose (x_1, x_2, x_3) is

$$y(x_1, x_2, x_3) = \alpha_1 + \beta\Bigg(\log(z_1) + \log((1-\rho_0)z_2 + \rho_0) + \log\left(\left(1 - \frac{\rho_1}{\rho_0}\right)(1 - z_3) + \frac{\rho_1}{\rho_0}\right)\Bigg),$$

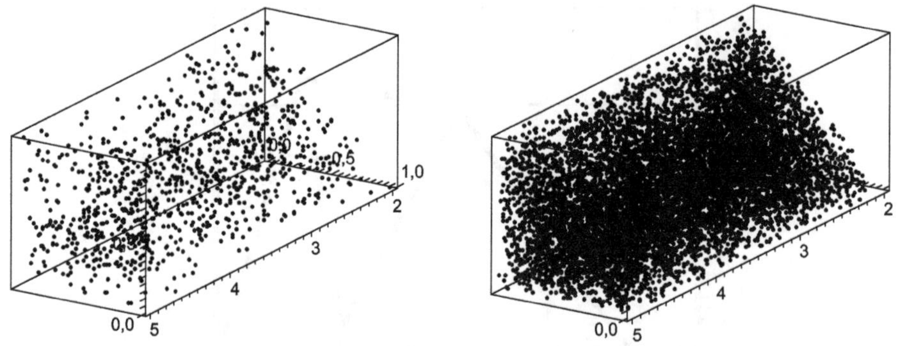

Fig. 9 Simulation of 10^3 (left) and 10^4 (right) uniformly distributed random points on $\mathbb{P}(2,5)$

where $\rho_0 = e^{(\alpha_2-\alpha_1)/\beta}$, $\rho_1 = e^{(\alpha_3-\alpha_1)/\beta}$, $z_1 = x_1 + x_2 + x_3$, $z_2 = x_1/(x_1 + x_2 + \rho_1 x_3/\rho_0)$ and $z_3 = x_3/(x_1 + x_2 + x_3)$. It is also shown there that experimental points (z_1, z_2, z_3) maximizing the statistical power in detecting synergy should be uniformly scattered in the domain

$$\mathbb{P}(z_l, z_u) = \left\{ \begin{pmatrix} z_1 \\ z_2 \\ z_3 \end{pmatrix} \in \mathbb{R}^3 : z_l < z_1 < z_u, 0 < z_2, 0 < z_3, 0 < z_2 + z_3 < 1 \right\}. \quad (10)$$

Since $\mathbb{P}(z_l, z_u)$ is a triangular prism, the methods described in Sections 4.2.1 and 4.2.2 apply to simulating uniformly distributed random points on $\mathbb{P}(z_l, z_u)$. Alternatively, we can use representation (10) to calculate A, b and Q for the application of Algorithm 1, since

$$\begin{aligned}\mathbb{P}(z_l, z_u) &= \left\{ \begin{pmatrix} z_1 \\ z_2 \\ z_3 \end{pmatrix} \in \mathbb{R}^3 : z_1 < z_u, -z_1 < -z_l, -z_2 < 0, -z_3 < 0, z_2 + z_3 < 1, \right. \\ &\quad \left. - z_2 - z_3 < 0 \right\} \\ &= \left\{ \begin{pmatrix} z_1 \\ z_2 \\ z_3 \end{pmatrix} \in \mathbb{R}^3 : \underbrace{\begin{pmatrix} 1 & 0 & 0 \\ -1 & 0 & 0 \\ 0 & -1 & 0 \\ 0 & 0 & -1 \\ 0 & 1 & 1 \\ 0 & -1 & -1 \end{pmatrix}}_{=A} \cdot \begin{pmatrix} z_1 \\ z_2 \\ z_3 \end{pmatrix} < \underbrace{\begin{pmatrix} z_u \\ -z_l \\ 0 \\ 0 \\ 1 \\ 0 \end{pmatrix}}_{=b} \right\}.\end{aligned}$$

Furthermore, the boundaries of $\mathbb{P}(z_l, z_u)$ can be read out off representation (10). Thus Q can be chosen as $Q = [z_l, z_u] \times [0, 1] \times [0, 1]$. For an illustration, we refer to Fig. 9, where 1000 and 10000 uniformly distributed random points are simulated on $\mathbb{P}(2, 5)$.

Acknowledgement
The authors are grateful to a referee who provided valuable suggestions on improving the paper and an additional recent reference, and declare that they have no competing interests in the paper.

References

Arnold, B, Castillo, E, Sarabia, JM: Multivariate distributions defined in terms of contoures. J. Stat. Plan. Infer. **138**, 4158–4171 (2008)
Avis, D: lrs: A Revised Implementation of the Reverse Search Vertex Enumeration Algorithm. Birkhäuser Verlag, Basel (2000). doi:10.1007/978-3-0348-8438-9_9
Balkema, G, Embrechts, P, Nolde, N: Meta densities and the shape of their sample clouds. J Multiv Analysis. **101**(7), 1738–1754 (2010)
Berenbaum, MC: What is synergy? Pharmacol. Rev. **41**(2), 93–141 (1989)
Böhm, J, Hertel, E: Mathematische Monographien: Polyedergeometrie in n- dimensionalen Räumen konstanter Krümmung, vol 14. VEB Deutscher Verlag der Wissenschaften, Berlin (1980)
Cutler, B, Dorsey, J, McMillan, L: Simplification and improvement of tetrahedral models for simulation. In: The 2004 Eurographics/ACM SIGGRAPH Symposium on Geometry Processing, pp. 93–102. ACM, New York, (2004)
Fang, KT, Yang, ZH: On uniform design of experiments with restricted mixtures and generation of uniform distribution on some domains. Stat. Probab. Lett. **46**, 113–120 (2000)
Fang, KT, Kotz, S, Ng, KW: Monographs on Statistics and Applied Probability: Symmetric Multivariate and Related Distributions, vol 36. Chapman & Hall, New York (1990)
Fernandez, C, Osiewalski, J: Steel, MFJ: Modeling and inference with ν-spherical distributions. J Am. Stat. Assoc. **90**, 1331–1340 (1995)
Fukuda, K, Prodon, A: Double Description Method Revisited. In: Combinatorics and Computer Science. Springer- Verlag, Berlin, Heidelberg, (1996). doi:10.1007/3-540-61576-8_77
Gupta, A, Song, D: l_p-norm spherical distribution. J. Stat. Plann. Infer. **60**(2), 241–260 (1997)
Henschel, V, Richter, W-D: Geometric generalization of the exponential law. J. Multivar. Anal. **81**, 189–204 (2002). doi:10.1006/jmva.2001.2001
Kalke, S, Richter, W-D: Simulation of the p- generalized gaussian distribution. J. Stat. Comput. Simul. **83**(4), 641–667 (2013). doi:10.1080/00949655.2011.631187
Kamiya, H, Takemura, A, Kuriki, S: Star-shaped distributions and their generalizations. J. Stat. Plan. Infer. **138**, 3429–3447 (2008)
von Neumann, J: Various techniques used in connection with random digits.monte carlo methods. Nat. Bureau Standards. **12**, 36–38 (1951)
Nolan, J: An R package for modeling and simulating generalized spherical and related distributions. J. Stat. Distrib. Appl. **3:14**, 11 (2016)
Richter, W-D: Eine geometrische Methode in der Stochastik. Rostocker Mathematisches Kolloquium. **44**, 63–72 (1991)
Richter, W-D: Continuous $l_{n,p}$- symmetric distributions. Lith. Math. J. **49**(1), 93–108 (2009). doi:10.1007/s10986-009-9030-3
Richter, W-D: Geometric and stochastic representations for elliptically contoured distributions. Commun. Stat. Theory Methods. **42**, 579–602 (2013). doi:10.1080/03610926.2011.611320
Richter, W-D: Geometric disintegration and star- shaped distributions. J. Stat. Distrib. Appl. **1:20**, 24 (2014). doi:10.1186/s40488-014-0020-6
Richter, W-D: Convex and radially concave contoured distributions. J. Probab. Stat. **2015**, 12 (2015a). doi:10.1155/2015/165468
Richter, W-D: Norm contoured distributions in $\mathbb{R}^2$. Lect. Notes Seminario Interdisciplinare di Matematica. **12**, 179–199 (2015b)
Richter, W-D, Schicker, K: Ball numbers of platonic bodies. J. Math. Anal. Appl. **416**(2), 783–799 (2014). doi:10.1016/j.jmaa.2014.03.007
Richter, W-D, Schicker, K: Polyhedral star-shaped distributions. J. Probab. Stat. **2017**, 35 (2017)
Rocchini, C, Cignoni, P: Generating random points in a tetrahedron. J. Graphics Tools. **5**(4), 9–12 (2000)
Sarabia, JM, Gomez-Deniz, E: Construction of multivariate distributions: a review of some results. SORT. **32**(1), 3–36 (2008). With discussion by Pardo, M.C., 39-42 and Navarro, J., 43-46
Tan, M, Fang, HB, Tian, GL, Houghton, PJ: Experimental design and sample size determination for testing synergism in drug combination studies based on uniform measures. Stat. Med. **22**, 2091–2100 (2003). doi:10.1002/sim.1467
Tian, GL, Fang, HB, Tan, M, Qin, H, Tang, ML: Uniform distributions in a class of convex polyhedrons with applications to drug combination studies. J. Multivariate Anal. **100**, 1854–1865 (2009). doi:10.1016/j.jmva.2009.02.011
Ziegler, GM: Lectures on Polytopes. Springer-Verlag, Berlin, Heidelberg (1995)

2

On p-generalized elliptical random processes

Klaus Müller and Wolf-Dieter Richter*

*Correspondence:
wolf-dieter.richter@uni-rostock.de
University of Rostock, Institute of Mathematics, Ulmenstraße 69, Rostock, Germany

Abstract

We introduce rank-k-continuous axis-aligned p-generalized elliptically contoured distributions and study their properties such as stochastic representations, moments, and density-like representations. Applying the Kolmogorov existence theorem, we prove the existence of random processes having axis-aligned p-generalized elliptically contoured finite dimensional distributions with arbitrary location and scale functions and a consistent sequence of density generators of p-generalized spherical invariant distributions. Particularly, we consider scale mixtures of rank-k-continuous axis-aligned p-generalized elliptically contoured Gaussian distributions and answer the question when an n-dimensional rank-k-continuous axis-aligned p-generalized elliptically contoured distribution is representable as a scale mixture of n-dimensional rank-k-continuous p-generalized Gaussian distribution for a suitable mixture distribution of a positive random variable. Based on this class of multivariate probability distributions, we introduce scale mixed p-generalized Gaussian processes having axis-aligned finite dimensional distributions being p-generalizations of elliptical random processes. Additionally, some of their characteristic properties are discussed and approximates of trajectories of several examples such as p-generalized Student-t and p-generalized Slash processes having axis-aligned finite dimensional distributions are simulated with the help of algorithms to simulate rank-k-continuous axis-aligned p-generalized elliptically contoured distributions.

Keywords: Axis-aligned p-generalized elliptically contoured distributions, Density-like representation, Kolmogorov consistency conditions, p-generalized spherically invariant random processes, Scale mixtures of multivariate axis-aligned p-generalized elliptically contoured Gaussian distributions, Simulation

1 Introduction

Random processes may be constructed and characterized in different ways. Apart from constructions via families of random variables whose members satisfy, e.g., specific autoregressive relations or are coefficients of specific series representations, the existence of random processes can be studied following the fundamental existence theorem due to Kolmogorov (1933). The explicit knowledge of the family of finite dimensional distributions (fdds) can be used then to establish some of the properties of the random process by proving corresponding ones of the fdds. Basic technical problems to be solved this way belong to multivariate distribution theory. In the present paper, Kolmogorov's theorem is used to prove the existence of real valued random processes having axis-aligned p-generalized elliptically contoured (apec) fdds, thus being p-generalizations of elliptical random processes having axis-aligned fdds.

Well studied examples of random processes which can be constructed via Kolmogorov's existence theorem are real valued Gaussian processes with emphasis on the Brownian motion, see Shiryaev (1996) and Schilling and Partzsch (2014). Apart from further examples as random processes with independent values, random processes with independent increments as well as Markov processes, spherically invariant random processes being also known as elliptical random processes can be constructed this way. The latter are introduced in Vershik (1964) as random processes consisting of quadratically integrable random variables such that if two of them have the same variance, they follow the same distribution. Corresponding characteristic functions and densities are determined in Blake and Thomas (1968). Yao (1973) and Kano (1994) characterize spherically invariant random processes by establishing that their families of fdds are what is called now scale mixtures of Gaussian distributions having one and the same mixture distribution. The notion of a scale mixture but is first introduced in Andrews and Mallows (1974) and, independently, Wise and jun Gallagher (1978) show that an elliptical random process can be represented as a product of a Gaussian process and a positive random variable being independent of it. Additionally, in Huang and Cambanis (1979), the structure of the space of all second order spherically invariant random processes is studied and used to solve nonlinear estimation problems. Finally, based on the concepts of expansive and semi-expansive sequences of elliptically contoured distributions and apart from analogue representation theorems in Yao (1973) and Kano (1994), a formula to determine the corresponding mixture distribution of the family of fdds of a spherically invariant random process is determined in Gómez-Sánchez-Manzano et al. (2006).

Besides a thematically assorted summary of several articles on the theory of spherically invariant random processes, numerous applications of these random processes such as modelings of bandlimited speech waveform, of radar clutters, of radio propagation disturbances and of equalization and array processing are dealt with in Yao (2003). Furthermore, the author discusses simulations of trajectories of spherically invariant random processes based on the work in Brehm and Stammler (1987), Conte et al. (1991), and Rangaswamy et al. (1995). More recent applications deal with fading models from spherically invariant random processes in Biglieri et al. (2015) and with MIMO radar target localization and performance evaluation under spherically invariant random process clutter in Zhang et al. (2017).

The notion of a scale mixture of Gaussian distributions is introduced in Andrews and Mallows (1974) as the distribution of the product of a Gaussian variable and an independent positive random variable. A multivariate generalization is given in Lange and Sinsheimer (1993). Using numerous equivalent definitions, scale mixtures of Gaussian distributions are also studied in West (1987), Gneiting (1997), Eltoft et al. (2006), Gómez-Sánchez-Manzano et al. (2006, 2008), and Hashorva (2012). According to Andrews and Mallows (1974), Lange and Sinsheimer (1993), and Gómez-Sánchez-Manzano et al. (2006), scale mixtures of Gaussian distributions are special cases of elliptically contoured distributions and an elliptically contoured distribution is a scale mixture of a Gaussian distribution if and only if the composition of its density generator and the square root function is completely monotone. Moreover, examples of scale mixtures of Gaussian distributions are Pearson type VII distributions, power exponential distributions as well as Slash distributions.

Applications of scale mixtures of Gaussian distributions are given in the fields of natural images, insurances and quantitative genetic in Wainwright and Simoncelli (2000), Choy and Chan (2003), and Gómez-Sánchez-Manzano et al. (2008). More recent applications are Gaussian scale mixture models for robust multivariate linear regression with missing data in Ala-Luhtala and Piché (2016), testing homogeneity in a scale mixture of Gaussian distributions in Niu et al. (2016), and adaptive robust regression with continuous Gaussian scale mixture errors in Seo et al. (2017).

For any choice of $p > 0$, introducing the notion of a p-generalization of a spherically invariant random process means the transition from spherically contoured to $l_{n,p}$-symmetric fdds, the transition from regular elliptically contoured to suitably introduced p-generalized elliptically contoured distributions and the associated consideration of suitable non-Euclidean instead of Euclidean geometries, respectively. To be more specific, a well-known example is the n-dimensional p-generalized (spherical) Gaussian distribution being introduced already in Subbotin (1923) and having the probability density function (pdf)

$$f(x) = \left(\frac{p^{1-\frac{1}{p}}}{2\Gamma\left(\frac{1}{p}\right)} \right)^n \exp\left\{ -\frac{1}{p} \sum_{i=1}^{n} |x_i|^p \right\}, \quad x = (x_1, \ldots, x_n)^{\mathrm{T}} \in \mathbb{R}^n,$$

and p-generalized Weibull, Pearson type II and Pearson type VII distributions are dealt with in Gupta and Song (1997). Additionally, a p-generalized spherical coordinate transformation, a p-generalized surface content measure as well as numerous p-generalized probability distributions and statistics such as p-generalized versions of the χ^2-, Student and Fisher distributions are considered in Richter (2007); Richter (2009).

The more general class of continuous $l_{n,p}$-symmetric distributions is studied in Arellano-Valle and Richter (2012), Kalke and Richter (2013), Müller and Richter (2016a, b, 2017a, b) as well as several references given there. In the present paper, we introduce a class of multivariate apec distributions containing both regular and singular distributions and covering the classes of continuous $l_{n,p}$-symmetric and common axis-aligned elliptically contoured distributions.

For a nonempty index set $I \subseteq \mathbb{R}$, a Polish space (E, ρ) and a family Q of probability measures on the product spaces $\left(E^J, \mathcal{B}^J\right)$ for nonempty finite subsets $J \subseteq I$ and the Borel sigma field $\mathcal{B}$ on E with respect to ρ, if Q is projective on E, Kolmogorov's existence theorem states the existence of a random process having time set I and state space E such that its family of fdds is equal to Q. The projectivity of Q on E can be shown by checking the consistency conditions in Kolmogorov (1956). This will be discussed for the particular case $E = \mathbb{R}$ in "Sketch of proof" section. This way, we prove the existence of real valued random processes having apec fdds. Such random processes are p-generalizations of elliptical random processes having axis-aligned fdds. Moreover, for the special case of scale mixed p-generalized Gaussian processes having axis-aligned fdds, basic properties such as characteristic representations, stationary properties and specific closedness properties are studied and certain approximates of their trajectories are simulated. Preparing for these results, we prove firstly that an apec distribution can be represented by a scale mixture of the apec Gaussian distribution if and only if its density-like generator composed with the pth root function, is completely monotone and secondly that the corresponding mixture distribution is in a well defined way closely connected to the inverse Laplace-Stieltjes transform of this composition.

The paper is structured as follows. In "The class of n-dimensional rank-k-continuous axis-aligned p-generalized elliptically contoured distributions" section, n-dimensional apec distributions are introduced as location-scale generalizations of continuous $l_{n,p}$-symmetric distributions, and some of their properties such as stochastic representations, moments and pdf-like representations are discussed. Furthermore, the pdfs of bivariate p-generalized spherical as well as of bivariate apec Gaussian distributions are visualized for several values of $p > 0$. Our main result on the existence of p-generalizations of elliptical random processes is presented in "Main result" section. A sketch of its proof consisting of four basic steps is given in "Sketch of proof" section, and an approximate simulation of the trajectories of the new random processes is discussed in "Simulation" section. Examples illustrating the developed theory are studied in the fourth section. In "Scale mixtures of apec Gaussian distributions" section, scale mixtures of multivariate apec Gaussian distributions are introduced and some of their characteristic properties such as stochastic representations, moments, specific conditional distributions, and their connections to completely monotone functions are discussed. Random processes whose families of fdds are families of scale mixtures of multivariate apec Gaussian distributions with one and the same mixture distribution as well as some of their basic properties are studied in "Scale mixed p-generalized Gaussian processes having axis-aligned fdds" section. All proofs are given in "Proofs" section. For the sake of a better readability, the proofs of certain results are prepared by proving certain particular cases first. An algorithm to simulate arbitrary apec distributions and another one to particularly simulate scale mixtures of apec Gaussian distributions with an explicitly known mixture distribution are presented in Appendix 7.1. The latter one is used in Appendix 7.2 to simulate approximations of trajectories of p-generalized Student-t as well as p-generalized Slash processes having axis-aligned fdds. Finally, we remark that all figures presented here are made using the program MATLAB.

2 The class of n-dimensional rank-k-continuous axis-aligned p-generalized elliptically contoured distributions

For each $p > 0$ and $n \in \mathbb{N}$, we denote the p-functional in $\mathbb{R}^n$ by $|x|_p = \left(\sum\limits_{i=1}^{n} |x_i|^p\right)^{\frac{1}{p}}$, $x = (x_1, \ldots, x_n)^{\mathrm{T}} \in \mathbb{R}^n$, and the $l_{n,p}$-generalized surface content of the $l_{n,p}$-unit sphere $S_{n,p} = \{x \in \mathbb{R}^n \colon |x|_p = 1\}$ by $\omega_{n,p}$,

$$\omega_{n,p} = \frac{\left(2\Gamma\left(\frac{1}{p}\right)\right)^n}{p^{n-1}\Gamma\left(\frac{n}{p}\right)}.$$

Furthermore, a function $g\colon [0, \infty) \to [0, \infty)$ satisfying $0 < I_n(g) < \infty$ is called a density generating function of an n-variate distribution where $I_n(g) = \int\limits_0^\infty r^{n-1} g(r)\, dr$. An n-dimensional random vector $X\colon \Omega \to \mathbb{R}^n$ on a probability space $(\Omega, \mathfrak{A}, P)$ having the pdf $\frac{g(|x|_p)}{\omega_{n,p} I_n(g)}$, $x \in \mathbb{R}^n$, is called continuous $l_{n,p}$-symmetrically distributed with density generating function g. A density generating function g of a continuous $l_{n,p}$-symmetric distribution satisfying $I_n(g) = \frac{1}{\omega_{n,p}}$ is called a density generator (dg) and denoted by $g^{(n,p)}$. The pdf of the continuous $l_{n,p}$-symmetric distribution with dg $g^{(n,p)}$ is $g^{(n,p)}(|x|_p)$, $x \in \mathbb{R}^n$, and the corresponding probability law is denoted by $\Phi_{g^{(n,p)}}$. With a view to the special cases listed below, $\Phi_{g^{(n,p)}}$ may also be called n-dimensional continuous p-generalized spherical distribution with dg $g^{(n,p)}$.

A well-known example of the latter type of probability distributions is the n-dimensional p-generalized (spherical) Gaussian distribution $N_{n,p} = \Phi_{g_{PE}^{(n,p)}}$ where

$$g_{PE}^{(n,p)}(r) = \left(\frac{p^{1-\frac{1}{p}}}{2\Gamma\left(\frac{1}{p}\right)}\right)^n \exp\left\{-\frac{1}{p}r^p\right\}, \quad r \geq 0.$$

For visualizations of the pdf of this distribution for $n \in \{1, 2\}$ and several $p > 0$, we refer to Kalke and Richter (2013) and Müller and Richter (2015). The class of continuous $l_{n,2}$-symmetric distributions coincides with the class of n-variate continuous spherical distributions and $N_{n,2}$ is the n-dimensional standard Gaussian distribution. Numerous properties such as stochastic representations, moments, and marginal distributions and several types of dgs are discussed in Gupta and Song (1997), Richter (2009), Arellano-Valle and Richter (2012), and Müller and Richter (2016a).

Let $\mu \in \mathbb{R}^n$ be a constant vector and $D = \operatorname{diag}(d_1, \ldots, d_n)$ an $n \times n$ diagonal matrix having nonnegative diagonal entries and positive rank $\operatorname{rk}(\Sigma) = k$. Moreover, let $I_1 = \{i_1, \ldots, i_k\} \subseteq \{1, \ldots, n\}$ with $|I_1| = k$ and $i_1 < i_2 < \ldots < i_k$ be the set of indices such that $d_i > 0$ if $i \in I_1$ and $d_i = 0$ if $i \in I_2 = \{1, \ldots, n\} \backslash I_1$. Let $e_i^{(n)}$ denote the ith unit vector in $\mathbb{R}^n$ $0_{n\times n}$ the $n \times n$ zero matrix, $S_1 = \operatorname{diag}\left(d_{i_1}, \ldots, d_{i_k}\right) \in \mathbb{R}^{k\times k}$, $W_1 = \left(e_{i_1}^{(n)} \cdots e_{i_k}^{(n)}\right) \in \mathbb{R}^{n\times k}$ and $W_2 \in \mathbb{R}^{n\times(n-k)}$ a matrix having columns $e_i^{(n)}$ for all $i \in I_2$, then,

$$W_1^{\mathrm{T}} D W_1 = S_1 \qquad \text{and} \qquad W_2^{\mathrm{T}} D W_2 = 0_{(n-k)\times(n-k)}.$$

Let $\sqrt{S_1} = \operatorname{diag}\left(\sqrt{d_{i_1}}, \ldots, \sqrt{d_{i_k}}\right)$. The distribution of a random vector X satisfying the stochastic representation

$$X \stackrel{d}{=} \mu + W_1\sqrt{S_1}Y \qquad \text{where } Y \sim \Phi_{g^{(k,p)}} \tag{1}$$

is called an n-dimensional rank-k-continuous axis-aligned p-generalized elliptically contoured (kapec) distribution with location parameter μ, scaling matrix D and dg $g^{(k,p)}$ and is denoted by $AEC_{n,p}(\mu, D, g^{(k,p)})$. For simplicity, the distribution of such random vector X is just called apec distribution if its continuity and dimension as well as the rank of the diagonal matrix parameter D are contextually clear or play only a minor role.

Here and in what follows, $X \stackrel{d}{=} Z$ and $X \sim \Psi$ mean that the random vectors X and Z follow the same distribution law and that the random vector X follows the distribution law $\mathfrak{L}(X) = \Psi$, respectively. In particular, for the special choice of μ and D to be the zero vector 0_n and identity matrix $I_{n\times n}$ in $\mathbb{R}^n$, respectively, we have $AEC_{n,p}\left(0_n, I_{n\times n}, g^{(n,p)}\right) = \Phi_{g^{(n,p)}}$. For the special case of $p = 2$, the class of $AEC_{n,2}\left(\mu, D, g^{(k,2)}\right)$-distributions is identical with the class of common n-variate axis-aligned elliptically contoured distributions. Furthermore, $AEC_{n,p}\left(\mu, D, g_{PE}^{(k,p)}\right)$ is called n-dimensional kapec Gaussian distribution and is denoted $AN_{n,p}(\mu, D)$. The family of apec distributions with full rank scaling matrices as well as their star-shaped extensions and certain aspects of their inferential applications are studied in Richter (2014, 2016, 2017).

Because of relation (1), a stochastic representation and properties of moments of n-dimensional kapec distributions stated in Lemmata 2.1 and 2.2 follow immediately from corresponding results of $l_{k,p}$-symmetric distributions in Richter (2009) and Arellano-Valle and Richter (2012).

Lemma 2.1 *Let $X \sim AEC_{n,p}(\mu, D, g^{(k,p)})$ where* rk(D) $= k$. *Then, the random vector X satisfies the stochastic representation*

$$X \stackrel{d}{=} \mu + R \cdot W_1\sqrt{S_1}U_p^{(k)}$$

where the random vector $U_p^{(k)}$ is k-dimensional p-generalized uniformly distributed on $S_{k,p}$, R and $U_p^{(k)}$ are stochastically independent and R is a nonnegative random variable with pdf

$$f_R(r) = \omega_{k,p}\, r^{k-1} g^{(k,p)}(r)\, \mathbb{1}_{[0,\infty)}(r), \quad r \in \mathbb{R}.$$

Lemma 2.2 *Let $X \sim AEC_{n,p}(\mu, D, g^{(k,p)})$ where* rk(D) $= k$. *Then,* $\mathbb{E}(X) = \mu$ *if $I_{k+1}(g^{(k,p)})$ is finite and* Cov(X) $= \sigma^2_{g^{(k,p)}}D$ *if $I_{k+2}(g^{(k,p)})$ is finite where the univariate variance component $\sigma^2_{g^{(k,p)}}$ of $\Phi_{g^{(k,p)}}$ satisfies $\sigma^2_{g^{(k,p)}} = \frac{\Gamma\left(\frac{3}{p}\right)\Gamma\left(\frac{k}{p}\right)}{\Gamma\left(\frac{1}{p}\right)\Gamma\left(\frac{k+2}{p}\right)}\omega_{k,p} I_{k+2}(g^{(k,p)})$. The components of X are independent if and only if $g^{(k,p)} = g_{PE}^{(k,p)}$.*

The justification for calling $\sigma^2_{g^{(n,p)}}$ the univariate variance component of $\Phi_{g^{(n,p)}}$ is given by the following lemma with $k = 1$. Examples of $\sigma^2_{g^{(n,p)}}$ are given in Müller and Richter (2016b). Let us remark that, according to Arellano-Valle and Richter (2012), for $k = 1, \ldots, n-1$, the marginal dg $g_{(n)}^{(k,p)}$ of an arbitrary k-dimensional marginal distribution of $\Phi_{g^{(n,p)}}$ is

$$g_{(n)}^{(k,p)}(r) = \frac{\omega_{n-k,p}}{p}\int_{r^p}^{\infty}\left(y - r^p\right)^{\frac{n-k}{p}-1} g^{(n,p)}\left(\sqrt[p]{y}\right)\, dy, \quad r \in [0,\infty),$$

where the variability of the choice of the k marginal variables is established by the permutation invariance of $\Phi_{g^{(n,p)}}$, see Müller and Richter (2016b).

Lemma 2.3 *For $k = 1, \ldots, n-1$,*

$$\sigma^2_{g_{(n)}^{(k,p)}} = \sigma^2_{g^{(n,p)}}.$$

Denoting $M_n^* = [0,\pi)^{\times(n-2)} \times [0, 2\pi)$ and $M_n = [0,\infty) \times M_n^*$ for $n \geq 2$, let the $l_{n,p}$-spherical coordinate transformation $SPH_p^{(n)} : M_n \to \mathbb{R}^n$ be defined as in Richter (2007). Note that $SPH_p^{(n)}$ is bijective a.e. in M_n and its inverse mapping as well as its Jacobian are explicitly known. The next lemma combines and states more precisely some earlier results and introduces a second stochastic representation of random vectors following the distribution $AEC_{n,p}(\mu, D, g^{(k,p)})$.

Lemma 2.4 *Let $X \sim AEC_{n,p}(\mu, D, g^{(k,p)})$ where* rk(D) $= k$. *Then, the random vector X satisfies the stochastic representation*

$$X \stackrel{d}{=} \mu + W_1\sqrt{S_1} \cdot SPH_p^{(k)}(R, \Psi_1, \ldots, \Psi_{k-1})$$

where the nonnegative random variables $R, \Psi_1, \ldots, \Psi_{k-1}$ are mutually stochastic independent having pdfs

$$f_R(r) = \omega_{k,p}\, r^{k-1} g^{(k,p)}(r)\, \mathbb{1}_{[0,\infty)}(r), \quad r \in \mathbb{R},$$

$$f_{\Psi_i}(\psi_i) = \frac{\omega_{k-i,p}}{\omega_{k-i+1,p}} \frac{(\sin(\psi_i))^{k-i-1}}{\left(N_p(\psi_i)\right)^{k-i+1}} \mathbb{1}_{[0,\pi)}(\psi_i), \quad \psi_i \in \mathbb{R}, i = 1, \ldots, k-2,$$

$$f_{\Psi_{k-1}}(\psi_{k-1}) = \frac{1}{\omega_{2,p}} \frac{1}{\left(N_p(\psi_{k-1})\right)^2} \mathbb{1}_{[0,2\pi)}(\psi_{k-1}), \quad \psi_{k-1} \in \mathbb{R}.$$

Here, $N_p(\psi) = \left(|\sin(\psi)|^p + |\cos(\psi)|^p\right)^{1/p}$ and f_Z denotes the pdf of Z.

While the distribution $AEC_{n,p}\big(\mu, D, g^{(n,p)}\big)$ is regular and has a pdf, the distribution $AEC_{n,p}\big(\mu, D, g^{(k,p)}\big)$ is singular if rk(D) $= k < n$ and may be characterized by a pdf-like representation as it was done in Khatri (1968) and Rao (1973, pp. 527-528) in case of singular normal distributions and in Arellano-Valle and Azzalini (2006, Appendix C) in case of singular unified skew-normal distributions. To this end, let $\mathcal{U}_{W_2^T}(\mu) = \{x \in \mathbb{R}^n \colon W_2^T x = W_2^T \mu\}$ be a k-dimensional affine subspace in $\mathbb{R}^n$ and $\lambda^{(k)}_{\mathcal{U}_{W_2^T}(\mu)}$ the k-dimensional Lebesgue measure defined on $\mathcal{U}_{W_2^T}(\mu)$.

Lemma 2.5 *Let $X \sim AEC_{n,p}\big(\mu, D, g^{(k,p)}\big)$ where* rk(D) $= k$. *Then, the distribution of X has pdf-like representation*

$$\frac{1}{\sqrt{d_{i_1} \cdot \ldots \cdot d_{i_k}}} g^{(k,p)}\left(\left|\sqrt{S_1}^{-1} W_1^T (x - \mu)\right|_p\right), \quad x \in \mathbb{R}^n, \tag{2}$$

and

$$W_2^T X = W_2^T \mu \qquad P - a.s. \tag{3}$$

where the function given in (2) is interpreted as pdf in the space $\mathcal{U}_{V_2^T}(\mu)$ in which the whole probability mass of X is concentrated according to Eq. 3.

Lemma 2.5 can be read as follows. For $X \sim AEC_{n,p}\big(\mu, D, g^{(k,p)}\big)$, the orthogonal projection $Y = \Pi_{\mathcal{U}_{W_2^T}(\mu)}(X)$ of X into the subspace $\mathcal{U}_{W_2^T}(\mu)$, and any event $B \in \mathfrak{B}^n$,

$$\begin{aligned} P(X \in B) &= P\left(Y \in \left(B \cap \mathcal{U}_{W_2^T}(\mu)\right)\right) \\ &= \frac{1}{\sqrt{d_{i_1} \cdot \ldots \cdot d_{i_k}}} \int_B g^{(k,p)}\left(\left|\sqrt{S_1}^{-1} W_1^T (x - \mu)\right|_p\right) \lambda^{(k)}_{\mathcal{U}_{W_2^T}(\mu)}(dx) \end{aligned} \tag{4}$$

meaning that the probability measure induced by the random vector X, $P^X = AEC_{n,p}\big(\mu, D, g^{(k,p)}\big)$, is absolutely continuous with respect to $\lambda^{(k)}_{\mathcal{U}_{V_2^T}(\mu)}$. Thus, (2) is the Radon-Nikodym derivative of P^X with respect to the Lebesgue measure $\lambda^{(k)}_{\mathcal{U}_{V_2^T}(\mu)}$ on the subspace $\mathcal{U}_{W_2^T}(\mu)$ of $\mathbb{R}^n$. Because of (4), $g^{(k,p)}$ might be called density-like generator of $AEC_{n,p}\big(\mu, D, g^{(k,p)}\big)$ if $k < n$. In particular, if rk(D) $= n$, then $W_1 = I_{n\times n}$ and W_2 is not defined. Hence, Eq. 3 is not applicable and the function in (2) is the common pdf of the distribution $AEC_{n,p}\big(\mu, D, g^{(n,p)}\big)$. An example is illustrated in Fig. 1.

At the end of this section, our consideration will be slightly extended in order to cover the case $k =$ rk(D) $= 0$ or, equivalently, $D = 0_{n\times n}$. To this end, $AEC_{n,p}\big(\mu, 0_{n\times n}, g^{(0,p)}\big)$ is

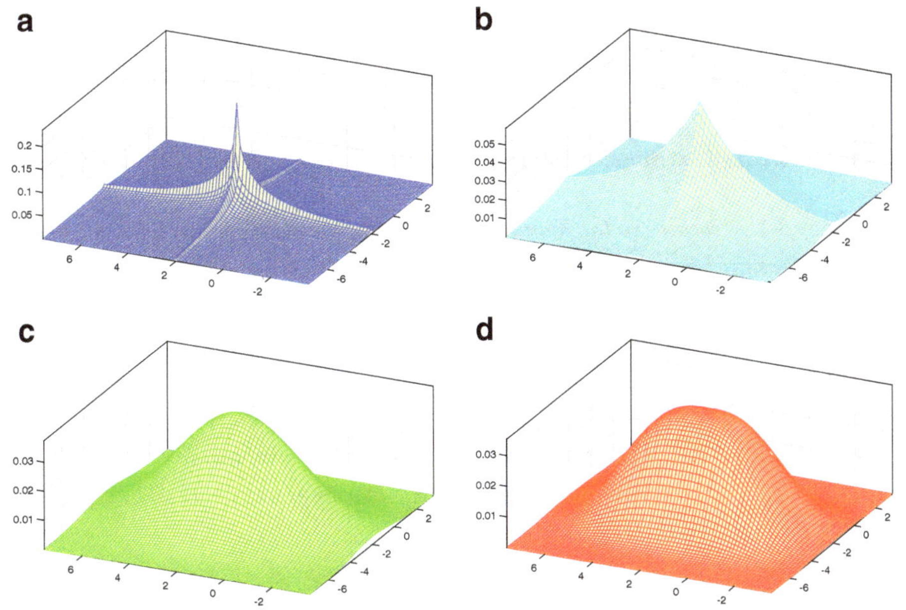

Fig. 1 Pdf of $AN_{2,p}\left(\begin{pmatrix} -2 \\ 2 \end{pmatrix}, \begin{pmatrix} 3 & 0 \\ 0 & 6 \end{pmatrix}\right)$. **a** $p = \frac{1}{2}$ **b** $p = 1$ **c** $p = 2$ **d** $p = 3$

defined to be the Dirac distribution at $\mu \in \mathbb{R}^n$ where $g^{(0,p)}$ is just a symbol to maintain previous notations.

While each finite dimensional distribution (fdd) of an elliptical process is elliptically contoured, in the next section the existence of random processes will be shown whose families of fdds consist of apec distributions.

3 Generalized elliptical random processes

3.1 Main result

In order to state our main result, we call a sequence $g^{(p)} = \left(g^{(k,p)}\right)_{k\in\mathbb{N}}$ of dgs of continuous $l_{k,p}$-symmetric distributions consistent if the following condition is satisfied for any $k \in \mathbb{N}$ and almost all $(x_1, \ldots, x_k)^{\mathrm{T}} \in \mathbb{R}^k$,

$$\int_{-\infty}^{\infty} g^{(k+1,p)}\left(\left|(x_1, \ldots, x_k, x_{k+1})^{\mathrm{T}}\right|_p\right)\, dx_{k+1} = g^{(k,p)}\left(\left|(x_1, \ldots, x_k)^{\mathrm{T}}\right|_p\right). \tag{5}$$

For the particular case of this definition if $p = 2$, we refer to Kano (1994). Moreover, for any nonempty subset I of $\mathbb{R}$, any functions $m\colon I \to \mathbb{R}$ and $S\colon I \to [0, \infty)$, and any sequence $g^{(p)} = \left(g^{(k,p)}\right)_{k\in\mathbb{N}}$ of dgs of continuous $l_{k,p}$-symmetric distributions, let the family

$$\bigcup_{n\in\mathbb{N}} \bigcup_{\substack{\{t_1,\ldots,t_n\}\subseteq I \\ |\{t_1,\ldots,t_n\}|=n}} \Big\{AEC_{n,p}\left(\mu, D, g^{(k,p)}\right) : \mu = (m(t_1), \ldots, m(t_n))^{\mathrm{T}}, \; D = \operatorname{diag}\left(S(t_1), \ldots, S(t_n)\right), k = \operatorname{rk}(\mathrm{D})\Big\}$$

of apec distributions having dgs from $g^{(p)}$ and location and scale functions m and S, respectively, be denoted by $\mathcal{AEC}^I_{g^{(p)}}(m, S)$. Note that strict positivity of S yields a family $\mathcal{AEC}^I_{g^{(p)}}(m, S)$ containing only regular distributions. In difference to this, allowing S to be nonnegative, the family $\mathcal{AEC}^{I}_{g^{(p)}}(m,S)$ consists both of regular and singular distributions. In particular, the univariate member of this family corresponding to $t \in I$ such that $S(t) = 0$ is $AEC_{1,p}\left(m(t), 0, g^{(0,p)}\right)$, i.e. an univariate kapec distribution with $k = 0$.

Theorem 4.1 *If $g^{(p)}$ is consistent, then $\mathcal{AEC}^{I}_{g^{(p)}}(m,S)$ is projective on $\mathbb{R}$.*

Corollary 3.1 *According to the Kolmogorov existence theorem, for any nonempty subset I of $\mathbb{R}$, functions $m\colon I \to \mathbb{R}$ and $S\colon I \to [0,\infty)$, and consistent sequence $g^{(p)}$, Theorem 4.1 yields the existence of a real-valued random process having $\mathcal{AEC}^{I}_{g^{(p)}}(m,S)$ as its family of fdds.*

A random process defined according to Theorem 4.1 and Corollary 3.1 is called random process having apec fdds with location and scale functions m and S, respectively, and sequence $g^{(p)}$ of dgs of continuous $l_{k,p}$-symmetric distributions. Such random process is denoted by $AEC\mathcal{P}_p\left(m, S; g^{(p)}\right)$.

3.2 Sketch of proof

Because of the complexity of the proof of Theorem 4.1, we first give a sketch of its principal ideas. For the outline of details of proof, we refer to "Proof of Theorem 4.1" section. The first step and fundamental argument to prove Theorem 4.1 and thus the existence of the random processes according to Corollary 3.1 is to show that the family $\mathcal{AEC}^{I}_{g^{(p)}}(m,S)$ satisfies Kolmogorov's consistency conditions. Let the set of all finite and nonempty subsets of I be denoted by $\mathcal{H}(I)$, $\mathcal{H}(I) = \{J \subseteq I\colon J \neq \emptyset,\ |J| < \infty\}$. According to Kolmogorov (1956), a family $Q = \{Q_J\}_{\{J \in \mathcal{H}(I)\}}$ of probability measures on $\left(\mathbb{R}^{|J|}, \mathfrak{B}^{|J|}\right)$, $J \in \mathcal{H}(I)$, is projective on $\mathbb{R}$ if the following two conditions are satisfied:

1) For all $t_1, \dots, t_n, t_{n+1} \in I$ being pairwise distinct and $A^{(n)} \in \mathfrak{B}^n$,

$$Q_{\{t_1,\dots,t_n,t_{n+1}\}}\left(A^{(n)} \times E\right) = Q_{\{t_1,\dots,t_n\}}\left(A^{(n)}\right). \tag{6}$$

2) For all $t_1, \dots, t_n \in I$, $A^{(n)} \in \mathfrak{B}^n$ being pairwise distinct and every permutation π of $\{1, \dots, n\}$,

$$Q_{\{t_1,\dots,t_n\}}\left(A^{(n)}\right) = Q_{\{t_{\pi(1)},\dots,t_{\pi(n)}\}}\left(A^{(n)}_{\pi}\right) \tag{7}$$

where $A^{(n)}_{\pi} = \left\{(x_{\pi(1)}, \dots, x_{\pi(n)})^{\mathrm{T}}\colon\ (x_1, \dots, x_n)^{\mathrm{T}} \in A^{(n)}\right\}$.

These two conditions are traditionally formulated using the notion of ordered sets which are assumed to have different elements, i.e. the sets $\{t_1, t_2\}$ and $\{t_2, t_1\}$ differ from each other if $t_1 \neq t_2$, whereas (7) is not required in case of considering unordered sets, see Shiryaev (1996, p. 168).

Condition (6) ensures that specific marginal distributions of elements of the family Q are elements of this family, too. Proving (6) for the family given in Theorem 4.1 will be done in steps two and three. Since both of them are connected with transitions from joint to marginal distributions, we will use the notion of marginal dgs $g^{(m,p)}_{(k)}$, $m = 1, \dots, k-1$, according to "The class of n-dimensional rank-k-continuous axis-aligned p-generalized elliptically contoured distributions" section. Additionally, let $g^{(k,p)}_{(k)} = g^{(k,p)}$. Making use of the marginal dg, in step two an equivalent formulation of (5) is given in the next lemma.

Lemma 3.1 *A sequence $g^{(p)} = \left(g^{(k,p)}\right)_{k\in\mathbb{N}}$ of dgs of continuous $l_{k,p}$-symmetric distributions is consistent if and only if for any $k \in \mathbb{N}$*

$$g^{(k,p)}_{(k+1)} = g^{(k,p)} \qquad a.e.\ in [0,\infty).$$

As a consequence, a sequence $g^{(p)}$ of dgs of continuous $l_{k,p}$-symmetric distributions is consistent if and only if for any $k \in \mathbb{N}$ the marginal dg $g^{(k,p)}_{(k+1)}$ corresponding to the $(k+1)$th element $g^{(k+1,p)}$ of $g^{(p)}$ coincides with the kth element $g^{(k,p)}$. In the third step, for $m \leq n$, m-dimensional marginal distributions of n-dimensional apec distributions are shown to be m-dimensional apec distributions with suitably modified vector and matrix parameters and transitions to marginal dgs.

Lemma 3.2 *For $\mu = (\mu_1,\ldots,\mu_n)^{\mathrm{T}} \in \mathbb{R}^n$ and $D = \mathit{diag}\,(d_1,\ldots,d_n)$ having nonnegative diagonal entries and rank $k \geq 0$, let $X = (X_1,\ldots,X_n)^{\mathrm{T}} \sim AEC_{n,p}(\mu, D, g^{(k,p)})$. Further, let $m \in \mathbb{N}$ with $m \leq n$, $J = \{j_1,\ldots,j_m\} \subseteq \{1,\ldots,n\}$ with $j_1 < \ldots < j_m$, and $X_J = \left(X_{j_1},\ldots,X_{j_m}\right)^{\mathrm{T}}$ the corresponding m-dimensional subvector of X. Then,*

$$X_J \sim AEC_{m,p}\left(\mu_J, D_J, g^{(k_J,p)}_{(k)}\right)$$

where $\mu_J = \left(\mu_{j_1},\ldots,\mu_{j_m}\right)^{\mathrm{T}}$, $D_J = \mathit{diag}\,\left(d_{j_1},\ldots,d_{j_m}\right)$, and $k_J = \mathrm{rk}(\mathrm{D_J}) \geq 0$.

In the final step four, condition (7) ensures that the considered family of probability distributions is big enough in a suitable sense. Its proof in case of $Q = \mathcal{AEC}^I_{g^{(p)}}(m,S)$ is based on the next lemma on distributions of specific linear transformations of random vectors following an apec distribution.

Lemma 3.3 *Let $X \sim AEC_{n,p}(\mu, D, g^{(k,p)})$ with $\mathrm{rk(D)} = k \geq 0$. Then, for every $(n \times n)$-permutation matrix M and every $b \in \mathbb{R}^n$,*

$$\mathfrak{L}(MX+b) = AEC_{n,p}\left(M\mu + b, MDM^{\mathrm{T}}, g^{(k,p)}\right).$$

These sketched four steps to prove Theorem 4.1 are outlined in detail in "Proof of Theorem 4.1" section in reverse order. At the end of the present section, we consider an example of random processes being defined by Theorem 4.1 and Corollary 3.1. More general examples are studied in "Scale mixtures and particular p-generalizations of elliptical random processes" section.

Example 3.1 *Let $g^{(p)}_{PE} = \left(g^{(k,p)}_{PE}\right)_{k\in\mathbb{N}}$ be the sequence of all dgs of multivariate p-generalized Gaussian distributions. Then, the consistency of $g^{(p)}_{PE}$ is immediately seen and for any nonempty subset I of $\mathbb{R}$ and any functions $m\colon I \to \mathbb{R}$ and $S\colon I \to [0,\infty)$, Theorem 4.1 yields the existence of the real-valued random process $AGP_p(m,S)$ having $\mathcal{AEC}^I_{g^{(p)}_{PE}}(m,S)$ as its family of fdds. Such stochastic process is called p-generalized Gaussian process having axis-aligned fdds.*

3.3 Simulation

In order to simulate a random process X having apec fdds, we consider $I = [0,1]$, simulate the marginal vector of X regarding to the equidistant partition $\left\{\frac{i}{200} : i = 0,\ldots,200\right\}$ of $[0,1]$ to get a realization of the random vector $\left(X_0, X_{\frac{1}{200}},\ldots,X_{\frac{199}{200}},X_1\right)^{\mathrm{T}}$. Then, we connect the components of this realization in ascending order by linear functions to get an approximate realization of a trajectory of X. Since components of apec Gaussian distributed random vectors are independent, simulation of the random process $AGP_p(m, S)$ according to the method described above is just the simulation of 201 univariate p-generalized Gaussian variables having specific location and scale parameters. We denote functions on $[0, 1]$ taking constant values 0 and 1 by $0_{[0,1]}$ and $1_{[0,1]}$, respectively. Results of the simulation of the random process $AGP_p(0_{[0,1]}, 1_{[0,1]})$ are shown for $p \in \{\frac{1}{2}, 1, 2, 3\}$ in Fig. 2. Note that scales of axes are highly dependent on the value of p, but also on the specific realization of a trajectory of the process. Moreover, in Fig. 3, the effect different location and scale functions m and S have on simulations of $AGP_3(m, S)$ are shown. See also Appendix 7.2 for several other simulations of random processes having apec fdds.

4 Scale mixtures and particular p-generalizations of elliptical random processes

4.1 Scale mixtures of apec Gaussian distributions

Let be $\mu \in \mathbb{R}^n$, $D \in \mathbb{R}^{n \times n}$ a diagonal matrix having nonnegative diagonal elements and rank $k \geq 0$, V a positive random variable, and $Z \sim AN_{n,p}(0_n, D)$ independent of V.

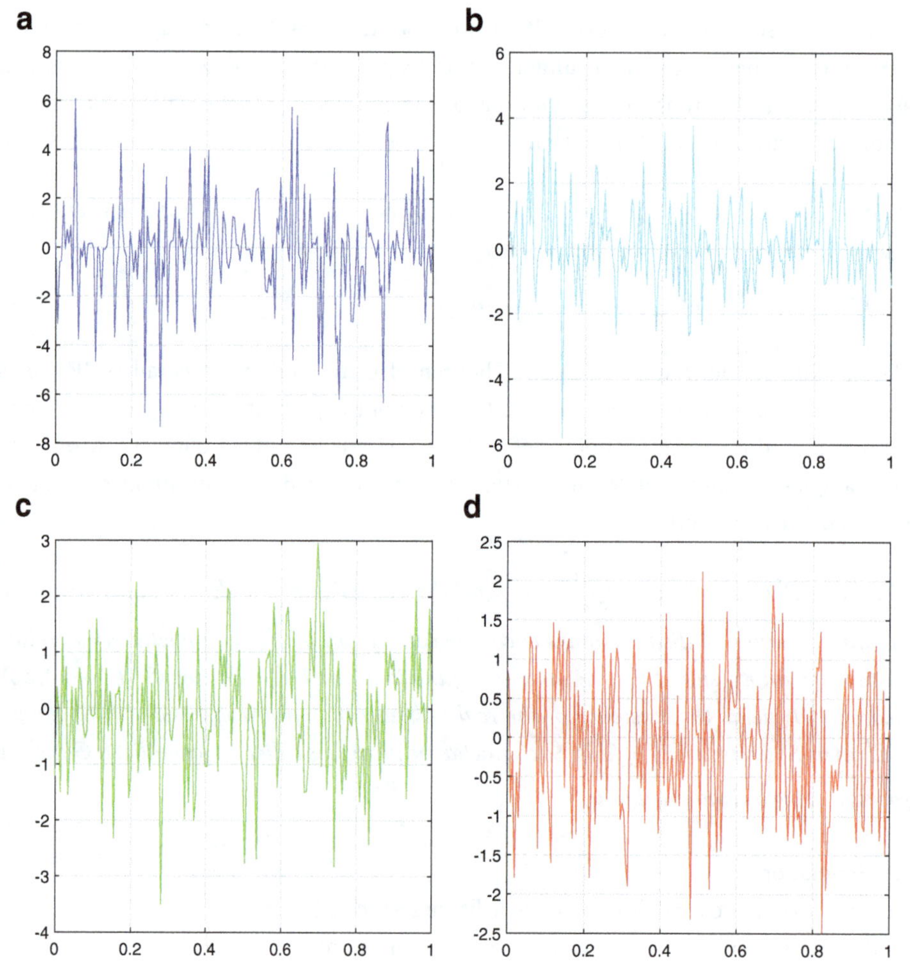

Fig. 2 Simulation of $AGP_p(0_{[0,1]}, 1_{[0,1]})$. **a** $p = \frac{1}{2}$ **b** $p = 1$ **c** $p = 2$ **d** $p = 3$

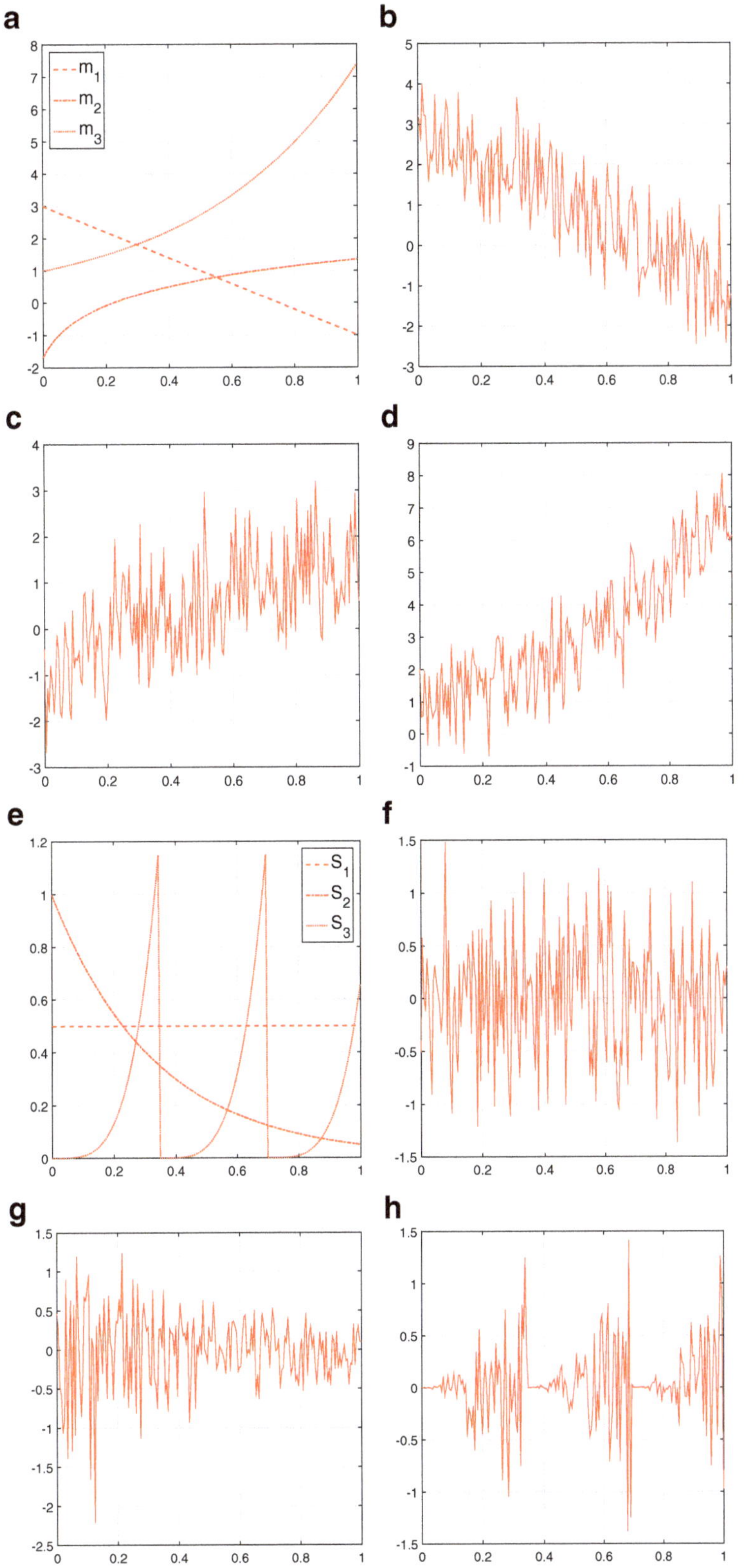

Fig. 3 **a** Location functions. Simulation of $AGP_3(m, 1_{[0,1]})$ **b** $m = m_1$ **c** $m = m_2$ **d** $m = m_3$ **e** Scale functions. Simulation of $AGP_3(0_{[0,1]}, S)$ **f** $S = S_1$ **g** $S = S_2$ **h** $S = S_3$

Furthermore, let G denote the cumulative distribution function (cdf) of V. Then, the distribution of an n-dimensional random vector X satisfying the stochastic representation

$$X \stackrel{d}{=} \mu + V^{-\frac{1}{p}} \cdot Z \tag{8}$$

is called scale mixture of the n-dimensional kapec Gaussian distribution with parameters μ and D and with mixture cdf G and is denoted by $SMAN_{n,p}(\mu, D, G)$.

The particular cases $SMAN_{1,2}(0,1,G)$, $SMAN_{n,2}(\mu, D, G)$ with full rank matrix D, and $SMN_{n,p}(G) = SMAN_{n,p}(0_n, I_{n\times n}, G)$ are introduced in Andrews and Mallows (1974), Lange and Sinsheimer (1993), and Arellano-Valle and Richter (2012), respectively, where numerous equivalent parameterizations of scale mixtures of the common multivariate Gaussian distribution and different notions such as normal/independent distributions or variance mixtures of Gaussian distribution are used. As a first characterization of the class of $SMAN_{n,p}(\mu, D, G)$-distributions, its connections to the classes of $SMN_{n,p}(G)$- and $AEC_{n,p}\left(\mu, D, g^{(k,p)}\right)$-distributions are studied next.

Lemma 4.1 *A random vector $X\colon \Omega \to \mathbb{R}^n$ satisfies $X \sim SMAN_{n,p}(\mu, D, G)$ with $k = \mathrm{rk(D)} \geq 1$ if and only if*

$$X \stackrel{d}{=} \mu + W_1\sqrt{S_1}\tilde{X} \qquad \text{where } \tilde{X} \sim SMN_{k,p}(G).$$

Corollary 4.1 *There holds $SMAN_{n,p}(\mu, D, G) = AEC_{n,p}\left(\mu, D, g^{(k,p)}_{SMN;G}\right)$ with $k = \mathrm{rk(D)}$ and*

$$g^{(k,p)}_{SMN;G}(r) = \left(\frac{p^{1-\frac{1}{p}}}{2\Gamma\left(\frac{1}{p}\right)}\right)^k \int_0^\infty v^{\frac{k}{p}} e^{-\frac{r^p}{p}v}\, dG(v), \quad r \geq 0.$$

As a result, scale mixtures of kapec Gaussian distributions are themselves kapec. Moreover, many properties of such scale mixtures (such as stochastic representations according to Lemmata 2.1 and 2.4) can be obtained from properties of n-dimensional kapec distributions by specializing dgs (according to that given in Corollary 4.1). Additionally, some properties as the first two moments of $SMAN_{n,p}(\mu, D, G)$ can be specialized as follows.

Corollary 4.2 *Let $X \sim SMAN_{n,p}(\mu, D, G)$ with $k = \mathrm{rk(D)} \geq 1$ and $V \sim G$. Then, $\mathbb{E}(X) = \mu$ if $\mathbb{E}\left(V^{-\frac{1}{p}}\right)$ is finite, and $\mathrm{Cov(X)} = \sigma^2_{g^{(k,p)}_{SMN;G}} D$ if $\mathbb{E}\left(V^{-\frac{2}{p}}\right)$ is finite where*

$$\sigma^2_{g^{(n,p)}_{SMN;G}} = p^{\frac{2}{p}} \frac{\Gamma\left(\frac{3}{p}\right)}{\Gamma\left(\frac{1}{p}\right)} \mathbb{E}\left(V^{-\frac{2}{p}}\right).$$

Because of the assertion of the following lemma, $SMAN_{n,p}(\mu, D, G)$ can be called a variance mixture of $AN_{n,p}(\mu, D)$. In the special case of $\mu = 0_n$, $D = I_{n\times n}$ and $p = 2$, the following lemma is covered by the main theorem in Kingman (1972).

Lemma 4.2 *Let $X \sim SMAN_{n,p}(\mu, D, G)$ with $k = \mathrm{rk(D)} \geq 1$ and $V \sim G$ a positive random variable. Then, the conditional distribution of X given $V = v$ satisfies*

$$\mathfrak{L}(X \mid V = v) = AN_{n,p}\left(\mu, v^{-\frac{2}{p}}D\right), \quad v > 0.$$

According to Corollary 4.1, each scale-mixture of the n-dimensional apec Gaussian distribution is an n-dimensional apec distribution with a specific dg. Now, we are interested in which $AEC_{n,p}(\mu, D, g^{(k,p)})$-distributions can be represented by scale mixtures of the n-dimensional apec Gaussian distribution. This question is answered by the following theorem using the notion of completely monotone functions on $[0, \infty)$. A function $f\colon (0, \infty) \to \mathbb{R}$ is called completely monotone if its restriction $f^* = f_{|(0,\infty)}$ to $(0, \infty)$ is completely monotone, i.e. f^* is infinitely often differentiable and satisfies the inequality $(-1)^m \frac{d^m f}{dx^m}(z) \geq 0$ for all $z \in (0, \infty)$ and all $m \in \mathbb{N}_0 = \mathbb{N} \cup \{0\}$, see Sasvári (2013).

Theorem 4.1 *Let $X \sim AEC_{n,p}(\mu, D, g^{(k,p)})$ with D having positive rank k. Then, $X \sim SMAN_{n,p}(\mu, D, G)$ for the cdf G of a suitable positive random variable if and only if the function h defined by $h(y) = g^{(k,p)}(\sqrt[p]{y})$, $y \in [0, \infty)$, is completely monotone.*

For the special case of $n = 1$ and $p = 2$, this theorem is proven in Andrews and Mallows (1974). Subsequently, the Euclidean case $p = 2$ of Theorem 4.1 in arbitrary dimensions ($n \in \mathbb{N}$) is proven in Lange and Sinsheimer (1993) and Gómez-Sánchez-Manzano et al. (2006). Particularly, the proof of Theorem 4.1 given in "Proofs regarding to "Scale mixtures of apec Gaussian distributions" section" section has analogies to that in Andrews and Mallows (1974) and the cdf G of the corresponding mixture distribution can be determined with the help of the inverse Laplace-Stieltjes transform of h.

Corollary 4.3 *Let $X \sim AEC_{n,p}(\mu, D, g^{(k,p)})$ with $k = \mathrm{rk}(D) \geq 1$ and assume that the function $y \mapsto g^{(k,p)}(\sqrt[p]{y})$ is completely monotone in $(0, \infty)$ and has the inverse Laplace-Stieltjes transform α, that is*

$$g^{(k,p)}(\sqrt[p]{y}) = \int_0^\infty e^{-yt}\, d\alpha(t), \quad y > 0.$$

Then, $X \sim SMAN_{n,p}(\mu, D, G)$ and the cdf G of the mixture distribution satisfies the representation

$$\alpha(t) = \frac{p}{\omega_{k,p}\,\Gamma\left(\frac{k}{p}\right)} \int_1^t z^{\frac{k}{p}}\, dG(pz), \quad t > 0.$$

Moreover, the probability law corresponding to G is regular and has pdf f_G if and only if α is absolutely continuous with pdf f_α and both pdfs are connected by the equation

$$f_G(s) = \omega_{k,p}\,\Gamma\left(\frac{k}{p}\right) p^{\frac{k}{p}-2} \cdot s^{-\frac{k}{p}} f_\alpha\left(\frac{s}{p}\right) \mathbb{1}_{(0,\infty)}(s), \quad s \in \mathbb{R}.$$

Example 4.1 *An n-dimensional apec Gaussian distribution is a scale mixture of itself with the Dirac distribution in 1 being the mixture distribution. The cdf of this Dirac distribution is the indicator function $s \mapsto \mathbb{1}_{(1,\infty)}(s)$.*

Example 4.2 *The n-dimensional kapec Pearson-type VII distribution with parameters M and ν, $M > \frac{k}{p}$ and $\nu > 0$, and dg*

$$g^{(k,p)}_{PT7;M,\nu}(r) = \left(\frac{p}{2\Gamma\left(\frac{1}{p}\right)}\right)^k \frac{\Gamma(M)}{\nu^{\frac{k}{p}}\Gamma\left(M - \frac{k}{p}\right)} \left(1 + \frac{r^p}{\nu}\right)^{-M}, \quad r \geq 0,$$

is the scale mixture of the n-dimensional kapec Gaussian distribution where the mixture distribution is the Gamma distribution $\Gamma_{M-\frac{k}{p},\frac{\nu}{p}}$ *having pdf*

$$f_G(s) = \frac{\left(\frac{\nu}{p}\right)^{M-\frac{k}{p}}}{\Gamma\left(M-\frac{k}{p}\right)} s^{M-\frac{k}{p}-1} e^{-\frac{\nu}{p}s} \mathbb{1}_{(0,\infty)}(s), \quad s \in \mathbb{R}.$$

Example 4.3 *A special case of the preceding one is the n-dimensional kapec Student-t distribution with parameter* $\nu > 0$ *and dg* $g_{St;\nu}^{(k,p)} = g_{PT7;\frac{\nu+k}{p},\nu}^{(k,p)}$ *being that of the scale mixture of the n-dimensional kapec Gaussian distribution with mixture distribution* $\Gamma_{\frac{\nu}{p},\frac{\nu}{p}}$.

Example 4.4 *The n-dimensional kapec Slash distribution with parameter* $\nu > 0$ *is defined as the scale mixture of the n-dimensional kapec Gaussian distribution with mixture distribution having pdf* $f_\nu^{Sl}(y) = \nu y^{\nu-1}\mathbb{1}_{(0,1)}(y)$, $y \in \mathbb{R}$.

4.2 Scale mixed *p*-generalized Gaussian processes having axis-aligned fdds

Let $g_{SMN;G}^{(p)} = \left(g_{SMN;G}^{(k,p)}\right)_{k\in\mathbb{N}}$ denote the sequence of dgs of scale mixtures of k-dimensional p-generalized Gaussian distributions with one and the same mixture cdf G with

$$G \text{ is independent of the index variable} k \text{ in } g_{SMN;G}^{(k,p)}. \tag{9}$$

According to Examples 4.1-4.4, representatives of mixture cdfs satisfying (9) are the Dirac distribution in 1, $\Gamma_{\frac{\nu}{p},\frac{\nu}{p}}$ as well as the distribution with pdf f_ν^{Sl}, whereas the cdf of the distribution $\Gamma_{M-\frac{k}{p},\frac{\nu}{p}}$ does not generally satisfy (9).

Lemma 4.3 *For the cdf G of a positive random variable satisfying (9), the sequence* $g_{SMN;G}^{(p)}$ *is consistent.*

Throughout this section, again let I be a nonempty subset of $\mathbb{R}$, $m: I \to \mathbb{R}$ and $S: I \to [0,\infty)$ arbitrary functions, and G the cdf of a positive random variable satisfying (9). Then, a random process having apec fdds, location and scale functions m and S, respectively, and the sequence $g_{SMN;G}^{(p)}$ of dgs exists according to Theorem 4.1 and Corollary 3.1. Such process is called a scale mixed p-generalized Gaussian process having axis-aligned fdds with location function m, scale function S and mixture cdf G and is denoted by $SMAGP_p(m,S,G)$, thus $AECP_p\left(m,S;g_{SMN;G}^{(p)}\right) = SMAGP_p(m,S,G)$. The motivation and justification of this naming is given by a characterizing property of such processes in Theorem 4.2 below.

On the one hand, for the special case $p = 2$, the class of $SMAGP_p(m,S,G)$-processes is equal to the class of spherically invariant random processes having axis-aligned fdds which is defined in Vershik (1964). Moreover, it is shown implicitly in Yao (1973) and explicitly in Kano (1994) that a sequence $g^{(2)}$ is consistent if and only if all elements of $g^{(2)}$ are dgs of scale mixtures of multivariate Gaussian distributions regarding to one and the same mixture distribution. On the other hand, for general $p > 0$, if the mixture distribution is chosen to be the Dirac distribution in 1, then $SMAGP_p\left(m,S,\mathbb{1}_{(1,\infty)}\right) = AGP_p(m,S)$. Furthermore, for any $\nu > 0$, let us denote the cdf of $\Gamma_{\frac{\nu}{p},\frac{\nu}{p}}$ and of the distribution with pdf f_ν^{Sl} by $G_{\nu/p}^{St}$ and G_ν^{Sl}, respectively. Then, $SMAGP_p\left(m,S,G_{\nu/p}^{St}\right)$ and $SMAGP_p(m,S,G_\nu^{Sl})$ are called p-generalized Student-t and p-generalized Slash process having axis-aligned fdds with location function m, scale function S and parameter ν, and are denoted by $AStP_p(m,S,\nu)$ and $ASlP_p(m,S,\nu)$, respectively.

Because of its construction, a scale mixed p-generalized Gaussian process X having axis-aligned fdds with location function m, scale function S and mixture cdf G is uniquely determined except for equivalence and denoted $X \sim SMAGP_p(m,S,G)$. Next, we state a characteristic representation of the random process $SMAGP_p(m,S,G)$ with the help of a specific p-generalized Gaussian process providing the motivation for the naming of such process.

Theorem 4.2 *Let $X = \{X_t\}_{t\in I}$ be a scale mixed p-generalized Gaussian process having axis-aligned fdds, $X \sim SMAGP_p(m,S,G)$. Then, X and $Y = \left\{m(t) + V^{-\frac{1}{p}} Z_t\right\}_{t\in I}$ are equivalent where the p-generalized Gaussian process $Z = \{Z_t\}_{t\in I} \sim AGP_p(0_I, S)$ having axis-aligned fdds is independent of the random variable $V \sim G$.*

For $p = 2$ and $m = 0_I$, Theorem 4.2 is proven in Wise and jun Gallagher (1978). In the sequel, using the characteristic representation from Theorem 4.2, we determine expectation and covariance functions as well as stationarity properties of the random process $SMAGP_p(m,S,G)$. Since $SMAGP_p(m,0_I,G)$ equals a.s. the location function m, the results of Theorems 4.3 and 4.5 below are restricted to non-vanishing scale functions, i.e. $S \neq 0_I$. Let $g^{(p)}_{SMN;G} = \left(g^{(k,p)}_{SMN;G}\right)_{k\in\mathbb{N}}$ be the sequence of dgs of scale mixtures of multivariate p-generalized Gaussian distributions with one and the same mixture cdf G such that $\mathbb{E}\left(V^{-\frac{2}{p}}\right)$ is finite where $V \sim G$. Then, because of Corollary 4.2 and property (9) of G, the sequence $\left(\sigma^2_{g^{(k,p)}_{SMN;G}}\right)_{k\in\mathbb{N}}$ of the corresponding univariate variance components is constant and an arbitrary element of it is subsequently denoted by $\sigma^2_{g^{(p)}_{SMN;G}}$.

Theorem 4.3 *Let $X = \{X_t\}_{t\in I} \sim SMAGP_p(m,S,G)$ with $S \neq 0_I$ and $V \sim G$. Then, the expectation function of the random process X exists and is equal to the location function m if $\mathbb{E}\left(V^{-\frac{1}{p}}\right)$ is finite. If $\mathbb{E}\left(V^{-\frac{2}{p}}\right)$ is finite, X is a second order random process with covariance function $\Gamma : I \times I \to \mathbb{R}$ given by*

$$\Gamma(s,t) = \begin{cases} \sigma^2_{g^{(p)}_{SMN;G}} \cdot S(t) & \text{if } s = t \\ 0 & \text{else} \end{cases}.$$

As announced before, different stationarity properties of the random process $SMAGP_p(m,S,G)$ are studied now. We start with a result on strict stationarity.

Theorem 4.4 *Let $X = \{X_t\}_{t\in I} \sim SMAGP_p(m,S,G)$. Then, X is strictly stationary if and only if m and S are constant.*

In the following theorem, we additionally take the notions of weak stationarity and white noise into consideration.

Theorem 4.5 *Let $X = \{X_t\}_{t\in I} \sim SMAGP_p(m,S,G)$, $V \sim G$, $\mu \in \mathbb{R}$ and $\delta > 0$. Then, the following statements are equivalent:*

1) *There holds $m(t) = \mu$ and $S(t) = \delta$ for all $t \in I$ and $\mathbb{E}\left(V^{-\frac{2}{p}}\right)$ is finite.*

2) *X is strictly stationary,* $\mathbb{E}\left(V^{-\frac{2}{p}}\right)$ *is finite, the expectation function of X attains the constant value* μ *and the covariance function* Γ *of X satisfies* $\Gamma(t,t) = \sigma^2_{g^{(p)}_{SMN;G}}\delta$ *for all* $t \in I$ *and* $\Gamma(s,t) = 0$ *for all* $s, t \in I$ *with* $s \neq t$.
3) *X is weakly stationary with constant expectation* μ *and covariance function* Γ *given by* $\Gamma(s,t) = K(s-t)$ *where K satisfies* $K(0) = \sigma^2_{g^{(p)}_{SMN;G}}\delta$ *and* $K(h) = 0$ *for all* $h \in \{s-t\colon s,t \in I\}\backslash\{0\}$.
4) *X is white noise with expectation* μ *and variance* $\sigma^2_{g^{(p)}_{SMN;G}}\delta$.

Finally, we establish the closedness of the class of all scale mixed p-generalized Gaussian processes having axis-aligned fdds with respect to linear transformations.

Theorem 4.6 *Let* $\{X_t\}_{t\in I} \sim SMAGP_p(m,S,G)$, $b\colon I \to \mathbb{R}$ *and* $\gamma\colon I \to \mathbb{R}$. *Then,*

$$\{\gamma(t)X_t + b(t)\}_{t\in I} \sim SMAGP_p\left(\gamma m + b, \gamma^2 S, G\right),$$

where $[\gamma m+b] : I \to \mathbb{R}$ *and* $\left[\gamma^2 S\right] : I \to [0,\infty)$ *are defined by* $[\gamma m + b](t) = \gamma(t)m(t) + b(t)$, $t \in I$, *and* $\left[\gamma^2 S\right](t) = (\gamma(t))^2 S(t)$, $t \in I$, *respectively.*

5 Proofs

5.1 Proofs of Lemmata 2.3 and 2.5

Before proving Lemma 2.3, we state a part of its proof as the following remark on the p-generalized surface content of p-generalized spheres of different dimensions in relation with a certain integral.

Remark 5.1 *For every* $\nu \in \mathbb{N}$ *with* $\nu \geq 2$ *and every* $\kappa \in \{1,\ldots,\nu-1\}$,

$$\frac{\omega_{\kappa,p}\,\omega_{\nu-\kappa,p}}{\omega_{\nu,p}} \int\limits_0^{\frac{\pi}{2}} \frac{(\cos(\psi))^{\nu-\kappa-1}\,(\sin(\psi))^{\kappa-1}}{\left((\sin(\psi))^p + (\cos(\psi))^p\right)^{\frac{\nu}{p}}}\,d\varphi = 1.$$

According to Richter (2009), the left hand side of the above equation is the limit of the cdf of the p-generalized Fisher statistic $T_{\nu-\kappa,\kappa}(p)$ evaluated at t as $t \to \infty$. Hence, Remark 5.1 follows from the elementary fact that the cdf of a univariate random variable evaluated at t tends to one as $t \to \infty$.

Proof of Lemma 2.3 Let $k \in \{1,\ldots,n-1\}$ be fixed. Denoting $\tau_{k,p} = \frac{\Gamma\left(\frac{3}{p}\right)\Gamma\left(\frac{k}{p}\right)}{\Gamma\left(\frac{1}{p}\right)\Gamma\left(\frac{k+2}{p}\right)}$, using integral transformation $y = z^p + r^p$ with $\frac{dy}{dz} = pz^{p-1}$ and finally renaming r and z by x and y, respectively, we get

$$\begin{aligned}
\sigma^2_{g^{(k,p)}_{(n)}} &= \tau_{k,p}\,\omega_{k,p} \int\limits_0^\infty r^{k+1} g^{(k,p)}_{(n)}(r)\,dr \\
&= \tau_{k,p}\,\omega_{k,p}\,\omega_{n-k,p} \int\limits_0^\infty \int\limits_0^\infty x^{k+1} y^{n-k-1} g^{(n,p)}\left(\sqrt[p]{x^p + y^p}\right) dy\,dx.
\end{aligned}$$

Applying the $l_{2,p}$-spherical coordinate transformation $x = r\frac{\cos(\psi)}{N_p(\psi)}$ and $y = r\frac{\sin(\psi)}{N_p(\psi)}$ with $N_p(\psi) = \left(|\sin(\psi)|^p + |\cos(\psi)|^p\right)^{1/p}$ and $\frac{d(x,y)}{d(r,\psi)} = \frac{r}{N_p^2(\psi)}$, see Richter (2007), Fubini's theorem and Remark 5.1 for $\nu = n+2$ and $\kappa = n-k$, it follows

$$\begin{aligned}\sigma^2_{g^{(k,p)}_{(n)}} &= \tau_{k,p}\,\omega_{k,p}\,\omega_{n-k,p} \int_0^\infty \int_0^{\frac{\pi}{2}} r^{n+1} g^{(n,p)}(r) \frac{(\cos(\psi))^{k+1}(\sin(\psi))^{n-k-1}}{\left((\sin(\psi))^p + (\cos(\psi))^p\right)^{\frac{n+2}{p}}}\, d\psi\, dr \\ &= \sigma^2_{g^{(n,p)}}.\end{aligned}$$

□

Proof of Lemma 2.5 It follows from $D = \left(W_1\sqrt{S_1}\right)\left(W_1\sqrt{S_1}\right)^{\mathrm{T}}$ and Lemma 2.1 that

$$W_1^{\mathrm{T}} X \overset{d}{=} W_1^{\mathrm{T}}\left(\mu + R\cdot\left(W_1\sqrt{S_1}\right)U_p^{(k)}\right) = W_1^{\mathrm{T}}\mu + R\cdot\sqrt{S_1}U_p^{(k)}.$$

Since $\sqrt{S_1}$ has full rank k, $W_1^{\mathrm{T}} X$ is k-dimensional rank-k-continuous p-generalized elliptically contoured distributed with parameters $W_1^{\mathrm{T}}\mu$ and S_1 and with dg $g^{(k,p)}$. By definition of this distribution, for $Y \sim \Phi_{g^{(k,p)}}$, it follows $W_1^{\mathrm{T}} X \overset{d}{=} W_1^{\mathrm{T}}\mu + \sqrt{S_1}Y$. Thus, $W_1^{\mathrm{T}} X$ has pdf

$$\frac{1}{\sqrt{d_{i_1}\cdot \ldots \cdot d_{i_k}}}\, g^{(k,p)}\left(\left|\sqrt{S_1}^{-1}\left(z - W_1^{\mathrm{T}}\mu\right)\right|_p\right), \quad z \in \mathbb{R}^k.$$

Since the columns of W_1 and W_2 together build an orthonormal basis of $\mathbb{R}^n$, we have $W_2^{\mathrm{T}} W_1 = 0_{(n-k)\times k}$ and

$$W_2^{\mathrm{T}} X \overset{d}{=} W_2^{\mathrm{T}}\mu + R\cdot W_2^{\mathrm{T}} W_1\sqrt{S_1}U_p^{(k)} = W_2^{\mathrm{T}}\mu \qquad \text{a.s.}$$

Thus, the orthogonal projection $Y = \Pi_{U_{W_2^{\mathrm{T}}}(\mu)}(X)$ of X into the space $U_{W_2^{\mathrm{T}}}(\mu)$ has the pdf

$$\frac{1}{\sqrt{d_{i_1}\cdot \ldots \cdot d_{i_k}}}\, g^{(k,p)}\left(\left|\sqrt{S_1}^{-1} W_1^{\mathrm{T}}(x - \mu)\right|_p\right), \quad x \in \mathbb{R}^n,$$

and the orthogonal projection of X into the orthogonal complement of $U_{W_2^{\mathrm{T}}}(\mu)$ has probability mass zero. □

5.2 Proof of Theorem 4.1

We start with considering a particular case of Lemma 3.3.

Lemma 5.1 *Let* $X \sim AEC_{n,p}\left(\mu, D, g^{(k,p)}\right)$ *where* $\mathrm{rk(D)} = k \geq 1$. *Then, for every* $(n \times n)$*-permutation matrix* M *and every* $b \in \mathbb{R}^n$,

$$\mathfrak{L}(MX + b) = AEC_{n,p}\left(M\mu + b, MDM^{\mathrm{T}}, g^{(k,p)}\right).$$

Proof With notations from "The class of n-dimensional rank-k-continuous axis-aligned p-generalized elliptically contoured distributions" section,

$$MX + b \overset{d}{=} (M\mu + b) + MW_1\sqrt{S_1}Y \qquad \text{where } Y \sim \Phi_{g^{(k,p)}}.$$

Since $MW_1\sqrt{S_1} \in \mathbb{R}^{n\times k}$ arises from $W_1\sqrt{S_1}$ by interchanging rows, it has rank k. Thus,

$$\begin{aligned}\mathfrak{L}(MX+b) &= AEC_{n,p}\left(M\mu+b, \left(MW_1\sqrt{S_1}\right)\left(MW_1\sqrt{S_1}\right)^{\mathrm{T}}, g^{(k,p)}\right)\\ &= AEC_{n,p}\left(M\mu+b, MDM^{\mathrm{T}}, g^{(k,p)}\right).\end{aligned}$$

□

Proof of Lemma 3.3 Because of Lemma 5.1, only the case $k=0$ has to be considered. In this case, $X \sim AEC_{n,p}\left(\mu, 0_{n\times n}, g^{(0,p)}\right)$, i.e. X follows the Dirac distribution in μ. Thus,

$$MX+b \overset{d}{=} M\mu+b \qquad P-\text{a.s.}$$

and $\mathfrak{L}(MX+b) = AEC_{n,p}\left(M\mu+b, M0_{n\times n}M^{\mathrm{T}}, g^{(0,p)}\right)$ because of $0_{n\times n} = M0_{n\times n}M^{\mathrm{T}}$. □

Denoting the cardinality of the set A by $|A|$, we continue with studying a particular case of Lemma 3.2.

Lemma 5.2 *Let be* $X = (X_1,\dots,X_n)^{\mathrm{T}} \sim AEC_{n,p}\left(\mu, D, g^{(k,p)}\right)$ *where* $\mu = (\mu_1,\dots,\mu_n)^{\mathrm{T}} \in \mathbb{R}^n$ *and assume* $D = \operatorname{diag}(d_1,\dots,d_n)$ *has nonnegative diagonal entries and rank* $k \geq 1$*. Further, let* $m \in \mathbb{N}$ *with* $m \leq n$, $J = \{j_1,\dots,j_m\} \subseteq \{1,\dots,n\}$ *with* $j_1 < \dots < j_m$ *and* $\left|\{\eta \in \{1,\dots,m\}\colon d_{j_\eta} > 0\}\right| \geq 1$*. Then, the corresponding m-dimensional subvector* $X_J = \left(X_{j_1},\dots,X_{j_m}\right)^{\mathrm{T}}$ *of X satisfies*

$$X_J \sim AEC_{m,p}\left(\mu_J, D_J, g_{(k)}^{(k_J,p)}\right)$$

where $\mu_J = \left(\mu_{j_1},\dots,\mu_{j_m}\right)^{\mathrm{T}}$, $D_J = \operatorname{diag}\left(d_{j_1},\dots,d_{j_m}\right)$ *and* $k_J = \mathrm{rk(D_J)} \geq 1$.

Proof Starting from the equation $X_J = \Gamma X$ where

$$\Gamma = \begin{pmatrix} e_{j_1}^{(n)\mathrm{T}} \\ \vdots \\ e_{j_m}^{(n)\mathrm{T}} \end{pmatrix} \in \mathbb{R}^{m\times n}$$

and using notations from "The class of n-dimensional rank-k-continuous axis-aligned p-generalized elliptically contoured distributions" section, it follows that

$$\Gamma W_1\sqrt{S_1} = \begin{pmatrix} e_{j_1}^{(n)\mathrm{T}} \\ \vdots \\ e_{j_m}^{(n)\mathrm{T}} \end{pmatrix}\left(\sqrt{d_{i_1}}e_{i_1}^{(n)} \cdots \sqrt{d_{i_k}}e_{i_k}^{(n)}\right) = \begin{pmatrix} f(1) \\ \vdots \\ f(n) \end{pmatrix} \in \mathbb{R}^{m\times k}$$

where

$$f(\eta) = \begin{cases} \sqrt{d_{i_l}}e_l^{(k)\mathrm{T}} & \text{if } j_\eta = i_l \text{ for an } l \in \{1,\dots,k\} \\ 0_k^{\mathrm{T}} & \text{else} \end{cases}, \qquad \eta = 1,\dots,m.$$

Thus, for $Y = (Y_1,\dots,Y_k) \sim \Phi_{g^{(k,p)}}$, we get

$$\Gamma W_1\sqrt{S_1}Y = \begin{pmatrix} h(1) \\ \vdots \\ h(m) \end{pmatrix} \in \mathbb{R}^m$$

where

$$h(\eta) = \sum_{l=1}^{k} \sqrt{d_{i_l}} Y_l \delta_{i_l j_\eta} = \begin{cases} \sqrt{d_{i_l}} Y_l & \text{if } j_\eta = i_l \text{ for an } l \in \{1, \dots, k\} \\ 0 & \text{else} \end{cases},$$

$\eta = 1, \dots, m$. Now, let

$$K = \{l \in \{1, \dots, k\} \colon i_l = j_\eta \text{ for an } \eta \in \{1, \dots, m\}\}. \tag{10}$$

Then, $|K| = \left|\left\{\eta \in \{1, \dots, m\} \colon \sigma_{j_\eta}^2 > 0\right\}\right| \geq 1$ and the matrix $\Gamma W_1 \sqrt{S_1}$ has $k - |K|$ zero columns. Since each non-zero column is the product of a positive constant with a unit vector in $\mathbb{R}^m$, the vector $\Gamma W_1 \sqrt{S_1} Y$ consists of $|K|$ different components of Y multiplied by positive constants and of $m - |K|$ zeros. Subsequently, put $K = \{l_1, \dots, l_{|K|}\}$ where $l_1 < l_2 < \dots < l_{|K|}$ is an increasing enumeration of the elements of K and let

$$M = \begin{pmatrix} \psi(1) \\ \vdots \\ \psi(m) \end{pmatrix} \in \mathbb{R}^{m \times |K|}$$

be a matrix consisting of the row vectors

$$\psi(\eta) = \begin{cases} \sqrt{d_{i_{l_\kappa}}} e_\kappa^{(|K|)^{\mathrm{T}}} & \text{if } j_\eta = i_{l_\kappa} \text{ for a } \kappa \in \{1, \dots, |K|\} \\ 0_{|K|}^{\mathrm{T}} & \text{else} \end{cases}, \quad \eta = 1, \dots, m.$$

Then, for $B \in \mathfrak{B}^m$, $Y \sim \Phi_{g^{(k,p)}}$ and $Z \sim \Phi_{g_{(k)}^{(|K|,p)}}$, it follows that

$$P\left(\Gamma W_1 \sqrt{S_1} Y \in B\right) = P\left(\begin{pmatrix} h(1) \\ \vdots \\ h(m) \end{pmatrix} \in B\,,\ Y_l \in \mathbb{R} \text{ for all} l \in \{1, \dots, k\} \backslash K\right)$$
$$= P(MZ \in B)$$

and, because of (1) and $\mathrm{rk(M)} = |K|$,

$$X_J = \Gamma X \stackrel{d}{=} \mu_J + \Gamma W_1 \sqrt{S_1} Y \stackrel{d}{=} \mu_J + MZ$$
$$\sim EC_{m,p}\left(\mu_J, MM^{\mathrm{T}}, g_{(k)}^{(|K|,p)}\right).$$

Note that M can be extended to $\Gamma W_1 \sqrt{S_1}$ by adding zero columns without changing the rank. Moreover, $MM^{\mathrm{T}} = \left(\Gamma W_1 \sqrt{S_1}\right)\left(\Gamma W_1 \sqrt{S_1}\right)^{\mathrm{T}} = \Gamma D \Gamma = D_J$ and $|K| = \mathrm{rk(M)} = \mathrm{rk}\left(\mathrm{MM}^{\mathrm{T}}\right) = \mathrm{rk(D_J)} = k_J$. Summarizing all, we have

$$\mathfrak{L}(X_J) = AEC_{m,p}\left(\mu_J, D_J, g_{(k)}^{(k_J,p)}\right).$$

□

Proof of Lemma 3.2 If $k = 0$, $X \sim AEC_{n,p}\left(\mu, 0_{n \times n}, g^{(0,p)}\right)$ and $J = \{j_1, \dots, j_m\} \subseteq \{1, \dots, n\}$ with $j_1 < \dots < j_m$. In this case, $X_J = \mu_J$ P-a.s. and

$$X_J \sim EC_{m,p}\left(\mu_J, 0_{m \times m}, g_{(0)}^{(0,p)}\right)$$

because the symbols $g^{(0,p)}$ and $g_{(k)}^{(0,p)}$ can be switched for a $k \in \mathbb{N} \cup \{0\}$. Now, let $X \sim AEC_{n,p}\left(\mu, D, g^{(k,p)}\right)$ where $D = \operatorname{diag}(d_1, \dots, d_n)$ has nonnegative diagonal elements and positive rank k, and let $J = \{j_1, \dots, j_m\} \subseteq \{1, \dots, n\}$ be an index set such that $j_1 < \dots < j_m$ and $\left|\left\{\eta \in \{1, \dots, m\} \colon \sqrt{d_{j_\eta}} > 0\right\}\right| \geq 1$. Then, Lemma 5.2 yields the assertion. Finally, let $X \sim AEC_{n,p}\left(\mu, D, g^{(k,p)}\right)$ where $D = \operatorname{diag}(d_1, \dots, d_n)$ has nonnegative diagonal elements and positive rank k but, now, where $J = \{j_1, \dots, j_m\} \subseteq \{1, \dots, n\}$

is an index set such that $j_1 < \ldots < j_m$ and $\left|\left\{\eta \in \{1,\ldots,m\}\colon \sqrt{d_{j_\eta}} > 0\right\}\right| = 0$. Using the notation from the proof of Lemma 5.2, the set K defined in (10) has cardinality $|K| = \left|\left\{\eta \in \{1,\ldots,m\}\colon \sqrt{d_{j_\eta}} > 0\right\}\right| = 0$. Because of this, $\Gamma W_1\sqrt{S_1}$ is equal to the $(m \times k)$ zero matrix and the distribution of $\Gamma W_1\sqrt{S_1}Y$ for $Y \sim \Phi_{g^{(k,p)}}$ is concentrated in 0_m. Since $D_J = \operatorname{diag}\left(d_{j_1},\ldots,d_{j_m}\right) = 0_{m\times m}$, $k_J = \operatorname{rk}(D_J) = 0$ and $X_J = \Gamma X \stackrel{d}{=} \mu_J + \Gamma W_1\sqrt{S_1}Y$ for $Y \sim \Phi_{g^{(k,p)}}$, it follows

$$X_J = \mu_J \qquad P-\text{a.s.},$$

i.e. $X_J \sim AEC_{m,p}\left(\mu_J, 0_{m\times m}, g_{(k)}^{(0,p)}\right)$. □

Proof of Lemma 3.1 Starting from (5) and using the transformation $\tilde{y} = |x|_p^p + y^p$, for $x \in \mathbb{R}^k$, we get

$$\begin{aligned} g^{(k,p)}\left(|x|_p\right) &= \int_{-\infty}^{\infty} g^{(k+1,p)}\left(\sqrt[p]{|x|_p^p + |y|^p}\right) dy \\ &= 2\int_{0}^{\infty} g^{(k+1,p)}\left(\sqrt[p]{|x|_p^p + y^p}\right) dy \\ &= \frac{2}{p}\int_{|x|_p^p}^{\infty} \left(\tilde{y} - |x|_p^p\right)^{\frac{1}{p}-1} g^{(k+1,p)}\left(\sqrt[p]{\tilde{y}}\right) d\tilde{y}. \end{aligned}$$

Because of $\omega_{1,p} = 2$,

$$g^{(k,p)}\left(|x|_p\right) = g_{(k+1)}^{(k,p)}\left(|x|_p\right), \quad x \in \mathbb{R}^k.$$

□

Proof of Theorem 4.1 For $n \in \mathbb{N}$ and arbitrary elements $t_1,\ldots,t_n,t_{n+1}$ of I, let $\mu^{(n+1)} = (m(t_1),\ldots,m(t_n),m(t_{n+1}))^{\mathrm{T}} \in \mathbb{R}^{n+1}$ and assume $D^{(n+1)} = \operatorname{diag}\left(S(t_1),\ldots,S(t_n),S(t_{n+1})\right)$ to have rank k. Further, let $Q_{\{t_1,\ldots,t_n,t_{n+1}\}}(\cdot) = AEC_{n+1,p}\left(\cdot \mid \mu^{(n+1)}, D^{(n+1)}, g^{(k,p)}\right) \in \mathcal{AEC}^{I}_{g^{(p)}}(m,S)$ be the probability measure induced by a random vector following the $(n+1)$-dimensional kapec distribution with parameters $\mu^{(n+1)}$ and $D^{(n+1)}$ and dg $g^{(k,p)} \in g^{(p)}$ if $k > 0$ and symbol $g^{(0,p)}$ if $k = 0$, respectively. By Lemma 3.2, it follows

$$\begin{aligned} Q_{\{t_1,\ldots,t_n,t_{n+1}\}}(A \times \mathbb{R}) &= AEC_{n+1,p}\left(A \times \mathbb{R} \mid \mu^{(n+1)}, D^{(n+1)}, g^{(k,p)}\right) \\ &= AEC_{n,p}\left(A \mid \mu^{(n)}, D^{(n)}, g_{(k)}^{(\kappa,p)}\right), \quad A \in \mathfrak{B}^n, \end{aligned}$$

where $\mu^{(n)} = (m(t_1),\ldots,m(t_n))^{\mathrm{T}}$ and $D^{(n)} = \operatorname{diag}\left(S(t_1),\ldots,S(t_n)\right)$ with $\kappa = \operatorname{rk}\left(D^{(n)}\right) \in \{k-1,k\}$. Furthermore, using Lemma 3.1 if $\kappa > 0$ and recalling the exchangeability of symbols $g_{(k)}^{(0,p)}$ and $g^{(0,p)}$ (to maintain the notation as in the proof of Lemma 3.2) if $\kappa = 0$, we have

$$Q_{\{t_1,\ldots,t_n,t_{n+1}\}}(A \times \mathbb{R}) = AEC_{n,p}\left(A \mid \mu^{(n)}, D^{(n)}, g^{(\kappa,p)}\right) = Q_{\{t_1,\ldots,t_n\}}(A).$$

Therefore, the marginal probability measure $Q_{\{t_1,\ldots,t_n\}}$ of $Q_{\{t_1,\ldots,t_{n+1}\}}$ corresponds to the element $AEC_{n,p}\left(\mu^{(n)}, D^{(n)}, g^{(\kappa,p)}\right)$ of $\mathcal{AEC}^{I}_{g^{(p)}}(m,S)$ and, thus, the Kolmogorov consistency condition (6) is satisfied. Now, let π be a permutation of $\{1,\ldots,n\}$ and M the corresponding permutation matrix. Additionally, let $Q_{\{t_1,\ldots,t_n\}}(\cdot) = AEC_{n,p}\left(\cdot \mid \mu, D, g^{(\kappa,p)}\right) \in$

$\mathcal{AEC}^I_{g^{(p)}}(m,S)$ be the probability measure induced by a random vector X with $X \sim AEC_{n,p}\big(\mu, D, g^{(\kappa,p)}\big)$ where $\mu = \mu^{(n)} = (m(t_1),\dots,m(t_n))^{\mathrm{T}}$ and $D = D^{(n)} = \operatorname{diag}(S(t_1),\dots,S(t_n))$ with $\kappa = \operatorname{rk}(\mathrm{D})$. Then, $Q_{\{t_{\pi(1)},\dots,t_{\pi(n)}\}}$ is induced by MX and, according to Lemma 3.3,

$$Q_{\{t_{\pi(1)},\dots,t_{\pi(n)}\}}(\cdot) = AEC_{n,p}\Big(\cdot \mid M\mu, MDM^{\mathrm{T}}, g^{(\kappa,p)}\Big).$$

If $\kappa = 0$, then $D = MDM^{\mathrm{T}} = 0_{n\times n}$. In this case, $Q_{\{t_1,\dots,t_n\}}$ and $Q_{\{t_{\pi(1)},\dots,t_{\pi(n)}\}}$ are Dirac measures in μ and $M\mu$, respectively, and, for $A \in \mathfrak{B}^n$,

$$\begin{aligned}Q_{\{t_{\pi(1)},\dots,t_{\pi(n)}\}}(A_\pi) = Q_{\{t_{\pi(1)},\dots,t_{\pi(n)}\}}(MA) &= \mathbb{1}_{MA}(M\mu)\\ &= \mathbb{1}_A(\mu) = Q_{\{t_1,\dots,t_n\}}(A).\end{aligned}$$

Thus, the Kolmogorov consistency condition (7) is satisfied if $\kappa = 0$. Now, let $\kappa > 0$. Using the notations of matrices S_1, W_1 and W_2 from "The class of n-dimensional rank-k-continuous axis-aligned p-generalized elliptically contoured distributions" section, $\big(W_1\sqrt{S_1}\big)\big(W_1\sqrt{S_1}\big)^{\mathrm{T}}$ is a decomposition of D with $W_1\sqrt{S_1} \in \mathbb{R}^{n\times\kappa}$ and $\operatorname{rk}\big(\mathrm{W}_1\sqrt{\mathrm{S}_1}\big) = \kappa$ and the columns of W_2 are a basis of the kernel of D. Consequently, on the one hand, $\big(MW_1\sqrt{S_1}\big)\big(MW_1\sqrt{S_1}\big)^{\mathrm{T}}$ is a corresponding decomposition of MDM^{T} with $MW_1\sqrt{S_1} \in \mathbb{R}^{n\times\kappa}$ and $\operatorname{rk}\big(\mathrm{MW}_1\sqrt{\mathrm{S}_1}\big) = \kappa$ since left multiplication of $W_1\sqrt{S_1}$ by permutation matrix M only interchanges columns and leaves the rank unchanged. On the other hand, the columns of MW_2 build a basis of the kernel of MDM^{T},

$$\mathcal{U}_{(MW_2)^{\mathrm{T}}}(M\mu) = \big\{My \in \mathbb{R}^n\colon W_2^{\mathrm{T}}y = W_2^{\mathrm{T}}\mu\big\} = M\cdot\mathcal{U}_{W_2^{\mathrm{T}}}(\mu),$$

and

$$\lambda^{(\kappa)}_{\mathcal{U}_{(MW_2)^{\mathrm{T}}}(M\mu)}(\cdot) = \lambda^{(\kappa)}_{M\cdot\mathcal{U}_{W_2^{\mathrm{T}}}(\mu)}(\cdot) = \lambda^{(\kappa)}_{\mathcal{U}_{W_2^{\mathrm{T}}}(\mu)}\big(f_{M^{\mathrm{T}}}(\cdot)\big)$$

where $f_{M^{\mathrm{T}}}$ is defined by $f_{M^{\mathrm{T}}}(x) = M^{\mathrm{T}}x$, $x \in \mathbb{R}^n$. Finally, for $A \in \mathfrak{B}^n$, Eq. 4 resulting from the pdf-like representation of an n-dimensional apec distribution together with the transformation $y = f_{M^{\mathrm{T}}}(x)$ having the Jacobian $|\det(M)| = 1$ yield

$$\begin{aligned}&Q_{\{t_{\pi(1)},\dots,t_{\pi(n)}\}}(A_\pi)\\ &= AEC_{n,p}\Big(MA \mid M\mu, MDM^{\mathrm{T}}, g^{(\kappa,p)}\Big)\\ &= \frac{1}{\det\big(\sqrt{S_1}\big)}\int\limits_{MA} g^{(\kappa,p)}\Big(\Big|\sqrt{S_1}^{-1}(MW_1)^{\mathrm{T}}(x - M\mu)\Big|_p\Big)\,\lambda^{(\kappa)}_{\mathcal{U}_{(MW_2)^{\mathrm{T}}}(M\mu)}(dx)\\ &= \frac{1}{\det\big(\sqrt{S_1}\big)}\int\limits_{f_{M^{\mathrm{T}}}(A)} g^{(\kappa,p)}\Big(\Big|\sqrt{S_1}^{-1}W_1^{\mathrm{T}}\big(f_{M^{\mathrm{T}}}(x) - \mu\big)\Big|_p\Big)\,\lambda^{(\kappa)}_{\mathcal{U}_{W_2^{\mathrm{T}}}(\mu)}\big(f_{M^{\mathrm{T}}}(dx)\big)\\ &= \frac{1}{\det\big(\sqrt{S_1}\big)}\int\limits_{A} g^{(\kappa,p)}\Big(\Big|\sqrt{S_1}^{-1}W_1^{\mathrm{T}}(y - \mu)\Big|_p\Big)\,\lambda^{(\kappa)}_{\mathcal{U}_{W_2^{\mathrm{T}}}(\mu)}(dy)\\ &= Q_{\{t_1,\dots,t_n\}}(A).\end{aligned}$$

Thus, the Kolmogorov consistency condition (7) is satisfied in case $\kappa > 0$, too. □

5.3 Proofs regarding to "Scale mixtures of apec Gaussian distributions" section

Proof of Lemma 4.1 For a positive random variable $V \sim G$, because of (1) and (8), we have $X \overset{d}{=} \mu + V^{-\frac{1}{p}}\cdot Z$ where $Z \sim AN_{n,p}(0_n, \Sigma)$. It follows $X \overset{d}{=} \mu + V^{-\frac{1}{p}}\cdot W_1\sqrt{S_1}\tilde{Z}$ where

$\tilde{Z} \sim N_{k,p}$ and $X \stackrel{d}{=} \mu + W_1\sqrt{S_1}V^{-\frac{1}{p}} \cdot \tilde{Z}$ where $\tilde{Z} \sim N_{k,p}$. Thus, $X \stackrel{d}{=} \mu + W_1\sqrt{S_1}\tilde{X}$ where $\tilde{X} \sim SMN_{k,p}(G)$. □

Proof of Corollary 4.1 In the case $k \geq 1$, the assertion follows from Lemma 4.1, Eq. 1 and the identity $SMN_{k,p}(G) = \Phi_{g^{(k,p)}_{SMN;G}}$ from Arellano-Valle and Richter (2012). In the case $k = 0$, $Z = 0_n$ a.s. in (8). Therefore, $X \sim SMAN_{n,p}(\mu, 0_{n\times n}, G)$, that is X has Dirac distribution in μ. Thus, $X \sim AEC_{n,p}\left(\mu, 0_{n\times n}, g^{(0,p)}_{SMN;G}\right)$, where $g^{(0,p)}_{SMN;G}$ is just a symbol to maintain notations. □

Proof of Corollary 4.2 By Corollary 4.1, the assertion follows from Lemma 2.2 with the specific dg $g^{(k,p)}_{SMN;G}$. Particularly, for $m \in \{1, 2\}$, $I_{k+m}\left(g^{(k,p)}_{SMN;G}\right)$ is finite if and only if $\mathbb{E}\left(V^{-\frac{m}{p}}\right)$ is finite. To see this, consider

$$I_{k+m}\left(g^{(k,p)}_{SMN;G}\right) = C_p^k \int_0^\infty r^{k+m-1} \int_0^\infty v^{\frac{k}{p}} e^{-\frac{r^p}{p}v}\, dG(v)\, dr$$

$$= C_p^k p^{\frac{k+m}{p}-1}\Gamma\left(\frac{k+m}{p}\right)\int_0^\infty v^{-\frac{m}{p}}\, dG(v).$$

Here, we used notation $C_p = \frac{p^{1-\frac{1}{p}}}{2\Gamma\left(\frac{1}{p}\right)}$, two times Fubini's theorem and changed variables $s = \frac{r^p}{p}v$ with $\frac{dr}{ds} = p^{\frac{1}{p}-1}v^{-\frac{1}{p}}s^{\frac{1}{p}-1}$. Finally, by Lemma 2.2, the specific univariate variance component is

$$g^{(k,p)}_{SMN;G} = p^{\frac{2}{p}}\frac{\Gamma\left(\frac{3}{p}\right)}{\Gamma\left(\frac{1}{p}\right)}\mathbb{E}\left(V^{-\frac{2}{p}}\right).$$

□

Proof of Lemma 4.2 Let $Z \sim AN_{n,p}(0_n, D)$ and assume Z to be independent of V. Making use of Eq. 8 and exploiting the independence of Z and V, for all $B \in \mathfrak{B}^n$ and $v > 0$,

$$P(X \in B \mid V = v) = P\left(\left(\mu + V^{-\frac{1}{p}}Z\right) \in B \mid V = v\right) = P\left(\left(\mu + v^{-\frac{1}{p}}W_1\sqrt{S_1}\tilde{Z}\right) \in B\right)$$

where $\tilde{Z} \sim N_{k,p}$. Because of $\left(v^{-\frac{1}{p}}W_1\sqrt{S_1}\right)\left(v^{-\frac{1}{p}}W_1\sqrt{S_1}\right)^{\mathrm{T}} = v^{-\frac{2}{p}}D$ with $v^{-\frac{1}{p}}W_1\sqrt{S_1} \in \mathbb{R}^{n\times k}$ and $\mathrm{rk}\left(v^{-\frac{1}{p}}W_1\sqrt{S_1}\right) = k$, according to (1) with dg $g^{(k,p)}_{PE}$, the assertion follows from

$$\mathfrak{L}\left(\mu + v^{-\frac{1}{p}}W_1\sqrt{S_1}\tilde{Z}\right) = AN_{n,p}\left(\mu, v^{-\frac{2}{p}}D\right), \quad v > 0.$$

□

Before proving the general statement of Theorem 4.1, we prove the following particular one.

Lemma 5.3 *Let $X \sim \Phi_{g^{(k,p)}}$. Then, $X \sim SMN_{k,p}(G)$ for the cdf G of a suitable positive random variable if and only if the function h defined by $h(y) = g^{(k,p)}\left(\sqrt[p]{y}\right)$, $y \in [0, \infty)$, is completely monotone.*

Proof Throughout this proof, let $X \sim \Phi_{g^{(k,p)}}$. If $X \sim SMN_{k,p}(G)$ for the cdf G of a suitable positive random variable, according to Corollary 4.1, $g^{(k,p)} = g^{(k,p)}_{SMN;G}$ and

$$h(y) = g^{(k,p)}_{SMN;G}\left(y^{\frac{1}{p}}\right) = C_p^k \int_0^\infty v^{\frac{k}{p}} e^{-\frac{y}{p}v}\, dG(v), \quad y \geq 0,$$

where $C_p = \frac{p^{1-\frac{1}{p}}}{2\Gamma\left(\frac{1}{p}\right)}$. Because of

$$\frac{d^m h}{dy^m}(y) = (-1)^m \frac{C_p^k}{p^m} \int_0^\infty v^{\frac{k}{p}+m} e^{-\frac{y}{p}v}\, dG(v), \quad y > 0$$

for all $m \in \mathbb{N} \cup \{0\}$, h is completely monotone in $[0, \infty)$. Now, let $h = g^{(k,p)}(\sqrt[p]{\cdot})$ be completely monotone on $[0, \infty)$. According to Hausdorff-Bernstein-Widder theorem, see Widder (1946), h is representable as the Laplace-Stieltjes transform of a nondecreasing function α, i.e.

$$h(y) = \int_0^\infty e^{-yt}\, d\alpha(t), \quad 0 < y < \infty,$$

and the integral converges for all $0 < y < \infty$. Additionally, denoting,

$$\beta(t) = \int_1^t C_p^{-k} v^{-\frac{k}{p}}\, d\alpha\left(\frac{1}{p}v\right), \quad t > 0,$$

Stieltjes integral properties yield

$$h(y) = \int_0^\infty e^{-y\left(\frac{1}{p}v\right)}\, d\alpha\left(\frac{1}{p}v\right) = C_p^k \int_0^\infty v^{\frac{k}{p}} e^{-\frac{1}{p}yv}\, d\beta(v), \quad y > 0.$$

Thus,

$$g^{(k,p)}(r) = h(r^p) = C_p^k \int_0^\infty v^{\frac{k}{p}} e^{-\frac{r^p}{p}v}\, d\beta(v), \quad r > 0.$$

Consequently, it remains to show that G defined by $G(v) = \beta(v) - \lim_{t \searrow 0} \beta(t)$, $v > 0$, is the cdf of a positive random variable. Note that G is nondecreasing since α has this property. Hence,

$$G(v_2) - G(v_1) = \beta(v_2) - \beta(v_1) = \int_{v_1}^{v_2} C_p^{-k} v^{-\frac{k}{p}}\, d\alpha\left(\frac{1}{p}v\right) \geq 0, \quad 0 < v_1 \leq v_2.$$

It remains to show that $1 = \lim_{v \to \infty} G(v) - \lim_{t \searrow 0} G(t)$. To this end, let $\tilde{g}^{(k,p)}(z,r) = C_p^k \int_{z^{-1}}^{z} v^{\frac{k}{p}} e^{-\frac{r^p}{p}v}\, d\beta(v)$, $1 < z < \infty$, denote a left and right truncated version of $g^{(k,p)}$. Using Fubini's theorem, change of variables $s = r\sqrt[p]{v}$ with $\frac{dr}{ds} = v^{-\frac{1}{p}}$ and the equality $\omega_{k,p} I_k\left(g^{(k,p)}_{PE}\right) = 1$, we have

$$\begin{aligned}\int_0^\infty r^{k-1}\tilde{g}^{(k,p)}(z,r)\,dr &= \int_{z^{-1}}^{z} v^{\frac{k}{p}} \int_0^\infty C_p^k r^{k-1} e^{-\frac{r^p}{p}v}\,dr\,d\beta(v)\\ &= \int_{z^{-1}}^{z} I_k\left(g_{PE}^{(k,p)}\right)\,d\beta(v)\\ &= \frac{1}{\omega_{k,p}}\left(\beta(z)-\beta\left(z^{-1}\right)\right),\quad z>1.\end{aligned}$$

Because $\tilde{g}^{(k,p)}(z,r)$ is a nonnegative function and $g^{(k,p)}(r) = h(r^p)$, it follows $0 \le r^{k-1}\tilde{g}^{(k,p)}(z,r) \le r^{k-1}g^{(k,p)}(r)$ for all $z > 1$ and $r > 0$. Furthermore, because of its structure as well as its nonnegativity, for all $r > 0$, the function $r^{k-1}\tilde{g}^{(k,p)}(z,r)$ is monotonically increasing in variable z and converges to $r^{k-1}g^{(k,p)}(r)$ as $z \to \infty$. Thus, the monotone convergence theorem of Beppo Levi yields the desired

$$\begin{aligned}\lim_{v\to\infty} G(v) - \lim_{t\searrow 0} G(t) &= \lim_{z\to\infty}\left(G(z)-G\left(z^{-1}\right)\right)\\ &= \lim_{z\to\infty}\beta(z)-\beta\left(z^{-1}\right)\\ &= \lim_{z\to\infty}\omega_{k,p}\int_0^\infty r^{k-1}\tilde{g}^{(k,p)}(z,r)\,dr\\ &= \omega_{k,p} I_k\left(g_{PE}^{(k,p)}\right)\end{aligned}$$

Therefore, G defined by $G(v) = \beta(v) - \lim_{t\searrow 0}\beta(t)$, $v > 0$, is the cdf of a positive random variable. Finally, because of

$$\begin{aligned}g^{(k,p)}(r) = h\left(r^p\right) &= C_p^k \int_0^\infty v^{\frac{k}{p}} e^{-\frac{r^p}{p}v}\,d\beta(v)\\ &= C_p^k \int_0^\infty v^{\frac{k}{p}} e^{-\frac{r^p}{p}v}\,d\left(\beta(v)-\lim_{t\searrow 0}\beta(t)\right)\\ &= C_p^k \int_0^\infty v^{\frac{k}{p}} e^{-\frac{r^p}{p}v}\,dG(v),\quad r>0,\end{aligned}$$

we have $g^{(k,p)} = g_{SMN;G}^{(k,p)}$ a.e. in $[0,\infty)$ and $X \sim SMN_{k,p}(G)$. □

Before proving the general statement of Corollary 4.3, we prove the following particular one.

Corollary 5.1 *Let $X \sim \Phi_{g^{(k,p)}}$ and assume that $g^{(k,p)}\left(\sqrt[p]{\cdot}\right)$ is completely monotone in $(0,\infty)$ and has inverse Laplace-Stieltjes transform α, $g^{(k,p)}\left(\sqrt[p]{y}\right) = \int_0^\infty e^{-yt}\,d\alpha(t)$, $y > 0$. Then, $X \sim SMN_{k,p}(G)$ and the mixture cdf G satisfies the representation*

$$\alpha(t) = \frac{p}{\omega_{k,p}\Gamma\left(\frac{k}{p}\right)}\int_1^t z^{\frac{k}{p}}\,dG(pz),\quad t>0.$$

Moreover, the probability distribution corresponding to G is regular and has pdf f_G if and only if α is absolutely continuous and has pdf f_α where both pdfs are connected by

$$f_G(s) = \omega_{k,p}\,\Gamma\left(\frac{k}{p}\right)p^{\frac{k}{p}-2}\cdot s^{-\frac{k}{p}}f_\alpha\left(\frac{s}{p}\right)\mathbb{1}_{(0,\infty)}(s), \quad s\in\mathbb{R}.$$

Proof of Corollary 5.1 According to the second part of the proof of Lemma 5.3, on the one hand, there exists a nondecreasing function α satisfying $g^{(k,p)}(\sqrt[p]{y}) = \int_0^\infty e^{-yt}\,d\alpha(t)$, $y>0$. Since $X\sim SMN_{k,p}(G)$ for a suitable mixture cdf G, on the other hand, we have $g^{(k,p)}(\sqrt[p]{y}) = g^{(k,p)}_{SMN;G}(\sqrt[p]{y}) = C_p^k\int_0^\infty v^{\frac{k}{p}}e^{-\frac{1}{p}yv}\,dG(v), y>0$. Then, changing variables $z=\frac{1}{p}v$,

$$\int_0^\infty e^{-yt}\,d\alpha(t) = C_p^k\int_0^\infty v^{\frac{k}{p}}e^{-\frac{1}{p}yv}\,dG(v) = \frac{p}{\omega_{k,p}\,\Gamma\left(\frac{k}{p}\right)}\int_1^t z^{\frac{k}{p}}e^{-zv}\,dG(pz)$$

and using properties of Stieltjes integrals, it turns out that

$$\alpha(t) = \frac{p}{\omega_{k,p}\,\Gamma\left(\frac{k}{p}\right)}\int_1^t z^{\frac{k}{p}}\,dG(pz), \quad t>0.$$

Hence, regularity properties of probability distributions regarding to G and α are equivalent. Moreover, since f_G is the pdf of a positive random variable and there holds

$$f_\alpha(t) = \frac{p}{\omega_{k,p}\,\Gamma\left(\frac{k}{p}\right)}t^{\frac{k}{p}}\frac{dG(pt)}{dt} = \frac{p^2}{\omega_{k,p}\,\Gamma\left(\frac{k}{p}\right)}t^{\frac{k}{p}}\cdot f_G(pt),$$

$t>0$, according to the above equation involving f_G, it follows $f_G(s)=0$ for all $s\le 0$ and

$$f_G(s) = \omega_{k,p}\,\Gamma\left(\frac{k}{p}\right)p^{-2}\left(\frac{s}{p}\right)^{-\frac{k}{p}}f_\alpha\left(\frac{s}{p}\right), \quad s>0.$$

□

Proof of Theorem 4.1 Let $X\sim SMAN_{n,p}(\mu,D,G)$ for the cdf G of a positive random variable. Then, $g^{(k,p)} = g^{(k,p)}_{SMN;G}$ according to Corollary 4.1 and $g^{(k,p)}_{SMN;G}(\sqrt[p]{\cdot})$ is completely monotone in $[0,\infty)$ according to Lemma 5.3. Vice versa, let $X\sim AEC_{n,p}\big(\mu,D,g^{(k,p)}\big)$ with $k=\mathrm{rk(D)}$ and assume $h(\cdot)=g^{(k,p)}\big(\sqrt[p]{\cdot}\big)$ to be completely monotone in $[0,\infty)$. Then, according to Lemma 5.3, $g^{(k,p)}$ is the dg of a distribution from

$$\{SMN_{k,p}(G)\colon G \text{ is the cdf of a positive random variable}\},$$

i.e. $\Phi_{g^{(k,p)}} = SMN_{k,p}(G)$ for a suitable cdf G of a positive random variable. Thus, $X\stackrel{d}{=}\mu+W_1\sqrt{S_1}\tilde{X}$ where $\tilde{X}\sim SMN_{k,p}(G)$ because of (1) and, finally, $X\sim SMAN_{n,p}(\mu,\Sigma,G)$ because of Lemma 4.1. □

Proof of Corollary 4.3 According to (1), for $X\sim AEC_{n,p}\big(\mu,D,g^{(k,p)}\big)$ with $\mathrm{rk(D)}=k$, we have

$$X\stackrel{d}{=}\mu+W_1\sqrt{S_1}\tilde{X} \qquad \text{where } \tilde{X}\sim\Phi_{g^{(k,p)}}.$$

Because $g^{(k,p)}(\sqrt[p]{\cdot})$ is completely monotone in $(0,\infty)$, Corollary 5.1 yields $\tilde{X} \sim SMN_{k,p}(G)$ as well as

$$\alpha(t) = \frac{p}{\omega_{k,p}\Gamma\left(\frac{k}{p}\right)} \int_1^t z^{\frac{k}{p}}\, dG(pz), \quad t > 0,$$

where α is the inverse Laplace-Stieltjes transform of $g^{(k,p)}(\sqrt[p]{\cdot})$. The relationship between the pdfs f_G and f_α follows in analogy to the second part of the proof of Corollary 5.1. $\square$

5.4 Proofs regarding to "Scale mixed p-generalized Gaussian processes having axis-aligned fdds" section

Proof of Lemma 4.3 Using Fubini's theorem and changing variables $y = v^{-\frac{1}{p}}z$ with $\frac{dy}{dz} = v^{-\frac{1}{p}}$, for all $k \in \mathbb{N}$ and $r \geq 0$, there holds

$$\int_{-\infty}^{\infty} g_{SMN;G}^{(k+1,p)}\left(\sqrt[p]{r^p + |y|^p}\right) dy = 2\int_0^{\infty} g_{SMN;G}^{(k+1,p)}\left(\sqrt[p]{r^p + y^p}\right) dy$$
$$= \left(C_p^k \int_0^{\infty} v^{\frac{k}{p}} e^{-\frac{r^p}{p}v}\, dG(v)\right) \frac{p^{1-\frac{1}{p}}}{\Gamma\left(\frac{1}{p}\right)} \int_0^{\infty} e^{-\frac{z^p}{p}}\, dz.$$

Since G is independent of k, see (9), the first factor on the right hand side of the latter equation is equal to the value of the dg $g_{SMN;G}^{(k,p)}$ evaluated at r. Furthermore, the corresponding second factor equals 1. Thus, the assertion follows with

$$\int_{-\infty}^{\infty} g_{SMN;G}^{(k+1,p)}\left(\left|(x_1,\ldots,x_k,x_{k+1})^{\mathrm{T}}\right|_p\right) dx_{k+1} = \int_{-\infty}^{\infty} g_{SMN;G}^{(k+1,p)}\left(\sqrt[p]{r^p + |y|^p}\right) dy$$
$$= g_{SMN;G}^{(k,p)}(r)$$
$$= g_{SMN;G}^{(k,p)}\left(\left|(x_1,\ldots,x_k)^{\mathrm{T}}\right|_p\right)$$

for all $k \in \mathbb{N}$ and $(x_1,\ldots,x_k)^{\mathrm{T}} \in \mathbb{R}^k$ where $r = \left|(x_1,\ldots,x_k)^{\mathrm{T}}\right|_p$ and $y = x_{k+1}$. $\square$

Proof of Theorem 4.2 Let $n \in \mathbb{N}$ and $J = \{t_1,\ldots,t_n\}$ an arbitrary subset of I having n elements. Then, $J \in \mathcal{H}(I)$, and $AEC_{n,p}\left(\mu, D, g_{SMN;G}^{(k,p)}\right)$ with $\mu = (m(t_1),\ldots,m(t_n))^{\mathrm{T}}$ and $D = \operatorname{diag}(S(t_1),\ldots,S(t_n))$ where $k = \operatorname{rk}(D)$ is the fdd of the random process X corresponding to $X_J = \left(X_{t_1},\ldots,X_{t_n}\right)^{\mathrm{T}}$. Moreover, $AN_{n,p}(0_n, D)$ and $\mathfrak{L}\left(\mu^{(n)} + V^{-\frac{1}{p}}Z_J\right)$ are the fdds of Z regarding to $Z_J = \left(Z_{t_1},\ldots,Z_{t_n}\right)^{\mathrm{T}}$ and of Y regarding to $Y_J = \left(Y_{t_1},\ldots,Y_{t_n}\right)^{\mathrm{T}}$, respectively. By (8) and Corollary 4.1,

$$\mathfrak{L}\left(\mu^{(n)} + V^{-\frac{1}{p}}Z_J\right) = SMAN_{n,p}(\mu, D, G) = AEC_{n,p}\left(\mu, D, g_{SMN;G}^{(k,p)}\right)$$

for all $n \in \mathbb{N}$ and every set $J = \{t_1,\ldots,t_n\} \in \mathcal{H}(I)$ with $|J| = n$. Thus, the random processes X and Y are equivalent meaning that they have one and the same family of fdds. $\square$

Before we prove Theorem 4.3, we consider the following special case of it. To this end, notice that the sequence $\left(\sigma^2_{g_{PE}^{(k,p)}}\right)_{k\in\mathbb{N}}$ of all univariate variance components of multivariate p-generalized spherical Gaussian distributions equals the sequence $\left(\sigma^2_{g_{SMN;G}^{(p)}}\right)_{k\in\mathbb{N}}$ with

$G = \mathbb{1}_{(1,\infty)}$. Thus, according to the paragraph before Theorem 4.3, it is constant. Subsequently, an arbitrary element of it is denoted by $\sigma^2_{g^{(p)}_{PE}}$ and satisfies $\sigma^2_{g^{(p)}_{PE}} = \sigma^2_{g^{(p)}_{SMN;\mathbb{1}_{(1,\infty)}}} = p^{\frac{2}{p}} \frac{\Gamma\left(\frac{3}{p}\right)}{\Gamma\left(\frac{1}{p}\right)}$.

Lemma 5.4 *Let $Z = \{Z_t\}_{t\in I} \sim AGP_p(m, S)$. Then, Z is a second order random process, its expectation function is equal to m, and its covariance function $\Gamma : I \times I \to \mathbb{R}$ is given by*

$$\Gamma(s,t) = \begin{cases} \sigma^2_{g^{(p)}_{PE}} \cdot S(t) & \textit{if } s = t \\ 0 & \textit{else} \end{cases}.$$

The proof of this lemma follows immediately from Corollaries 4.1 and 4.2 and is therefore omitted, here.

Proof of Theorem 4.3 Let $Z = \{Z_t\}_{t\in I} \sim AGP_p(0_I, S)$ be independent of $V \sim G$. Then, according to Theorem 4.2, X is equivalent to the random process $Y = \left\{m(t) + V^{-\frac{1}{p}} Z_t\right\}_{t\in I}$ and $V^{-\frac{1}{p}}$ and Z_t as well as $V^{-\frac{2}{p}}$ and $Z_s Z_t$ are independent for all indices $s, t \in I$. Because of

$$\mathbb{E}(X_t) = \mathbb{E}\left(m(t) + V^{-\frac{1}{p}} Z_t\right) = m(t) + \mathbb{E}\left(V^{-\frac{1}{p}}\right)\mathbb{E}(Z_t)$$

and $\mathbb{E}(Z_t) = 0$ for all $t \in I$ according to Lemma 5.4, the value of expectation of X_t exists and is equal to $m(t)$ if $\mathbb{E}(V^{-\frac{1}{p}})$ is finite. Furthermore, for all $t \in I$, the independence $V^{-\frac{2}{p}}$ and $Z_t Z_t = Z_t^2$ yields

$$\mathbb{E}\left(X_t^2\right) = (m(t))^2 + 2m(t)\mathbb{E}\left(V^{-\frac{1}{p}}\right)\mathbb{E}(Z_t) + \mathbb{E}\left(V^{-\frac{2}{p}}\right)\mathbb{E}\left(Z_t^2\right).$$

As Z is a second order random process, X is a second order random process, too, if $\mathbb{E}\left(V^{-\frac{2}{p}}\right)$ is finite. In this case, for all $s, t \in I$, using the independence of $V^{-\frac{2}{p}}$ and $Z_s Z_t$ as well as the covariance function of a centered p-generalized Gaussian process Z having axis-aligned fdds with scale function S from Lemma 5.4, it follows

$$\begin{aligned}\Gamma(s,t) = CovX_s, X_t &= \mathbb{E}\left(V^{-\frac{2}{p}}\right)\mathbb{E}\left(Z_s Z_t\right) \\ &= \begin{cases} \mathbb{E}\left(V^{-\frac{2}{p}}\right)\sigma^2_{g^{(p)}_{PE}} \cdot S(t) & \text{if } s = t \\ 0 & \text{else} \end{cases}.\end{aligned}$$

The equation $\mathbb{E}\left(V^{-\frac{2}{p}}\right)\sigma^2_{g^{(p)}_{PE}} = \sigma^2_{g^{(p)}_{SMN;G}}$ yields the asserted result. □

Proof of Theorem 4.4 Let X be strictly stationary. Then, for all $t_1 \in I$ and $h \in H_{t_1} = \{h \in \mathbb{R}: t_1 + h \in I\}$, the distributions $SMAN_{1,p}(m(t_1), S(t_1), G)$ of X_{t_1} and $SMAN_{1,p}(m(t_1+h), S(t_1+h), G)$ of X_{t_1+h} are equal. If $S(t_1) = 0$, the distribution of X_{t_1} is the univariate Dirac distribution in $m(t_1)$ which can be considered to be the scale mixture of the univariate kapec Gaussian distribution with $k = 0$, location parameter $m(t_1)$ and scale parameter 0. Therefore, for all $h \in H_{t_1}$, $\mathfrak{L}\left(X_{t_1+h}\right)$ is the univariate Dirac distribution in $m(t_1)$, too, and it follows $S(t_1 + h) = 0 = S(t_1)$ for all $h \in H_{t_1}$. Thus, $S = 0_I$. Since $\mathfrak{L}\left(X_{t_1+h}\right) = SMAN_{1,p}(m(t_1+h), 0, G)$ is defined to be the Dirac distribution in $m(t_1+h)$, it follows $m(t_1 + h) = m(t_1)$ for all $h \in H_{t_1}$, i.e. m is constant on I. If $S(t_1) > 0$, according to "The class of n-dimensional rank-k-continuous axis-aligned p-generalized elliptically contoured distributions" section, for all $t_1 \in I$ and $h \in H_{t_1}$, $\mathfrak{L}(X_{t_1})$ and $\mathfrak{L}(X_{t_1+h})$ have pdfs

$$f_{X_{t_1}}(x) = \frac{C_p}{\sqrt{S(t_1)}} g^{(1,p)}_{SMN;G}\left(\left|\frac{x - m(t_1)}{\sqrt{S(t_1)}}\right|\right), \quad x \in \mathbb{R},$$
$$f_{X_{t_1+h}}(x) = \frac{C_p}{\sqrt{S(t_1+h)}} g^{(1,p)}_{SMN;G}\left(\left|\frac{x - m(t_1+h)}{\sqrt{S(t_1+h)}}\right|\right), \quad x \in \mathbb{R},$$

respectively, where $C_p = \frac{p^{1-\frac{1}{p}}}{2\Gamma\left(\frac{1}{p}\right)}$. Because $\mathfrak{L}(X_{t_1}) = \mathfrak{L}(X_{t_1+h})$, we have $f_{X_{t_1}} = f_{X_{t_1+h}}$, too. As $f_{X_{t_1}}$ and $f_{X_{t_1+h}}$, $h \in H_{t_1}$, are symmetric with respect to the straight lines $x = m(t_1)$ and $x = m(t_1+h)$, respectively, being parallel to the ordinate axis, it follows $m(t_1) = m(t_1+h)$ for all $t_1 \in I$ and $h \in H_{t_1}$. Thus, m is constant on I. Furthermore, since $f_{X_{t_1}}(m(t_1)) = \frac{C_p}{\sqrt{S(t_1)}}$ and $f_{X_{t_1+h}}(m(t_1+h)) = \frac{C_p}{\sqrt{S(t_1+h)}}$, the identity of these pdfs implies $S(t_1) = S(t_1+h)$ for all $t_1 \in I$ and $h \in H_{t_1}$. Thus, the constancy of S on I is shown. The other direction of this proof is omitted, here. □

Proof of Theorem 4.5 Let assume 1). According to Theorem 4.4, the constancy of m and S yields strict stationarity of X. Moreover, according to Theorem 4.3, it follows by the existence of expectation of $V^{-\frac{2}{p}}$ that X is a second order random process having expectation function m and covariance function Γ given by $\Gamma(t,t) = \sigma^2_{g^{(p)}_{SMN;G}} S(t)$ for all $t \in I$ and $\Gamma(s,t) = 0$ for all $s,t \in I$ with $s \neq t$. Because of $m(t) = \mu$ and $S(t) = \delta$ for all $t \in I$, the expectation function of X is constantly equal to μ and the covariance function of X satisfies $\Gamma(t,t) = \sigma^2_{g^{(p)}_{SMN;G}} \delta$ for all $t \in I$ and $\Gamma(s,t) = 0$ for all $s,t \in I$ with $s \neq t$. Thus, 1) implies 2). Further, every strictly stationary second order random process is weakly stationary and the covariance function Γ of X from 2) evaluated in $(s,t) \in I \times I$ is representable as a function only depending on the difference $s - t$ since it follows from the property 3) of function K that $\Gamma(t,t) = \sigma^2_{g^{(p)}_{SMN;G}} \delta = K(0) = K(t-t)$ for all $t \in I$ as well as $\Gamma(s,t) = 0 = K(s-t)$ for all $s,t \in I$ with $s \neq t$. Thus, the implication from 2) to 3) is shown. Additionally, it follows from 3) that $CovX_s, X_t = \Gamma(s,t) = 0$ for all $s,t \in I$ with $s \neq t$ and $\mathbb{E}(X_t) = m(t) = \mu$ as well as $Var(X_t) = \Gamma(t,t) = \sigma^2_{g^{(p)}_{SMN;G}} \delta$ for all $t \in I$. Hence, assuming 3), random variables X_t, $t \in I$, are uncorrelated and have constant expectation μ and variance $\sigma^2_{g^{(p)}_{SMN;G}} \delta$. Thus, 4) follows from 3). Finally, let us assume 4) to hold. According to Theorem 4.3, X is a second order random process if $\mathbb{E}\left(V^{-\frac{2}{p}}\right)$ is finite. Furthermore, because of the definition of white noise as in 4), it holds $m(t) = \mathbb{E}(X_t) = \mu$ as well as $\sigma^2_{g^{(p)}_{SMN;G}} S(t) = Cov(X_t, X_t) = Var(X_t) = \sigma^2_{g^{(p)}_{SMN;G}} \delta$ for all $t \in I$. Then, m and S are constantly equal to μ and δ, respectively. Thus, the implication from 4) to 1) is shown. □

Finally, the proof of Theorem 4.6 is based on Lemma 5.6. In preparation for the proof of this lemma, we establish the following special case.

Lemma 5.5 *Let $X \sim AEC_{n,p}\left(\mu, D, g^{(k,p)}\right)$ with $D = diag\,(d_1, \ldots, d_n)$ having non-negative diagonal elements and positive rank k. Further, let be $b \in \mathbb{R}^n$ and $\Gamma = diag\,(\gamma_1, \ldots, \gamma_n) \in \mathbb{R}^{n \times n}$ such that $\Gamma D \Gamma = diag\left(\gamma_1^2 d_1, \ldots, \gamma_n^2 d_n\right)$ has positive rank $k_\Gamma \geq 1$. Then,*

$$\mathfrak{L}(\Gamma X + b) = AEC_{n,p}\left(\Gamma\mu + b, \Gamma D\Gamma, g_{(k)}^{(k_\Gamma,p)}\right).$$

Proof Assuming $\gamma_{m_\epsilon} \neq 0$ for $\epsilon = 1, \ldots, l$ and $\gamma_{m_\epsilon} = 0$ for $\epsilon = l+1, \ldots, n$ where $m_1 < m_2 < \ldots < m_l$ and $m_{l+1} < m_{l+2} < \ldots < m_n$, and using notations from "The class of n-dimensional rank-k-continuous axis-aligned p-generalized elliptically contoured distributions" section, it follows that

$$\Gamma W_1\sqrt{S_1} = \begin{pmatrix} \gamma_1 e_1^{(n)\mathrm{T}} \\ \vdots \\ \gamma_n e_n^{(n)\mathrm{T}} \end{pmatrix} \left(\sqrt{d_{i_1}} e_{i_1}^{(n)} \cdots \sqrt{d_{i_k}} e_{i_k}^{(n)} \right) = \begin{pmatrix} \gamma_1 f(1) \\ \vdots \\ \gamma_n f(n) \end{pmatrix} \in \mathbb{R}^{n\times k}$$

where

$$f(\eta) = \begin{cases} \sqrt{d_{i_j}} e_j^{(k)\mathrm{T}} & \text{if } \eta = i_j \text{ for a } j \in \{1, \ldots, k\} \\ 0_k^{\mathrm{T}} & \text{else} \end{cases}, \quad \eta = 1, \ldots, n.$$

Since $\gamma_\eta = 0$ for $\eta \in \{m_{l+1}, \ldots, m_n\}$, there holds

$$\Gamma W_1\sqrt{S_1} = \begin{pmatrix} h(1) \\ \vdots \\ h(n) \end{pmatrix} \in \mathbb{R}^{n\times k}$$

where

$$h(\eta) = \begin{cases} \gamma_\eta \sqrt{d_\eta} e_j^{(k)\mathrm{T}} & \text{if } \eta \in K \text{and } \eta = i_j \text{ for a } j \in \{1, \ldots, k\} \\ 0_k^{\mathrm{T}} & \text{else} \end{cases},$$

$\eta = 1, \ldots, n$, and

$$K = \left\{\eta\colon \eta = i_j \text{ for a } j \in \{1, \ldots, k\} \text{ and } \eta = m_\epsilon \text{ for a } \epsilon \in \{1, \ldots, l\}\right\}. \tag{11}$$

Then, $|K| \geq 1$ because of $\mathrm{rk}(\Gamma D\Gamma) \geq 1$, and $\Gamma W_1\sqrt{S_1}$ has $|K|$ columns being the product of a positive constant, a constant from $\mathbb{R}\backslash\{0\}$ and a unit vector of $\mathbb{R}^k$. Particularly, all these unit vectors differ from each other and, using the notation δ_{im} of Kronecker's Delta, we have

$$|K| = \sum_{j=1}^{k}\sum_{\epsilon=1}^{l} \delta_{i_j m_\epsilon}.$$

Hence, $\Gamma W_1\sqrt{S_1}$ has $k - |K|$ columns being 0_n. For $Y = (Y_1, \ldots, Y_k)^{\mathrm{T}} \sim \Phi_{g^{(k,p)}}$, it follows that

$$\Gamma W_1\sqrt{S_1}Y = \begin{pmatrix} \theta(1) \\ \vdots \\ \theta(n) \end{pmatrix} \in \mathbb{R}^n$$

where

$$\theta(\eta) = \begin{cases} \gamma_\eta \sqrt{d_\eta} Y_j & \text{if } \eta \in K \text{ and } \eta = i_j \text{for a } j \in \{1, \ldots, k\} \\ 0 & \text{else} \end{cases},$$

$\eta = 1, \ldots, n$, and the vector $\Gamma W_1 \sqrt{S_1} Y$ consists of $|K|$ different components of Y. Thus, for $B \in \mathfrak{B}^n$, we have

$$P\left(\Gamma W_1 \sqrt{S_1} Y \in B\right) = P\left(\begin{pmatrix} \theta(1) \\ \vdots \\ \theta(n) \end{pmatrix} \in B\,,\ Y_j \in \mathbb{R} \text{ for all } j \in \{1, \ldots, k\} \backslash J\right)$$

where $J = \left\{j \in \{1, \ldots, k\} : i_j \in K\right\}$. Now, let

$$J = \left\{j_1, \ldots, j_{|K|}\right\} \qquad \text{with } j_1 < j_2 < \ldots < j_{|K|}$$

be an enumeration of the elements of J and

$$M = \begin{pmatrix} \psi(1) \\ \vdots \\ \psi(n) \end{pmatrix} \in \mathbb{R}^{n \times |K|}$$

where

$$\psi(\eta) = \begin{cases} \gamma_\eta \sqrt{d_\eta}\, e_\kappa^{(|K|)^{\mathrm{T}}} & \text{if } \eta \in K \text{and } \eta = i_{j_\kappa} \text{ for a } \kappa \in \{1, \ldots, |K|\} \\ 0_{|K|}^{\mathrm{T}} & \text{else} \end{cases}$$

for $\eta = 1, \ldots, n$. Then, $|J| = |K|$ and $\Gamma W_1 \sqrt{S_1} Y \overset{d}{=} MZ$ for $Z \sim \Phi_{g_{(k)}^{(|K|,p)}}$. Thus, because of $\mathrm{rk(M)} = |K|$, it follows

$$\begin{aligned} \Gamma X + b &\overset{d}{=} (\Gamma\mu + b) + \Gamma W_1 \sqrt{S_1} Y, \quad Y \sim \Phi_{g^{(k,p)}} \\ &\overset{d}{=} (\Gamma\mu + b) + MZ, \quad Z \sim \Phi_{g_{(k)}^{(|K|,p)}} \\ &= AEC_{n,p}\left(\Gamma\mu + b, MM^{\mathrm{T}}, g_{(k)}^{(|K|,p)}\right). \end{aligned}$$

Note that M can be extended to $\Gamma W_1 \sqrt{S_1}$ by adding $k - |K|$ zero columns. Therefore,

$$MM^{\mathrm{T}} = \left(\Gamma W_1 \sqrt{S_1}\right)\left(\Gamma W_1 \sqrt{S_1}\right)^{\mathrm{T}} = \Gamma W_1 S_1 W_1^{\mathrm{T}} \Gamma = \Gamma D \Gamma,$$

and $|K| = \mathrm{rk(M)} = \mathrm{rk}\left(\mathrm{MM^T}\right) = \mathrm{rk}(\Gamma \mathrm{D} \Gamma)$. Finally, this yields

$$\mathfrak{L}(\Gamma X + b) = AEC_{n,p}\left(\Gamma\mu + b, \Gamma D \Gamma, g_{(k)}^{(k_\Gamma,p)}\right).$$

□

Using this particular result, we prove the following more general one.

Lemma 5.6 *Let $X \sim AEC_{n,p}\left(\mu, D, g^{(k,p)}\right)$ with $D = \mathrm{diag}\ (d_1, \ldots, d_n)$ having nonnegative diagonal elements and rank $k \geq 0$. Further, let be $b \in \mathbb{R}^n$ and $\Gamma = \mathrm{diag}\ (\gamma_1, \ldots, \gamma_n) \in \mathbb{R}^{n \times n}$. Then,*

$$\mathfrak{L}(\Gamma X + b) = AEC_{n,p}\left(\Gamma\mu + b, \Gamma D \Gamma, g_{(k)}^{(k_\Gamma,p)}\right),$$

where $\Gamma D \Gamma = \mathrm{diag}\left(\gamma_1^2 d_1, \ldots, \gamma_n^2 d_n\right)$ and $k_\Gamma = \mathrm{rk}(\Gamma \mathrm{D} \Gamma) \geq 0$.

Proof of Lemma 5.6 Let $k = 0$, that is $X \sim AEC_{n,p}\left(\mu, 0_{n \times n}, g^{(0,p)}\right)$. Then, $\Gamma X + b$ follows the Dirac distribution in $\Gamma\mu + b$. Using the exchangeability of $g^{(0,p)}$ and $g_{(0)}^{(0,p)}$, we have

$$\begin{aligned} \mathfrak{L}(\Gamma X + b) &= AEC_{n,p}\left(\Gamma\mu + b, 0_{n \times n}, g^{(0,p)}\right) \\ &= AEC_{n,p}\left(\Gamma\mu + b, \Gamma 0_{n \times n} \Gamma, g_{(0)}^{(0,p)}\right). \end{aligned}$$

If D has positive rank and Γ is assumed to satisfy $k_\Gamma = \mathrm{rk}(\Gamma D \Gamma) \geq 1$, the assertion coincides with the result of Lemma 5.5. Finally, let D have positive rank and Γ be assumed to satisfy $k_\Gamma = \mathrm{rk}(\Gamma D \Gamma) = 0$ and $\Gamma D \Gamma = 0_{n\times n}$, respectively. In Analogy to the proof of Lemma 5.5 and using the same notations, the set K in (11) is empty. Then, $|K| = 0$, $\Gamma W_1 \sqrt{S_1}$ consists only of zero columns, and, for $Y \sim \Phi_{g^{(k,p)}}$ and every $B \in \mathfrak{B}^n$, we have

$$P\Big(\Gamma W_1 \sqrt{S_1} Y \in B\Big) = P(0_n \in B) = \mathbb{1}_B(0_n).$$

Particularly, if $B = \{0_n\}$, it follows that

$$P\Big(\Gamma W_1 \sqrt{S_1} Y = 0_n\Big) = P\Big(\Gamma W_1 \sqrt{S_1} Y \in \{0_n\}\Big) = 1.$$

Thus, $\Gamma W_1 \sqrt{S_1} Y = 0_n$ P-a.s., and the stochastic representation $\Gamma X + b \overset{d}{=} (\Gamma\mu + b) + \Gamma W_1 \sqrt{S_1} Y$ where $Y \sim \Phi_{g^{(k,p)}}$ holds according to (1), yields

$$\Gamma X + b = \Gamma \mu + b \qquad P-\text{a.s.}$$

or, equivalently, $\mathfrak{L}(\Gamma X + b) = AEC_{n,p}\Big(\Gamma\mu + b, 0_{n\times n}, g_{(k)}^{(0,p)}\Big)$. □

Proof of Theorem 4.6 Let be $n \in \mathbb{N}$ and $J = \{t_1, \ldots, t_n\}$ an arbitrary subset of I. Moreover, let $Y_t = \gamma(t) X_t + b(t)$, $t \in I$, and $Y = \{Y_t\}_{t\in I}$. Then, for $Y_J = \big(Y_{t_1}, \ldots, Y_{t_n}\big)^{\mathrm{T}}$ and $X_J = \big(X_{t_1}, \ldots, X_{t_n}\big)^{\mathrm{T}}$, we have

$$Y_J = \begin{pmatrix} Y_{t_1} \\ \vdots \\ Y_{t_n} \end{pmatrix} = \begin{pmatrix} \gamma(t_1) X_{t_1} + b(t_1) \\ \vdots \\ \gamma(t_n) X_{t_n} + b(t_n) \end{pmatrix} = \Gamma X_J + b$$

where $b = (b(t_1), \ldots, b(t_n))^{\mathrm{T}}$ and $\Gamma = \mathrm{diag}\ (\gamma(t_1), \ldots, \gamma(t_n))$. Since

$$\mathfrak{L}(X_J) = AEC_{n,p}\Big(\mu, D, g_{SMN;G}^{(k,p)}\Big)$$

where $\mu = (m(t_1), \ldots, m(t_n))^{\mathrm{T}}$ and $D = \mathrm{diag}\ (S(t_1), \ldots, S(t_n))$ with $k = \mathrm{rk(D)}$, making use of Lemmata 4.3 and 5.5, it follows

$$\begin{aligned} \mathfrak{L}(Y_J) = \mathfrak{L}(\Gamma X_J + b) &= AEC_{n,p}\left(\Gamma\mu + b, \Gamma D \Gamma, \left(g_{SMN;G}^{(k,p)}\right)_{(k)}^{(k_\Gamma,p)}\right) \\ &= AEC_{n,p}\Big(\Gamma\mu + b, \Gamma D \Gamma, g_{SMN;G}^{(k_\Gamma,p)}\Big). \end{aligned}$$

Thus, $AEC_{n,p}\Big(\Gamma\mu + b, \Gamma D \Gamma, g_{SMN;G}^{(k_\Gamma,p)}\Big)$ is the fdd of Y corresponding to Y_J. Finally, because of $\Gamma\mu + b = ([\gamma m + b](t_1), \ldots, [\gamma m + b](t_n))^{\mathrm{T}}$ and $\Gamma D \Gamma = \mathrm{diag}\ \big(\big[\gamma^2 S\big](t_1), \ldots, \big[\gamma^2 S\big](t_n)\big)$, we get

$$Y \sim SMAGP_p\left(\gamma m + b, \gamma^2 S, G\right).$$

□

6 Discussion

In the present paper, first, kapec distributions are introduced and their properties such as stochastic representations, moments, and density-like representations are studied. Secondly, based on the Kolmogorov existence theorem, the existence of random processes having apec fdds with arbitrary location and scale functions and a consistent sequence of dgs of p-generalized spherical distributions is shown. Particularly, a sequence of dgs of scale mixtures of multivariate p-generalized Gaussian distributions with one and the same mixture distribution is consistent and the corresponding processes are p-generalizations of elliptical random processes having axis-aligned fdds, see Yao (1973) and Kano (1994) for the case of $p = 2$. Thirdly, the question is answered when an n-dimensional kapec distribution with dg $g^{(k,p)}$ is representable as a scale mixture of n-dimensional kapec Gaussian distribution for a suitable mixture distribution of a positive random variable. It is established that the complete monotony of the composition h of $g^{(k,p)}$ with the pth root function is a necessary and sufficient condition for such representation and that the inverse Laplace-Stieltjes transform of h is connected to the cdf of the mixture distribution. For the particular case $p = 2$, the univariate consideration is covered by Andrews and Mallows (1974) and the multivariate one by Lange and Sinsheimer (1993) and Gómez-Sánchez-Manzano et al. (2006), respectively.

7 Appendix 1: Further aspects of simulations

7.1 Algorithms to simulate apec distributions

The following two algorithms to simulate $X \sim AEC_{n,p}(\mu, D, g^{(k,p)})$ are based on the two stochastic representations of X, see Lemmata 2.1 and 2.4. In both cases, let $\tilde{X} \sim \Phi_{g^{(k,p)}}$ and use notations from "The class of n-dimensional rank-k-continuous axis-aligned p-generalized elliptically contoured distributions" section.

Algorithm 1 1) *Generation of a random vector $U_p^{(k)}$ following the k-dimensional p-generalized uniform distribution on $S_{k,p}$:*

a) *Generate $\tilde{Z} = \left(\tilde{Z}_1, \ldots, \tilde{Z}_k\right)^{\mathrm{T}}$ following the k-dimensional p-generalized Gaussian distribution by generating k independent and identically univariate p-generalized Gaussian distributed random variables $\tilde{Z}_1, \ldots, \tilde{Z}_k$.*

b) *Compute $R_{\tilde{Z}} = |\tilde{Z}|_p$ and $U_p^{(k)} = \frac{\tilde{Z}}{R_{\tilde{Z}}}$.*

2) *Generate $R_{\tilde{X}}$ having pdf $f_{R_{\tilde{X}}}(r) = \omega_{k,p}\, r^{k-1} g^{(k,p)}(r)\, \mathbb{1}_{[0,\infty)}(r)$, $r \in \mathbb{R}$, and being a univariate random radius variable.*

3) *Compute $\tilde{X} = R_{\tilde{X}}\, U_p^{(k)}$ and $X = \mu + W_1\sqrt{S_1}\tilde{X}$.*

Algorithm 2 1) *Generation of the random radius and angle variables according to Lemma 2.4: Generate*

a) *R with $f_R(r) = \omega_{k,p}\, r^{k-1} g^{(k,p)}(r)\, \mathbb{1}_{[0,\infty)}(r)$, $r \in \mathbb{R}$,*

b) *Ψ_i with $f_{\Psi_i}(\psi_i) = \frac{\omega_{k-i,p}}{\omega_{k-i+1,p}} \frac{(\sin(\psi_i))^{k-i-1}}{(N_p(\psi_i))^{k-i+1}}\, \mathbb{1}_{[0,\pi)}(\psi_i)$, $\psi_i \in \mathbb{R}$, for $i = 1, \ldots, k-2$,*

c) *Ψ_{k-1} with $f_{\Psi_{k-1}}(\psi_{k-1}) = \frac{1}{\omega_{2,p}} \frac{1}{(N_p(\psi_{k-1}))^2}\, \mathbb{1}_{[0,2\pi)}(\psi_{k-1})$, $\psi_{k-1} \in \mathbb{R}$.*

2) *Compute $\tilde{X} = SPH_p^{(k)}(R, \Psi_1, \ldots, \Psi_{k-1})$ and $X = \mu + W_1\sqrt{S_1}\tilde{X}$.*

For the particular case of simulating $X \sim SMAN_{n,p}(\mu, D, G)$ where the mixture cdf G is explicitly known in a closed form, the following algorithm can be used. This is based on (8) and Lemma 4.1 where $\tilde{X} \sim SMN_{k,p}(G)$ and notations from "The class of n-dimensional rank-k-continuous axis-aligned p-generalized elliptically contoured distributions" section are used.

Algorithm 3 *1) Generate $\tilde{Z} = \left(\tilde{Z}_1, \ldots, \tilde{Z}_k\right)^{\mathrm{T}}$ following the k-dimensional p-generalized spherical Gaussian distribution by generating k independent and identically univariate p-generalized Gaussian distributed random variables $\tilde{Z}_1, \ldots, \tilde{Z}_k$.*

2) Generate independently a univariate random variable V having cdf G.

3) Compute $\tilde{X} = V^{-\frac{1}{p}} \cdot \tilde{Z}$ and $X = \mu + W_1\sqrt{S_1}\tilde{X}$.

7.2 Simulation of *p*-generalized Student as well as *p*-generalized Slash processes

According to the method described in "Simulation" section, but simulating a 201-dimensional apec Student-*t* and Slash distributed random vector with the help of an algorithms from Appendix 7.1 instead of 201 independent univariate *p*-generalized Gaussian variables, we get approximates of trajectories of *p*-generalized Student-*t* and *p*-generalized Slash processes having axis-aligned fdds. Particularly, approximate realizations of $AStP_p\left(0_{[0,1]}, 1_{[0,1]}, \nu\right)$ as well as of $ASlP_p\left(0_{[0,1]}, 1_{[0,1]}, \nu\right)$ for $\nu \in \{1, 3, 10\}$ and $p = \frac{1}{2}$ and $p = 3$, respectively, are visualized in Figs. 4 and 5. Note that our considerations are restricted to location function $0_{[0,1]}$ and scale function $S = 1_{[0,1]}$ while the effects of varying location and scale functions are already shown in Fig. 3. Furthermore, on the one hand, notice that the height of amplitudes of the realizations of $AStP_p\left(0_{[0,1]}, 1_{[0,1]}, \nu\right)$ and $ASlP_p\left(0_{[0,1]}, 1_{[0,1]}, \nu\right)$, respectively, increases if $p > 0$ decreases or $\nu > 0$ increases. On the other hand, the effects that scales of axes are highly dependent on the specific realization of a trajectory of the process have to be in mind, too.

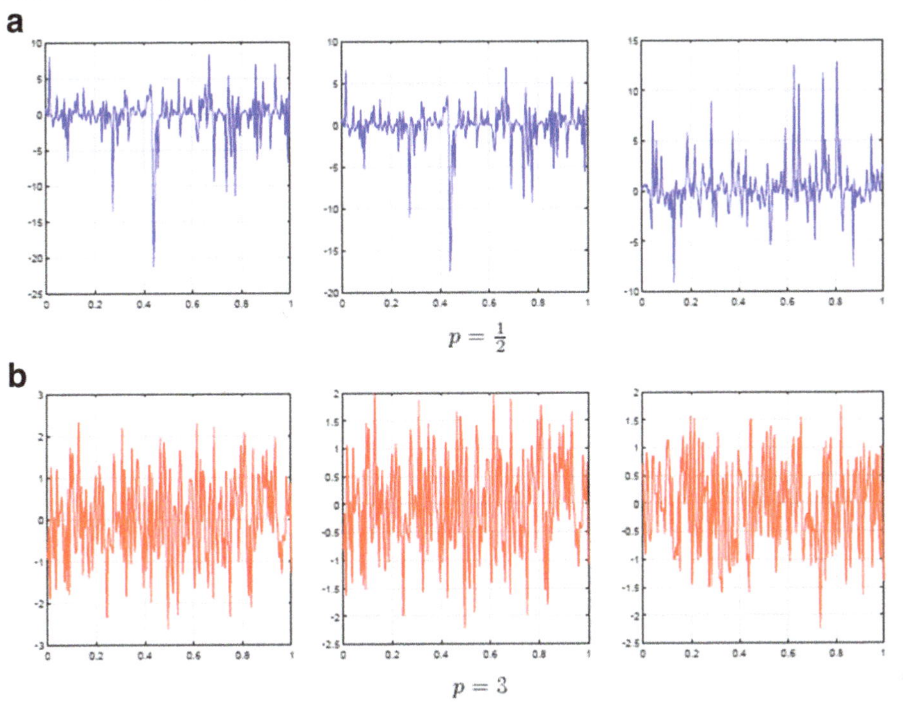

Fig. 4 Simulation of $AStP_p\left(0_{[0,1]}, 1_{[0,1]}, 1\right)$ (left), $AStP_p\left(0_{[0,1]}, 1_{[0,1]}, 3\right)$ (center), and $AStP_p\left(0_{[0,1]}, 1_{[0,1]}, 10\right)$ (right)

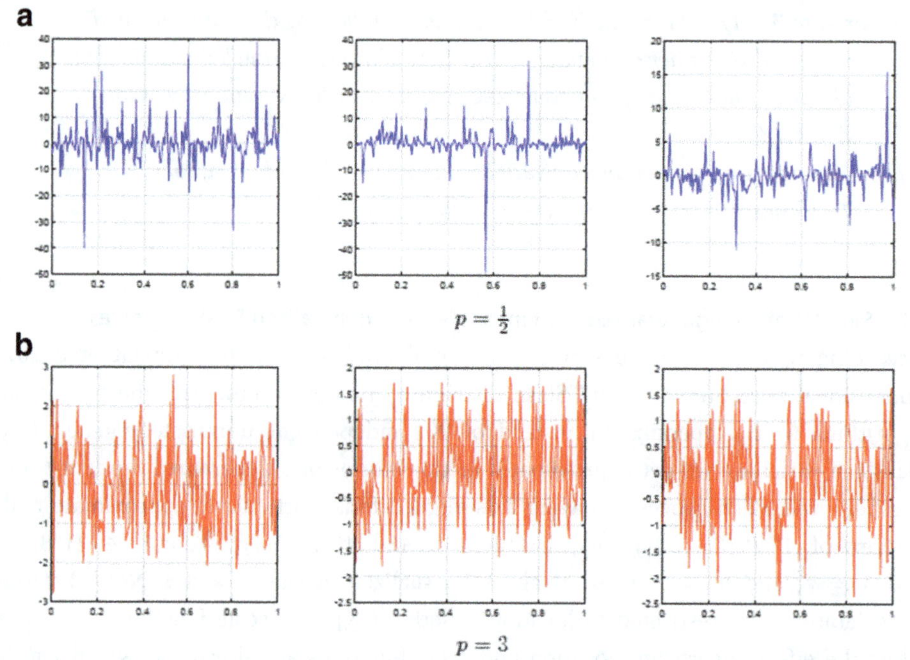

Fig. 5 Simulation of $ASIP_p(0_{[0,1]}, 1_{[0,1]}, 1)$ (left), $ASIP_p(0_{[0,1]}, 1_{[0,1]}, 3)$ (center), and $ASIP_p(0_{[0,1]}, 1_{[0,1]}, 10)$ (right)

Abbreviations

apec: Axis-aligned p-generalized elliptically contoured; cdf: Cumulative distribution function; dg: density generator; fdd: Finite dimensional distribution; fdds:family of finite dimensional distributions; kapec: Rank-k-continuous axis-aligned p-generalized elliptically contoured; pdf: Probability density function

Acknowledgements

We are grateful to the Reviewers and the Associate Editor for their valuable hints.

Funding

Not applicable.

Authors' contributions

Not applicable.

Authors' information (optional)

Not applicable.

Ethics approval and consent to participate

Not applicable.

Consent for publication

Not applicable.

Competing interests

The authors declare that they have no competing interests.

References

Ala-Luhtala, J, Piché, R: Gaussian scale mixture models for robust linear multivariate regression with missing data. Commun Stat Simul Comput. **45**(3), 791–813 (2016). https://doi.org/10.1080/03610918.2013.875565

Andrews, DF, Mallows, CL: Scale mixtures of normal distributions. J R Stat Soc Ser B Methodol. **36**(1), 99–102 (1974)

Arellano-Valle, RB, Azzalini, A: On the unification of families of skew-normal distributions. Scand J Stat. **33**(3), 561–574 (2006). https://doi.org/10.1111/j.1467-9469.2006.00503.x

Arellano-Valle, RB, Richter, W-D: On skewed continuous $l_{n,p}$-symmetric distributions. Chil J Stat. **3**(2), 193–212 (2012)

Biglieri, E, Yao, K, Yang, CA: Fading models from spherically invariant processes. IEEE Trans Wirel Commun. **14**(10), 5526–5538 (2015). https://doi.org/10.1109/TWC.2015.2439283

Blake, IF, Thomas, JB: On a class of processes arising in linear estimation theory. IEEE Trans Inf Theory. **14**, 12–16 (1968). https://doi.org/10.1109/TIT.1968.1054100

Brehm, H, Stammler, W: Description and generation of spherically invariant speech-model signals. Signal Process. **12**(2), 119–141 (1987). https://doi.org/10.1016/0165-1684(87)90001-6

Choy, BS, Chan, CM: Scale mixtures distributions in insurance applications. Astin Bull. **33**(1), 93–104 (2003). https://doi.org/10.2143/AST.33.1.1041

Conte, E, Longo, M, Lops, M: Modelling and simulation of non-Rayleigh radar clutter. IEE Proc. **138**(2), 121–130 (1991). https://doi.org/10.1049/ip-f-2.1991.0018

Eltoft, T, Kim, T, Lee, TW: Multivariate scale mixture of Gaussians modeling. In: Rosca, J, Erdogmus, D, Príncipe, JC, Haykin S (eds.) Independent component analysis and blind signal separation. 6th international conference, ICA 2006, Charleston, SC, USA, March 5–8, 2006, pp. 799–806. Proceedings, Springer, Berlin, chap 99, (2006). https://doi.org/10.1007/11679363

Gneiting, T: Normal scale mixtures and dual probability densities. J Stat Comput Simulation. **59**(4), 375–384 (1997). https://doi.org/10.1080/00949659708811867

Gómez-Sánchez-Manzano, E, Gómez-Villegas, MA, Marín, JM: Sequences of elliptical distributions and mixtures of normal distributions. J Multivariate Anal. **97**(2), 295–310 (2006). https://doi.org/10.1016/j.jmva.2005.03.008

Gómez-Sánchez-Manzano, E, Gómez-Villegas, MA, Marín, JM: Multivariate exponential power distributions as mixtures of normal distributions with Bayesian applications. Commun Stat, Theory Methods. **37**(6), 972–985 (2008). https://doi.org/10.1080/03610920701762754

Gupta, AK, Song, D: l_p-norm spherical distributions. J Stat Plann Inf. **60**(2), 241–260 (1997). https://doi.org/10.1016/S0378-3758(96)00129-2

Hashorva, E: Exact tail asymptotics in bivariate scale mixture models. Extremes. **15**(1), 109–128 (2012). https://doi.org/10.1007/s10687-011-0129-7

Huang, ST, Cambanis, S: Spherically invariant processes: Their nonlinear structure, discrimination, and estimation. J Multivariate Anal. **9**, 59–83 (1979). https://doi.org/10.1016/0047-259X(79)90067-8

Kalke, S, Richter, W-D: Simulation of the p-generalized Gaussian distribution. J Statist Comput Simul. **83**(4), 639–665 (2013). https://doi.org/10.1080/00949655.2011.631187

Kano, Y: Consistency property of elliptical probability density functions. J Multivar Anal. **51**(1), 139–147 (1994). https://doi.org/10.1006/jmva.1994.1054

Khatri, C: Some results for the singular normal multivariate regression models. Sankhyā Ser A. **30**, 267–280 (1968)

Kingman, JFC: On random sequences with spherical symmetry. Biometrika. **59**, 492–494 (1972). https://doi.org/10.1093/biomet/59.2.492

Kolmogorov, AN: Grundbegriffe der Wahrscheinlichkeitsrechnung, Ergebnisse der Mathematik und ihrer Grenzgebiete, vol 2. Springer, Berlin, Heidelberg (1933). https://doi.org/10.1007/978-3-642-49888-6

Kolmogorov, AN: Foundations of the theory of probability. Chelsea Publishing Company 1956, VIII, New York (1956)

Lange, K, Sinsheimer, JS: Normal/independent distributions and their applications in robust regression. J Comput Graph Stat. **2**(2), 175–198 (1993)

Müller, K, Richter, W-D: Exact extreme value, product, and ratio distributions under non-standard assumptions. AStA Adv stat Anal. **99**(1), 1–30 (2015). https://doi.org/10.1007/s10182-014-0228-2

Müller, K, Richter, W-D: Exact distributions of order statistics of dependent random variables from $l_{n,p}$-symmetric sample distributions, $n \in \{3, 4\}$. Depend Model. **4**, 1–29 (2016a). https://doi.org/10.1515/demo-2016-0001

Müller, K, Richter, W-D: Extreme value distributions for dependent jointly $l_{n,p}$-symmetrically distributed random variables. Depend Model. **4**, 30–62 (2016b). https://doi.org/10.1515/demo-2016-0002

Müller, K, Richter, W-D: Exact distributions of order statistics from $l_{n,p}$-symmetric sample distributions. Depend Model. **5**, 221–245 (2017a). https://doi.org/10.1515/demo-2017-0013

Müller, K, Richter, W-D: Modeling extremal data. J Stat Comput Simul. **87**(5), 933–955 (2017b). https://doi.org/10.1080/00949655.2016.1238089

Niu, X, Li, P, Zhang, P: Testing homogeneity in a scale mixture of normal distributions. Stat Pap. **57**(2), 499–516 (2016). https://doi.org/10.1007/s00362-015-0665-3

Rangaswamy, M, Weiner, D, Öztürk, A: Computer generation of correlated non-Gaussian radar clutter. IEEE Trans Aerosp Electron Syst. **31**(1), 106–116 (1995)

Rao, CR: Linear statistical inference and its applications, 2nd edn. Wiley, New York etc. (1973)

Richter, W-D: Generalized spherical and simplicial coordinates. J Math Anal Appl. **336**, 1187–1202 (2007). https://doi.org/10.1016/j.jmaa.2007.03.047

Richter, W-D: Continuous $l_{n,p}$-symmetric distributions. Lith Math J. **49**(1), 93–108 (2009). https://doi.org/10.1007/s10986-009-9030-3

Richter, W-D: Geometric disintegration and star-shaped distributions. J Stat Distrib Appl. **1:20**, 1–24 (2014). https://doi.org/10.1186/s40488-014-0020-6

Richter, W-D: Exact inference on scaling parameters in norm and antinorm contoured sample distributions. J Stat Distrib Appl. **3:13**, 1–16 (2016). https://doi.org/10.1186/s40488-016-0046-z

Richter, W-D: Statistical reasoning in dependent p-generalized elliptically contoured distributions and beyond. J Stat Distrib Appl. **4:21**, 1–25 (2017). https://doi.org/10.1186/s40488-017-0074-3

Sasvári, Z: Multivariate characteristic and correlation functions. de Gruyter, Berlin (2013). https://doi.org/10.1515/9783110223996

Schilling, RL, Partzsch, L: Brownian motion. An introduction to stochastic processes. With contributions by Björn Böttcher. 2nd revised and extended ed, 2nd edn. de Gruyter, Berlin (2014). https://doi.org/10.1515/9783110307306

Seo, B, Noh, J, Lee, T, Yoon, YJ: Adaptive robust regression with continuous Gaussian scale mixture errors. J Korean Stat Soc. **46**(1), 113–125 (2017). https://doi.org/10.1016/j.jkss.2016.08.002

Shiryaev, AN: Probability. Transl. from the Russian by R. P. Boas, 2nd edn. Graduate texts in mathematics, Springer-Verlag, New York (1996). https://doi.org/10.1007/978-1-4757-2539-1

Subbotin, M: On the law of frequency of error. Rec Math Moscou. **31**, 296–301 (1923)

Vershik, AM: Some characteristic properties of Gaussian stochastic processes. Theory Probab Appl. **9**(2), 353–356 (1964). https://doi.org/10.1137/1109053

Wainwright, MJ, Simoncelli, EP: Scale mixtures of Gaussians and the statistics of natural images. In: Solla, S, Leen, T, Müller K (eds.) Advances in Neural Information Processing Systems 12, pp. 855–861. MIT Press, Cambridge, (2000)

West, M: On scale mixtures of normal distributions. Biometrika. **74**(3), 646–648 (1987). https://doi.org/10.1093/biomet/74.3.646

Widder, DV: The Laplace transform. Princeton University Press, Princeton (1946)

Wise, GL, jun Gallagher, NC: On spherically invariant random processes. IEEE Trans Inf Theory. **24**, 118–120 (1978). https://doi.org/10.1109/TIT.1978.1055841

Yao, K: A representation theorem and its applications to spherically-invariant random processes. IEEE Trans Inf Theory. **19**, 600–608 (1973). https://doi.org/10.1109/TIT.1973.1055076

Yao, K: Spherically invariant random processes: Theory and Application. In: Bhargava VK, Poor, HV, Tarokh, V, Yoon S (eds.) Communications, Information and Network Security, pp. 315–331. Springer Science + Business Media, New York, (2003). chap 16

Zhang, X, El Korso, MN, Pesavento, M: MIMO radar target localization and performance evaluation under SIRP clutter. Signal Process. **130**, 217–232 (2017). https://doi.org/10.1016/j.sigpro.2016.06.031

3

Statistical reasoning in dependent *p*-generalized elliptically contoured distributions and beyond

Testing scaling parameters, the role semi-inner products play, and simulating star-shaped distributed random vectors

Wolf-Dieter Richter

Correspondence:
wolf-dieter.richter@uni.rostock.de
University of Rostock, Institute of Mathematics, Ulmenstraße 69, Haus 3, 18057 Rostock, Germany

Abstract

First, likelihood ratio statistics for checking the hypothesis of equal variances of two-dimensional Gaussian vectors are derived both under the standard $(\sigma_1^2, \sigma_2^2, \varrho)$-parametrization and under the geometric (a, b, α)-parametrization where a^2 and b^2 are the variances of the principle components and α is an angle of rotation. Then, the likelihood ratio statistics for checking the hypothesis of equal scaling parameters of principle components of p-power exponentially distributed two-dimensional vectors are considered both under independence and under rotational or correlation type dependence. Moreover, the role semi-inner products play when establishing various likelihood equations is demonstrated. Finally, the dependent p-generalized polar method and the dependent p-generalized rejection-acceptance method for simulating star-shaped distributed vectors are presented.

Keywords: Modeling with correlation and variances of Euclidean coordinates, Modeling with rotation and variances of principle components, Geometric parametrization, Likelihood ratio, p-generalized Fisher distribution, Semi-inner product, Dependent p-generalized polar method, Dependent p-generalized rejection-acceptance simulation, Star-shaped distributions

PACS: 02.50.-r, 02.50.Ng, 02.70.-Rr, 02.70.Uv

Mathematics Subject Classification: 62F03, 62F10, 62H15, 62E10

1 Introduction

One of the classical statistical problems deals with comparing variances. Results for quite arbitrary pairs of dependent random variables are due to Morgan (1939), Pitman (1939), Tiku and Balakrishnan (1986), McCulloch (1987), Wilcox (1990) and Mudholkar et al. (2003) and are recently reviewed in Wilcox (2015). For a survey of various practical applications in Snedecor and Cochran (1967), Lord and Novick (1968), Games et al. (1972), Levy (1976) and Rothstein et al. (1981), see again Wilcox (2015). A closely related but nevertheless rather different study in Richter (2016) compares scaling parameters of two-dimensional axes-aligned p-generalized elliptically contoured distributions. Such distributions show independence if the density generating function is that of the p-generalized

Gaussian law and l_p-dependence if the density generating function is of another type, but they show no rotational or correlation type dependence. The present paper is aimed now to study correlation type dependence modeling within the family of p-power exponential distribution laws.

It is well known that two jointly Gaussian distributed random variables are independent if and only if they are uncorrelated. The density level sets of such a vector are axes-aligned ellipses. If the components of a two-dimensional Gaussian vector are not independent then the vector may be constructed by rotating through its distribution center an axes-aligned elliptically contoured distributed Gaussian vector that has heteroscedastic components. The correlation coefficient may be expressed in such situation in terms of the angle of rotation and the ratio of variances, see Dietrich et al. (2013). This type of dependence between two random variables is called here a rotational or correlation type dependence. Basic facts on modeling two-dimensional Gaussian vectors with correlation and variances of Euclidean coordinates on the one hand and with rotation and variances of principle components on the other hand will be summarized in the presented paper. Considering these two models side by side demonstrates different aspects of 'standard' modeling with the more stochastically interpretable parameters and of 'flexible' modeling with the more geometrically motivated parameters.

It is outlined in Wilcox (2015) that "seemingly the best-known technique for testing $H_0 : \sigma_1^2 = \sigma_2^2$ is a method derived by Morgan (1939) and Pitman (1939). Letting $U = X + Y$ and $V = X - Y$, if the null hypothesis is true, then ϱ_{UV}, Pearsons correlation between U and V, is zero. So testing H_0 can be accomplished by testing $\varrho_{UV} = 0$." We do not consider here the test problem in the same full generality as in Wilcox (2015) where the joint distribution of X and Y is not basically restricted to belong to the families of p-generalized elliptically contoured or star-shaped distributions. While it is proved in Wilcox (2015) that certain heteroscedastic consistent estimators perform well in certain cases of heavy tailed distributions, here we use case sensitive estimators depending on the given value of the shape-tail parameter p, see Sections 3 and 4 and recognize the consequences drawn in Section 5. Note that cases of heavier and lighter than Gaussian distribution tails are observed here in dependence of whether $p \in (0, 2)$ or $p > 2$, respectively. A study demonstrating far and narrow tail effects when sampling from those distribution classes can be seen in Richter (2015a).

In the case of a two-dimensional Gaussian distribution $\Phi_{\mu,\Sigma}$, it turns out that the class of distributions satisfying H_0 is the union of the following two subsets. The elements of the first one are the spherical Gaussian distributions and the second one contains all elliptically contoured Gaussian distributions having the lines $y = +(-)x$ as the main axes of their density level ellipses. Any number from the interval $(-1, 1)$ is attained by the correlation coefficient of a suitably chosen element from the latter subset. Thus, H_0 covers two quite different cases of correlation and uncorrelation. We modify here the null hypothesis in a way that one of these two subsets is not included.

One of the likelihood equations needed to be solved for constructing the likelihood ratio statistic for testing the just mentioned modified hypothesis is formulated here on using a so-called semi-inner product in the sample space. This rises the question whether this analytical tool plays also a role in estimating location. We give a positive answer to this question in the case of axes-aligned p-power exponential distributions.

The paper is structured as follows. Gaussian correlation models and likelihood ratio tests for checking equality of variances of two dependent random variables are studied in Section 2. The content of this section is of some interest of its own although it might be partly known to the reader. Testing equality of scaling parameters of axes-aligned p-power exponential distributions is dealt with in Section 3. The more general results are presented in Sections 4-6. Section 4 deals with testing equality of scaling parameters of principal components of general, i.e. arbitrarily rotated, p-generalized elliptically contoured p-power exponential distributions. Derivations are omitted in the sections on Gaussian and axes-aligned p-generalized elliptically contoured distributions. They can be considered being standard and follow also from proving the more general results in Section 4. Throughout Sections 2-4, we restrict our consideration to the case of known expectations. Practical examples of this type are given in Richter (2016). Section 5 gives a new geometric-analytical insight into estimating the location parameter of the p-power exponential, or p-generalized Gaussian or Laplace, law using semi-inner products in the sample space.

Differently from the situation of statistics in Gaussian sample distributions, many statistical questions in p-generalized Gaussian and more general star-shaped sample distributions cannot yet fully be answered in a theoretical way. For intermittent empirical studies, and much beyond it, methods for simulating such distributions are needed. Generalizing the methods in Kalke and Richter (2013), Section 6 presents corresponding direct and acceptance-rejection methods and indicates how to extend to the dependent p-generalized multivariate case the classical and the rejecting polar methods in Box and Muller (1958) and Marsaglia and Bray (1964), respectively.

2 Likelihood ratio tests for scaling parameters in two-dimensional Gaussian distributions

Testing equality of scaling parameters can be interpreted in Gaussian models at least in two different ways. We deal here with equality of variances of the marginal variables or Euclidean coordinates if the Gaussian density is given in the classical $(\sigma_1^2, \sigma_2^2, \varrho)$ variances-correlation parametrization, and with equality of variances of principal components if the Gaussian density is given in the geometric (a, b, α)-parametrization from Dietrich et al. (2013) where a^2 and b^2 are the variances of the principle components and α is an angle of rotation.

2.1 The common $(\sigma_1^2, \sigma_2^2, \varrho)$-parametrization

In this section, we consider the marginal variables variances-correlation (mvv-c) model. Likelihood ratio tests with respect to the equality of two variances will be given separately for the cases of a known and an unknown correlation coefficient. Let $(X_i, Y_i)^T, i = 1, \ldots, n$ be independent Gaussian random vectors following the density $\varphi_{\mu,\Sigma}(.,.) = \varphi(.,.|\sigma_1, \sigma_2, \varrho)$ where $\mu = (\mu_1, \mu_2)^T$ is a vector from $\mathbb{R}^2$ and $\Sigma = \begin{pmatrix} \sigma_1^2 & \varrho\sigma_1\sigma_2 \\ \varrho\sigma_1\sigma_2 & \sigma_2^2 \end{pmatrix}$ is a positive definite matrix, and let $(x_i, y_i)^T, i = 1, \ldots, n$ be a corresponding concrete sample. We introduce the likelihood function

$$L(\sigma_1, \sigma_2, \varrho) = \prod_{i=1}^{n} \varphi(x_i, y_i|\sigma_1, \sigma_2, \varrho)$$

and its restriction to the case of equal variances, $\tilde{L}(\sigma,\varrho) = L(\sigma,\sigma,\varrho)$.

2.1.1 The case of an unknown correlation coefficient

We intend now to decide between the two hypotheses

$$H_0 : \sigma_1 = \sigma_2 \qquad \text{and} \qquad H_A : \sigma_1 \neq \sigma_2$$

using the likelihood ratio test statistic $Q = \tilde{L}(\tilde{\sigma},\tilde{\varrho})/L(\hat{\sigma}_1,\hat{\sigma}_2,\hat{\varrho})$ where $(\hat{\sigma}_1,\hat{\sigma}_2,\hat{\varrho}) = mle(\sigma_1,\sigma_2,\varrho)$ and $(\tilde{\sigma},\tilde{\varrho}) = mle|_{H_0}(\sigma,\varrho)$ are maximum likelihood and restricted to H_0 such estimators, respectively. Standard calculations show that Q allows the representation

$$\frac{Q^{2/n}}{4} = \frac{\Sigma_x^2\Sigma_y^2 - \Sigma_{xy}^2}{\left(\Sigma_x^2 + \Sigma_y^2\right)^2 - 4\Sigma_{xy}^2} \tag{1}$$

where

$$\Sigma_x^2 = \sum_1^n (x_i - \mu_1)^2, \Sigma_y^2 = \sum_1^n (y_i - \mu_2)^2, \Sigma_{xy} = \sum_1^n (x_i - \mu_1)(y_i - \mu_2).$$

Let $\alpha \in (0,1)$. According to the likelihood ratio rule, H_0 will be rejected if $Q < t_\alpha$ where t_α is chosen from the interval $(0,1)$ in a way such that

$$P(Q < t_\alpha)|_{H_0} = \alpha.$$

A restatement of this size α-test is based upon the following alternative representation of the likelihood ratio,

$$Q^{2/n} = \left(\frac{\hat{\sigma}_1\hat{\sigma}_2}{\tilde{\sigma}^2}\right)^2 \frac{1-\hat{\varrho}^2}{1-\tilde{\varrho}^2},$$

where

$$\hat{\sigma}_1^2 = \Sigma_x^2/n, \hat{\sigma}_2^2 = \Sigma_y^2/n, \hat{\varrho} = \Sigma_{xy}/\left(\Sigma_x\Sigma_y\right)$$

and

$$\tilde{\sigma}^2 = \left(\Sigma_x^2 + \Sigma_y^2\right)/(2n), \tilde{\varrho} = \Sigma_{xy}/\left(n\tilde{\sigma}^2\right).$$

Rewording the corresponding likelihood ratio decision rule, it is then a size $(\alpha_1 + \alpha_2)$-test to reject H_0 if

$$\frac{\Sigma_x^2}{\Sigma_y^2} \in (0, \lambda_{\alpha_1}] \cup [\lambda_{1-\alpha_2,\infty})$$

where $\lambda_q, q \in (0,1)$, is suitably chosen from $(0,\infty)$ such that

$$P\left(\frac{\Sigma_x^2}{\Sigma_y^2} < \lambda_q\right)|_{H_0} = q.$$

Here, Σ_x^2/Σ_y^2 is the ratio of two dependent Chi-squared distributed random variables. The distributions of all statistics considered here and in later sections may be simulated using the methods presented in Section 6. Alternatively, the geometric measure representation in Richter (2014) may be used to establish the exact distributions of several of these statistics, or at least to derive suitable approximations.

2.1.2 *The case of a known correlation coefficient*

Let $\varrho = \varrho_0$ be a known number and put $L(\sigma_1, \sigma_2) = L(\sigma_1, \sigma_2, \varrho_0)$ and $\tilde{L}(\sigma) = L(\sigma, \sigma, \varrho_0)$. The likelihood ratio $Q = \sup_{\sigma} \tilde{L}(\sigma) / \sup_{\sigma_1, \sigma_2} L(\sigma_1, \sigma_2)$ allows the representation

$$\frac{Q^{1/n}}{2} = \frac{\Sigma_x \Sigma_y - \varrho_0 \Sigma_{xy}}{\Sigma_x^2 + \Sigma_y^2 - 2\varrho_0 \Sigma_{xy}}.$$

Let $\alpha \in (0,1)$. The likelihood ratio decision rule leads to rejecting H_0 if $Q < t_\alpha, \alpha \in (0,1)$ satisfies $P(Q < t_\alpha)|_{H_0} = \alpha$ or, equivalently, if $Q^{1/n}/2 < z_\alpha$ for a suitably chosen $z_\alpha = z(t_\alpha) \in (0, 1/2)$, that is, if

$$1 - \varrho_0 (1 - 2z_\alpha) \frac{\Sigma_{xy}}{\Sigma_x \Sigma_y} < z_\alpha \left(\frac{\Sigma_x}{\Sigma_y} + \frac{\Sigma_y}{\Sigma_x} \right).$$

2.2 The geometric (a, b, α)-parametrization

We consider now the principal components variances-rotation (pcv-r) model. For simplicity, we assume that $\mu_1 = \mu_2 = 0$. The geometric parametrization of the Gaussian density is then

$$\varphi^*(x, y|a, b, \alpha) = \frac{1}{2ab\pi} \exp\left\{ -\frac{1}{2} \left[\left(\frac{x \cos\alpha + y \sin\alpha}{a} \right)^2 + \left(\frac{-x \sin\alpha + y \cos\alpha}{b} \right)^2 \right] \right\},$$

see Dietrich et al. (2013). Here,

$$a = (\sigma_1^2 \cos^2\alpha + \sigma_2^2 \sin^2\alpha + 2\varrho\sigma_1\sigma_2 \sin\alpha \cos\alpha)^{1/2},$$
$$b = \left(\sigma_2^2 \cos^2\alpha + \sigma_1^2 \sin^2\alpha - 2\varrho\sigma_1\sigma_2 \sin\alpha \cos\alpha\right)^{1/2},$$

and

$$\alpha = \begin{cases} 0 & \text{if } \varrho = 0 \\ \gamma + \begin{cases} 0 & \text{if } \varrho(\sigma_1 - \sigma_2) > 0 \\ \pi/2 & \text{if } \varrho(\sigma_1 - \sigma_2) < 0 \end{cases} & \text{if } \varrho \neq 0, \sigma_1 \neq \sigma_2 \\ \pi/4 & \text{if } \varrho \neq 0, \sigma_1 = \sigma_2 \end{cases}$$

with $\gamma = \frac{1}{2} \arctan\left(2\varrho\sigma_1\sigma_2 / \left(\sigma_1^2 - \sigma_2^2\right)\right)$. We put $\arctan(+(-)\infty) = +(-)\pi/2$ and remark that a^2 and b^2 are the variances of principal components of the related Gaussian random vector. The Euclidean coordinates of such a vector are correlated if $\varrho \neq 0$ and may then also be called rotational dependent because then $\alpha \neq 0$.

For testing equality of variances of principle components

$$H_0: \quad a = b \qquad \text{vs.} \quad H_A: \quad a \neq b,$$

we introduce the likelihood function

$$L^*(a, b, \alpha) = \prod \varphi^*(x_i, y_i|a, b, \alpha)$$

and its restriction to $H_0, \tilde{L}^*(a, \alpha) = L^*(a, a, \alpha)$.

2.2.1 *The case of an unknown α*

The likelihood ratio statistic $Q^* = \max_{a,\alpha} \tilde{L}^*(a, \alpha)|_{H_0} / \max_{a,b,\alpha} L^*(a, b, \alpha)$ in case α is to be estimated, allows the representation

$$\frac{(Q^*)^{2/n}}{4} = \frac{I(\hat{\alpha}) J(\hat{\alpha})}{\left(\Sigma_x^2 + \Sigma_y^2\right)^2}$$

where

$$I(\alpha) = (\cos\alpha)^2\,\Sigma_x^2 + (\sin\alpha)^2\,\Sigma_y^2 + 2\,(\sin\alpha\cos\alpha)\,\Sigma_{xy},$$

$$J(\alpha) = (\sin\alpha)^2\,\Sigma_x^2 + (\cos\alpha)^2\,\Sigma_y^2 - 2\,(\sin\alpha\cos\alpha)\,\Sigma_{xy}$$

and the maximum likelihood estimator $\hat{\alpha} = \text{mle}\,(\alpha)$ is

$$\hat{\alpha} = \frac{1}{2}\arctan\left(2\frac{\Sigma_{xy}}{\Sigma_y^2 - \Sigma_x^2}\right).$$

If H_0 is true then the correlation or rotational dependence of the Euclidean coordinates is zero. Correspondingly, no restricted under H_0 estimator of the angle of rotation α has any effect onto the statistic Q^*.

2.2.2 *The case of a known α*

If the angle of rotation α is known, the likelihood ratio allows the representation

$$\frac{(Q^*)^{2/n}}{4} = \frac{I(\alpha)J(\alpha)}{\left(\Sigma_x^2 + \Sigma_y^2\right)^2}.$$

The plug-in version of this statistic where, for unknown $\alpha, \alpha = \hat{\alpha} = \text{mle}\,(\alpha)$, is just the statistic from the previous section. Differently from this situation, the likelihood ratio statistic in Section 2.1.1 using both the unrestricted and the restricted maximum likelihood estimators of α, ist not such an immediate plug-in version of the statistic considered in Section 2.1.2.

3 Likelihood ratio test for scaling parameters in axes-aligned *p*-generalized elliptically contoured distributions

The present section is aimed to shortly summarize some results from the axes-aligned or independence case. To start with, we recall that the univariate p-power exponential distribution has the density

$$f_p(x;\mu,\sigma) = \frac{C_p}{\sigma}\exp\left\{-\frac{|x-\mu|^p}{p\sigma^p}\right\},\ x \in R,$$

which is also called p-generalized Gaussian or Laplace density, $p > 0$. The parameter p controls both the shape of the density and the tail behaviour of the distribution and may therefore be called a shape-tail parameter. Note that $C_p = p^{1-1/p}/(2\Gamma(1/p))$ and the first and second order moments of a correspondingly distributed random variable X are

$$\mathbb{E}X = \mu \in R \quad \text{and} \quad V(X) = \sigma^2\frac{\Gamma(3/p)}{\Gamma(1/p)}.$$

Moreover, such random variable X allows the stochastic representation

$$X \stackrel{d}{=} \mu + \sigma X_0$$

where X_0 follows the standard p-power exponential density, i.e. $X_0 \sim f_p(.;0,1)$. Because of this representation, σ is called a scaling parameter. Note that $E|X-\mu|^p = \sigma^p$. Two independent such variables follow the joint product density

$$f(x,y|\sigma_1,\sigma_2) = \frac{C_p^2}{\sigma_1\sigma_2}\exp\left\{-\frac{|x-\mu_1|^p}{p\sigma_1^p} - \frac{|y-\mu_2|^p}{p\sigma_2^p}\right\},\ (x,y) \in R^2$$

having the distribution center $(\mu_1,\mu_2)^T \in R^2$ and whose level sets are axes-aligned p-generalized ellipses.

Note that the axes-aligned p-generalized elliptically contoured p-power exponential densities introduced this way should not be confused with functions of the type $f_{(X,Y)}(x,y) = C\exp\left\{-Q(x,y)^{p/2}\right\}$ with Q being a quadratic form. The latter type of densities has been considered in Kuwana and Kariya (1991), Gómez et al. (1998), Gómez-Villegas et al. (2011) and Dang et al. (2015) and may also be called elliptically contoured p-power exponential densities. The corresponding type of distributions may be considered as a particular Kotz type distribution within the broad family of elliptically contoured distributions, see Fang et al. (1990) and Nadarajah (2003).

Testing

$$H_0 : \sigma_1 = \sigma_2 \quad \text{vs.} \quad H_A : \sigma_1 \neq \sigma_2$$

in the model of the present section means checking equality of scaling parameters. Let $(X_i, Y_i), i = 1, \ldots, n$ be independent random vectors following the density $f(.,.|\sigma_1, \sigma_2)$ and put $X_{(n)} = (X_1, \ldots, X_n)^T,\ Y_{(n)} = (Y_1, \ldots, Y_n)^T$. We still assume that the expectations μ_1 and μ_2 are known. In case of a true hypothesis H_0, the test statistic

$$T = \frac{(\text{mle}(\sigma_1)/\sigma_1)^p}{(\text{mle}(\sigma_2)/\sigma_2)^p} = \left(\frac{\sigma_2}{\sigma_1}\right)^p \frac{|X_{(n)} - \mu_1 1_n|_p^p}{|Y_{(n)} - \mu_2 1_n|_p^p}$$

follows the p-generalized Fisher distribution with (n, n) d.f., $T|_{H_0} \sim F_{n,n}(p)$. The latter distribution was derived in Richter (2009). It can be considered as the distribution of the ratio of independent p-generalized Chi-squared distributed variables that were introduced in Richter (2007). The density of the p-generalized Fisher distribution with (n, n) degrees of freedom is according to Richter (2009)

$$f_{n,n,p}(t) = \frac{\Gamma(2n/p)}{(\Gamma(n/p))^2} \frac{t^{n/p-1}}{(1+t)^{2n/p}}, t > 0,$$

see Fig. 1. With the notation

$$\Sigma_{p,X} = \left(\sum_{i=1}^{n} |X_i - \mu_1|^p\right)^{1/p} \text{and} \Sigma_{p,Y} = \left(\sum_{i=1}^{n} |Y_i - \mu_2|^p\right)^{1/p},$$

the statistic T can alternatively be represented as

$$T = \left(\frac{\sigma_2}{\sigma_1}\right)^p \frac{\Sigma_{p,X}^p}{\Sigma_{p,Y}^p}.$$

The decision rule according to which one rejects H_0 if $T < F_{n,n,\alpha_2}(p)$ or $T > F_{n,n,1-\alpha_1}(p)$ performs an exact size $(\alpha_1+\alpha_2)$-test. This test turns out to be the corresponding likelihood ratio test. Figures 2, 3 and 4 deal with the performance of this test showing histograms of simulation results for the test statistic T, under H_0. To this end, a random vector $(X, Y)^T$ following the distribution $\Phi_{(1,1),p,(0,0),I_2}$ was simulated $n \times N$-times, and the value of the statistic T was calculated N-times based upon this sample. The choices of the values of n and p allows direct comparisons with Fig. 1.

Figure 5 shows the influence an increasing simulation sample size N has onto the accuracy of the estimation of the density of the test statistic if the null hypothesis is true. In the case $n = 30, p = 2$ and for four different values of the simulation sample size N, Table 1 presents the correspondingly calculated percentiles of orders 5 and 95, respectively, and the exact Fisher quantiles $F_{30,30,q} = F_{30,30,q}(2), q \in \{0.05, 0.95\}$.

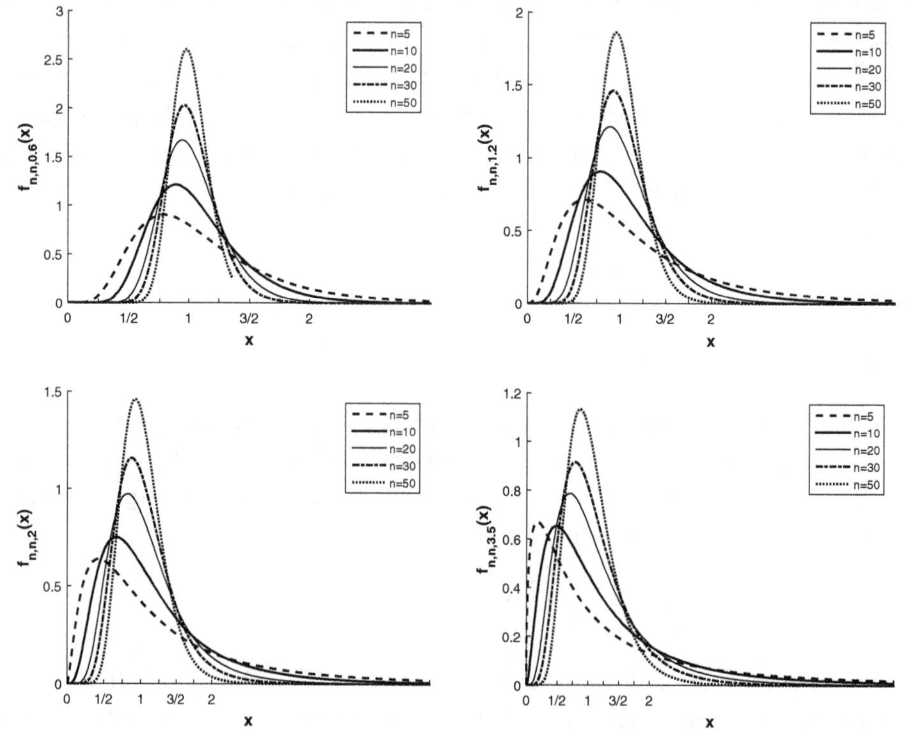

Fig. 1 The p-generalized Fisher densities $f_{n,n,p}$ for four values of the shape-tail parameter, $p \in \{0.6, 1.2, 2, 3.5\}$. Recognize different scalings on the ordinate axes

The likelihood ratio test can be equivalently reformulated as to reject H_0 if, for a suitably chosen c, the likelihood ratio Q satisfies $Q < c$. Let

$$L(\sigma_1, \sigma_2) = \prod_{i=1}^{n} f(x_i, y_i | \sigma_1, \sigma_2)$$

and $\tilde{L}(\sigma) = L(\sigma, \sigma)$, and denote unrestricted and restricted under H_0 mle's of σ_1, σ_2 and $\sigma_1 (= \sigma_2 = \sigma$, say) by $\hat{\sigma}_1, \hat{\sigma}_2$ and $\tilde{\sigma}$, respectively. The likelihood ratio statistic $Q = \tilde{L}(\tilde{\sigma})/L(\widehat{\sigma_1}, \widehat{\sigma_2})$ allows the representations

$$\frac{Q^{p/n}}{4} = \frac{\Sigma_{p,X}^{p} \Sigma_{p,Y}^{p}}{\left(\Sigma_{p,X}^{p} + \Sigma_{p,Y}^{p}\right)^2} = \frac{|X_{(n)} - \mu_1 1_n|_p^p |Y_{(n)} - \mu_2 1_n|_p^p}{\left(|X_{(n)} - \mu_1 1_n|_p^p + |Y_{(n)} - \mu_2 1_n|_p^p\right)^2}.$$

According to the general geometric measure-theoretical methodology of investigation in Richter (2014) and papers referred to there, the restricted distribution function of T if H_0 is true is

$$P(T < t)|_{H_0} = P\left(\left(X_{(n)}^T Y_{(n)}^T\right)^T \in C_{n,p}(t)\right)|_{H_0},\ t \in R$$

where

$$C_{n,p}(t) = \left\{(x^T y^T)^T \in R^n \times R^n : \frac{|x|_p^p}{|y|_p^p} < t\right\}$$

is a cone with vertex in $0 \in R^{2n}$ and $|z|_p = |z|_{(1,1),p}$. A geometric measure representation of the standardized p-power exponential law applies to show that this distribution is the p-generalized Fisher or $F_{n,n}(p)$-distribution. As mentioned before, this method may

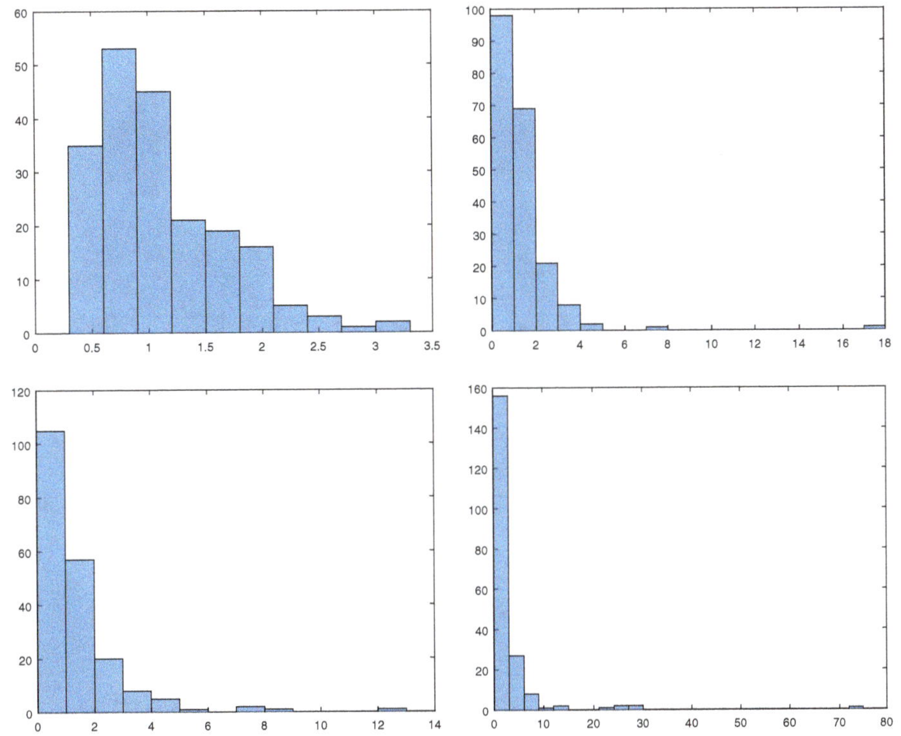

Fig. 2 Histograms of 200 simulated values of the test statistic T in case H_0 is true, $n = 5$ and shape-tail parameter attains in consecutive order from *upper left* to *lower right* picture the values $p = 0.6,\ 1.2,\ 2,\ 3.5$. Recognize different scalings on the abscissa and ordinate axes

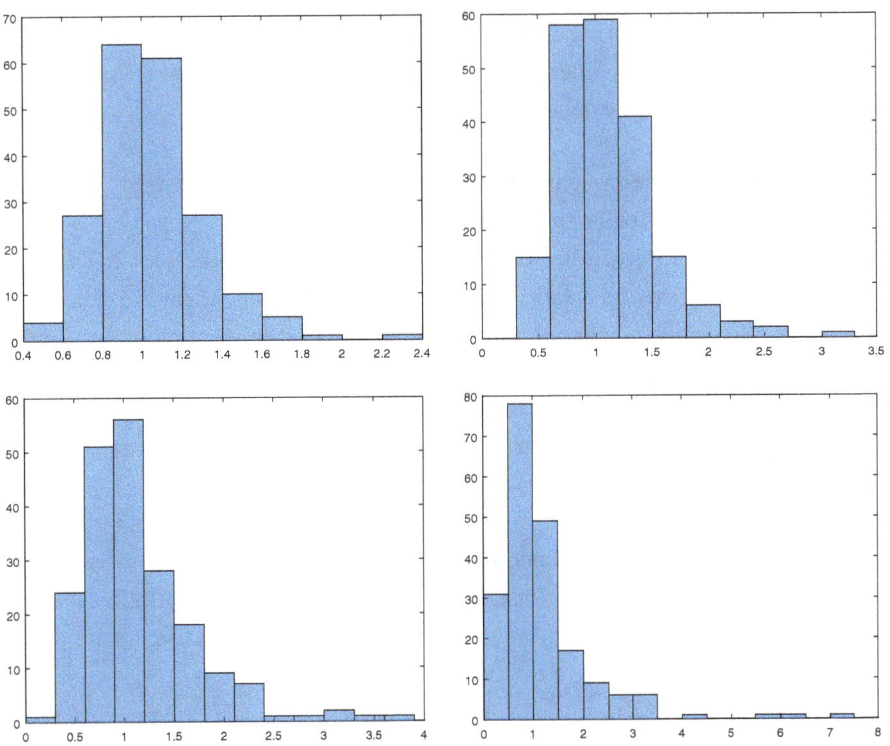

Fig. 3 Histograms as in Fig. 2 but for $n = 20$

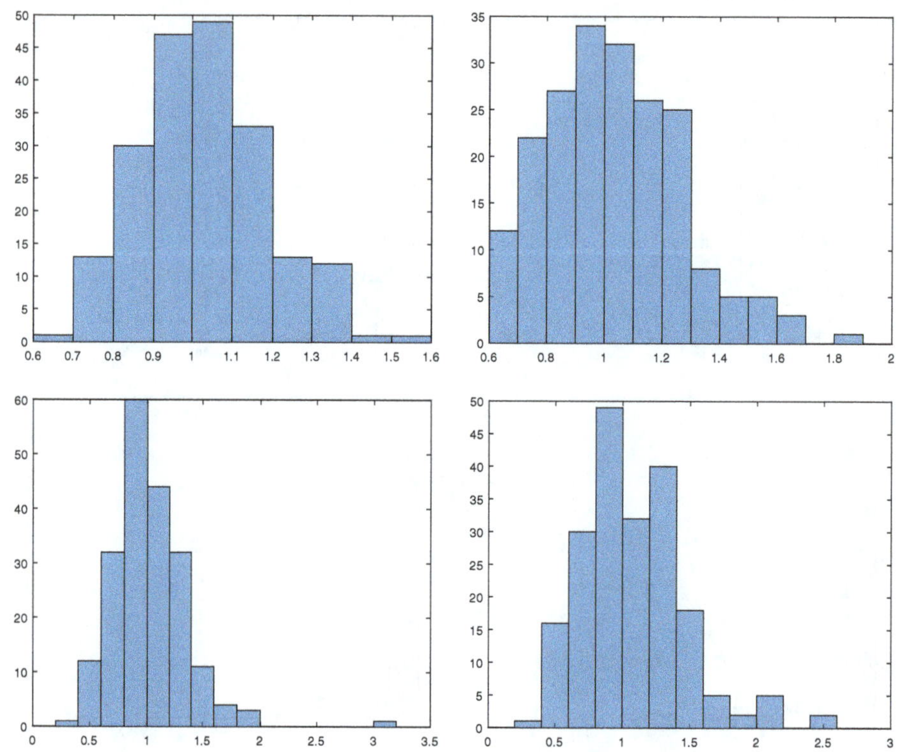

Fig. 4 Histograms as in Fig. 2 but for $n = 50$

also be used to derive exact distributions of other statistics. Differently from what was considered so far, throughout the following section the vectors $(X_i, Y_i)^T$ are allowed to be rotated through $(\mu_1, \mu_2)^T$, in other words, the variables X_i, Y_i are allowed to be rotational dependent or correlation type dependent, $i = 1, 2, \ldots$.

4 Tests for equal scaling parameters in correlational dependent *p*-generalized elliptically contoured distributions

This section is aimed to generalize the results presented in Section 3 for the case that two random variables may be rotational or correlation type dependent. To this end, we start in Section 4.1 with a p-generalization of the (a, b, α)-representation of the Gaussian law. Section 4.2 is aimed to give a geometric explanation of correlation in the particular case of a two-dimensional Gaussian distribution. Roughly spoken, correlation is interpreted by rotation under heteroscedasticity. Section 4.3 presents a test for checking homoscedasticity of principal components.

4.1 The geometric $((a, b), p, \alpha)$-parametrization

Let a random vector follow a rotational dependent p-generalized elliptically contoured p-power exponential distribution, $(X, Y)^T \sim \Phi_{(a,b),p,(\mu_1,\mu_2),D(\alpha)}$ where a, b, p are positive parameters, $(\mu_1, \mu_2)^T \in R^2$ and $D(\alpha) = \begin{pmatrix} \cos\alpha & \sin\alpha \\ -\sin\alpha & \cos\alpha \end{pmatrix}$ with $0 \leq \alpha < 2\pi$ is an orthogonal matrix causing a clockwise rotation around the origin through an angle of size α. Such random vector has according to Richter (2014) and Richter (2015a), (33), the density

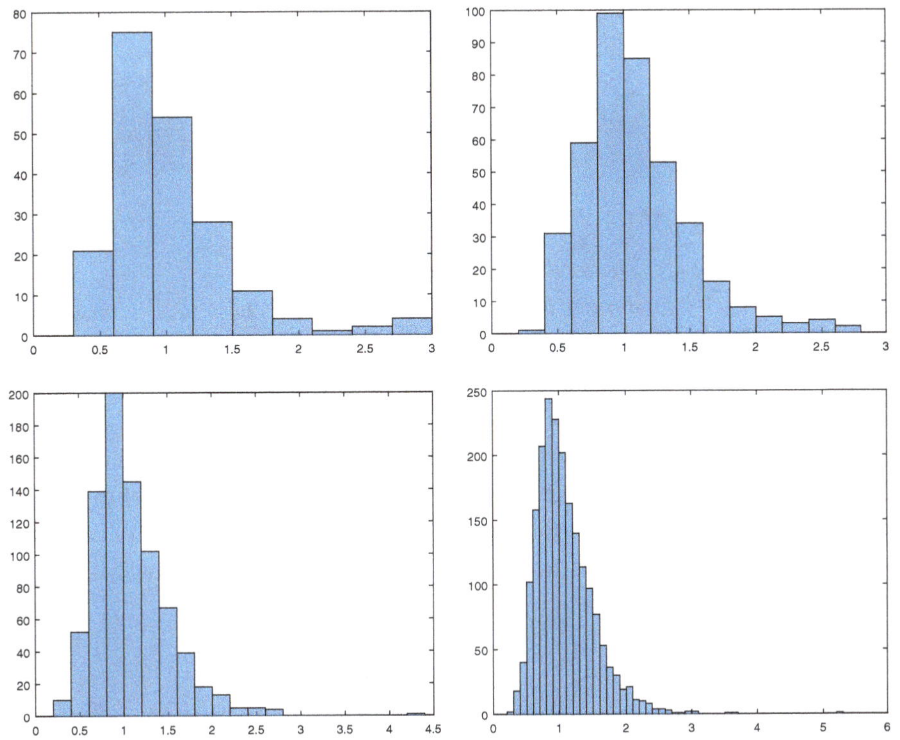

Fig. 5 Histograms of simulated values of the test statistic T where H_0 is true, $n = 30, p = 2$ and the simulation sample size attains in consecutive order from *upper left* to *lower right* picture the values $N = 200, 400, 800, 2000$

$$f_{(X,Y)}(x,y) = \frac{C_p^2}{ab} \exp\left\{ -\frac{1}{p} \left| D(\alpha) \begin{pmatrix} x-\mu_1 \\ y-\mu_2 \end{pmatrix} \right|_{(a,b),p}^p \right\}$$

where the functional

$$|z|_{(a,b),p} = (|\frac{z_1}{a}|^p + |\frac{z_2}{b}|^p)^{1/p}, (x,y)^T \in \mathbb{R}^2$$

is a norm if $p \geq 1$ and an antinorm if $0 < p \leq 1$. For the latter notion, see Moszyńska and Richter (2012). The level sets of the density $f_{(X,Y)}$ are p-generalized ellipses being not necessarily axes-aligned but centered at the point $(\mu_1, \mu_2)^T$.

Moreover, the stochastic representation

$$D(\alpha) \begin{pmatrix} X-\mu_1 \\ Y-\mu_2 \end{pmatrix} \stackrel{d}{=} RU$$

holds true where $R \geq 0$ and $U \sim \omega_{E_{(a,b),p}}$ are independent, R follows the density

$$f_R(r) = re^{-\frac{r^p}{p}} / \int_0^\infty re^{-\frac{r^p}{p}} dr, \; r > 0$$

Table 1 Simulating quantiles $F_{30,30,q}(2), q = 0.05, q = 0.95$

Simulation sample size N:	200	400	800	2000	$F_{30,30,q}(2)$
5-percentile	0.541	0.523	0.546	0.545	0.543
95-percentile	1.933	1.984	1.864	1.858	1.841=1/0.543

and U follows the $E_{(a,b),p}$-star-generalized uniform distribution on the Borel σ-field of the p-generalized axes-aligned ellipse having main axes of lengths $2a, 2b$,

$$E_{(a,b),p} = \left\{ (x,y)^T \in R^2 : \left| (x,y)^T \right|_{(a,b),p} = 1 \right\}.$$

Thus,

$$P(U \in A) = \omega_{E_{(a,b),p}}(A) = \mathfrak{U}(A)/\mathfrak{U}(E_{(a,b),p}),\ A \in \mathfrak{B}^2 \cap E_{(a,b),p}$$

where $\mathfrak{U}$ denotes the $E_{(a,b),p}$-generalized arc-length measure.
Let us finally remark that another definition of a bivariate p-generalized error density is given in Taguchi (1978).

4.2 Geometry of variance homogeneity

In this section, we exploit the fact that under heteroscedasticity a rotation causes a particular type of dependence, and give a new geometric interpretation of the hypothesis of variance homogeneity. To this end, we restrict our consideration once again to the Gaussian case. Let us assume that $(X,Y)^T$ is an anti-clockwise rotated axes-aligned Gaussian vector

$$\begin{pmatrix} X \\ Y \end{pmatrix} = D^T(\alpha) \begin{pmatrix} \xi \\ \eta \end{pmatrix} \sim N\left(\begin{pmatrix} \mu_1 \\ \mu_2 \end{pmatrix}, \Sigma \right), \Sigma = \begin{pmatrix} \sigma_1^2 & \varrho\sigma_1\sigma_2 \\ \varrho\sigma_1\sigma_2 & \sigma_2^2 \end{pmatrix}$$

where

$$\begin{pmatrix} \xi \\ \eta \end{pmatrix} \sim N\left(\begin{pmatrix} \nu_1 \\ \nu_2 \end{pmatrix}, \begin{pmatrix} a^2 & 0 \\ 0 & b^2 \end{pmatrix} \right).$$

Then,

$$\begin{aligned} f_{(X,Y)}(x,y) &= f_{(\xi,\eta)}(D(\alpha)(x,y)^T) \\ &= \frac{1}{2\pi ab} \exp\left\{ -\frac{1}{2} \begin{pmatrix} x-\mu_1 \\ y-\mu_2 \end{pmatrix}^T D(\alpha)^T \begin{pmatrix} a^2 & 0 \\ 0 & b^2 \end{pmatrix}^{-1} D(\alpha) \begin{pmatrix} x-\mu_1 \\ y-\mu_2 \end{pmatrix} \right\} \\ &= \frac{1}{2\pi ab} \exp\left\{ -\frac{1}{2} \left(\frac{(x-\mu_1)\cos\alpha + (y-\mu_2)\sin\alpha}{a} \right)^2 \right. \\ &\qquad \left. -\frac{1}{2} \left(\frac{-(x-\mu_1)\sin\alpha + (y-\mu_2)\cos\alpha}{b} \right)^2 \right\}. \end{aligned}$$

According to Dietrich et al. (2013), one can represent the parameters $\sigma_1, \sigma_2, \varrho$ in terms of the parameters a, b, α as follows:

$$\sigma_i = \sigma_i(\alpha) = [V((\xi,\eta)\theta_i)]^{1/2} > 0,\ i = 1,2$$

and

$$\varrho = \varrho(\alpha) = \frac{\varrho_*}{2} \sin(2\alpha) \left(\frac{\sigma_1(\alpha)}{\sigma_2(\alpha)} + \frac{\sigma_2(\alpha)}{\sigma_1(\alpha)} \right)$$

where

$$\theta_1 = \theta_1(\alpha) = \begin{pmatrix} \cos\alpha \\ \sin\alpha \end{pmatrix}, \theta_2 = \theta_2(\alpha) = \begin{pmatrix} \cos(\alpha + \frac{\pi}{2}) \\ \sin(\alpha + \frac{\pi}{2}) \end{pmatrix}$$

and

$$\varrho_* = \frac{a^2 - b^2}{a^2 + b^2}.$$

Let us try now to geometrically understand the hypothesis $H_0 : \sigma_1^2 = \sigma_2^2$ that is often supposed to hold in the statistical literature. It follows from the representations

$$\sigma_1^2 = a^2 \cos^2 \alpha + b^2 \sin^2 \alpha \quad \text{and} \quad \sigma_2^2 = a^2 \sin^2 \alpha + b^2 \cos^2 \alpha$$

that hypothesis H_0 means that

- either $\alpha \in \left\{\frac{\pi}{4}, \frac{3\pi}{4}\right\}$ with arbitrary a,b
- or $\alpha \notin \left\{\frac{\pi}{4}, \frac{3\pi}{4}\right\}$ and $a = b$.

If $a = b$ then $\varrho = 0$ and $\Sigma = \sigma^2 I_{2n}$ thus the density level sets of $(\xi, \eta)^T$ and $(X, Y)^T$ are Euclidean circles. If $\alpha \in \left\{\frac{\pi}{4}, \frac{3\pi}{4}\right\}$ then these level sets are arbitrary ellipses with main axes belonging to the lines $\left\{(x, y)^T \in \mathbb{R}^2 : y = x\right\}$ and $\left\{(x, y)^T \in \mathbb{R}^2 : y = -x\right\}$, and the correlation attains any value from the interval $(-1, 1)$. Thus, H_0 is not focussing, or is wavering, with respect to the parameters a, b, α and the shape of the density level ellipses and might therefore not always being primarily of interest, from this geometric point of view. If one presumes just the hypothesis

$$H_0^* : a = b$$

then the sample distribution and with it the distributions of all statistics derived from this sample vector are the same as in the axes-aligned and homoscedastic case.

Let us finally consider the following well known particular case of homoscedasticity. If $(X, Y)^T \sim \Phi_{\mu,\Sigma}$ then the random vector

$$(\xi, \eta)^T = D\left(\frac{\pi}{4}\right)(X, Y)^T = \frac{1}{2}(X + Y, X - Y)^T = \frac{1}{2}(U, V)^T$$

has the covariance matrix

$$\Psi = D\left(\frac{\pi}{4}\right)\Sigma D^T\left(\frac{\pi}{4}\right) = \frac{1}{2}\begin{pmatrix} \sigma_1^2 + \sigma_2^2 + 2\varrho\sigma_1\sigma_2 & \sigma_2^2 - \sigma_1^2 \\ \sigma_2^2 - \sigma_1^2 & \sigma_1^2 + \sigma_2^2 - 2\varrho\sigma_1\sigma_2 \end{pmatrix}.$$

Thus, if $\sigma_1 = \sigma_2 = \sigma$, say, then $\Psi = 2\sigma^2 \begin{pmatrix} 1 + \varrho & 0 \\ 0 & 1 - \varrho \end{pmatrix}$.

4.3 Testing homoscedasticity of principal components

We are well motivated now for testing equality of scaling parameters in the $((a, b), p, \alpha)$-parameterized model by checking the hypothesis $H_0 : a = b$ vs. the alternative $H_A : a \neq b$. Throughout this section, let

$$\hat{\alpha} = \begin{cases} \alpha & \text{if } \alpha \text{ is known} \\ \text{mle}(\alpha) & \text{if } \alpha \text{ is unknown.} \end{cases}$$

Let us further be given a concrete sample $\mathfrak{x}_i = (x_i, y_i)^T, i = 1, \ldots, n$ from independent identically and according to $\Phi_{(a,b),p,(0,0),D(\alpha)}$ distributed random vectors. Then, the likelihood function $L(a, b, \alpha)$ satisfies the equation

$$(ab)^n L(a, b, \alpha)/C_p^{2n}$$

$$= \exp\left\{-\frac{1}{pa^p}\sum_1^n |x_i \cos\alpha + y_i \sin\alpha|^p - \frac{1}{pb^p}\sum_1^n |-x_i \sin\alpha + y_i \cos\alpha|^p\right\}.$$

Let us consider the first two of the three likelihood equations. The partial derivatives of $\ln L$ with respect to a and b attain the value zero if $a = \hat{a}(\hat{\alpha})$ and $b = \hat{b}(\hat{\alpha})$, respectively,

where

$$\widehat{a}(\alpha)^p = \frac{1}{n}\sum_1^n |x_i \cos\alpha + y_i \sin\alpha|^p \text{and } \widehat{b}(\alpha)^p = \frac{1}{n}\sum_1^n |y_i \cos\alpha - x_i \sin\alpha|^p,$$

what ever the value of α is. The resulting equation

$$L(\hat{a}, \hat{b}, \hat{\alpha}) = \frac{C_p^{2n} e^{-2n/p}}{\left(\hat{a}(\hat{\alpha})\hat{b}(\hat{\alpha})\right)^n}$$

will be used later for constructing the likelihood ratio statistic. An angle $\hat{\alpha}$ solves the third likelihood equation if

$$\frac{\sum_{i=1}^n |\mathfrak{x}_i^T\theta_1(\hat{\alpha})|^{p-2}\left(\mathfrak{x}_i^T\theta_1(\hat{\alpha})\right)\left(\mathfrak{x}_i^T\theta_2(\hat{\alpha})\right)}{\sum_{i=1}^n |\mathfrak{x}_i^T\theta_1(\hat{\alpha})|^p} - \frac{\sum_{i=1}^n |\mathfrak{x}_i^T\theta_2(\hat{\alpha})|^{p-2}\left(\mathfrak{x}_i^T\theta_2(\hat{\alpha})\right)\left(\mathfrak{x}_i^T\theta_1(\hat{\alpha})\right)}{\sum_{i=1}^n |\mathfrak{x}_i^T\theta_2(\hat{\alpha})|^p} = 0.$$

An n-dimensional vector-algebraic reformulation of this equation is

$$\frac{\left[\begin{pmatrix}\mathfrak{x}_1^T\theta_2(\hat{\alpha})\\ \vdots \\ \mathfrak{x}_n^T\theta_2(\hat{\alpha})\end{pmatrix}, \begin{pmatrix}\mathfrak{x}_1^T\theta_1(\hat{\alpha})\\ \vdots \\ \mathfrak{x}_n^T\theta_1(\hat{\alpha})\end{pmatrix}\right]_p}{\left\|\begin{pmatrix}\mathfrak{x}_1^T\theta_1(\hat{\alpha})\\ \vdots \\ \mathfrak{x}_n^T\theta_1(\hat{\alpha})\end{pmatrix}\right\|_p^2} = \frac{\left[\begin{pmatrix}\mathfrak{x}_1^T\theta_1(\hat{\alpha})\\ \vdots \\ \mathfrak{x}_n^T\theta_1(\hat{\alpha})\end{pmatrix}, \begin{pmatrix}\mathfrak{x}_1^T\theta_2(\hat{\alpha})\\ \vdots \\ \mathfrak{x}_n^T\theta_2(\hat{\alpha})\end{pmatrix}\right]_p}{\left\|\begin{pmatrix}\mathfrak{x}_1^T\theta_2(\hat{\alpha})\\ \vdots \\ \mathfrak{x}_n^T\theta_2(\hat{\alpha})\end{pmatrix}\right\|_p^2}$$

where $[.,.]_p$ denotes a semi-inner-product defined by

$$[x, y]_p = \frac{\sum_{i=1}^n x_i y_i |y_i|^{p-2}}{|y|_p^{p-2}}, x, y \in \mathbb{R}^n.$$

For the theory and applications of semi-inner products we refer to Lumer (1961), Giles (1967), Dragomir (2004) and Horváth et al. (2015). We just mention here that, for all x, y, z from $R^n, a \in R$,

$$[x+z, y]_p = [x, y]_p + [z, y]_p, \ [ax, y]_p = a[x, y]_p, \ [x, ay]_p = a[x, y]_p,$$

$$[x, x]_p \geq 0, [x, x]_p = 0 \text{ iff } x = 0, \text{and } [x, y]_p \leq [x, x]_p^{1/2}[y, y]_p^{1/2}.$$

In general, a semi-inner product is not symmetric and non-linear in the second argument. With the notations $\xi_i(\alpha) = \left(\mathfrak{x}_1^T\theta_i(\alpha), \ldots, \mathfrak{x}_n^T\theta_i(\alpha)\right)^T, i = 1, 2$ and $x_{(n)} = (x_1, \ldots, x_n)$, $y_{(n)} = (y_1, \ldots, y_n)$, $\hat{\alpha}$ solves the equation

$$\frac{[\xi_2(\hat{\alpha}), \xi_1(\hat{\alpha})]_p}{|\xi_1(\hat{\alpha})|_p^2} = \frac{[\xi_1(\hat{\alpha}), \xi_2(\hat{\alpha})]_p}{|\xi_2(\hat{\alpha})|_p^2}$$

or

$$\frac{\left[(\cos\hat{\alpha})y_{(n)} - (\sin\hat{\alpha})x_{(n)}, (\cos\hat{\alpha})x_{(n)} + (\sin\hat{\alpha})y_{(n)}\right]_p}{|(\cos\hat{\alpha})x_{(n)} + (\sin\hat{\alpha})y_{(n)}|_p^2}$$

$$= \frac{\left[(\cos\hat{\alpha})x_{(n)} + (\sin\hat{\alpha})y_{(n)}, (\cos\hat{\alpha})y_{(n)} - (\sin\hat{\alpha})x_{(n)}\right]_p}{|(\cos\hat{\alpha})y_{(n)} - (\sin\hat{\alpha})x_{(n)}|_p^2}. \quad (2)$$

The Hessian matrix of $\ln L(a,b,\alpha)$ at the critical point $(a,b,\alpha)=(\hat{a},\hat{b},\hat{\alpha})$ is

$$HM=\begin{pmatrix} -\frac{np}{\hat{a}^2(\hat{\alpha})} & 0 & \epsilon_1 \\ 0 & -\frac{np}{\hat{b}^2(\hat{\alpha})} & \epsilon_2 \\ \epsilon_1 & \epsilon_2 & \epsilon_3 \end{pmatrix}$$

with

$$\epsilon_1=\epsilon_1(\hat{\alpha})=pn^{1+1/p}\frac{[\xi_2(\hat{\alpha}),\xi_1(\hat{\alpha})]_p}{|\xi_1(\hat{\alpha})|_p^3},\epsilon_2=\epsilon_2(\hat{\alpha})=-pn^{1+1/p}\frac{[\xi_1(\hat{\alpha}),\xi_2(\hat{\alpha})]_p}{|\xi_2(\hat{\alpha})|_p^3}$$

and

$$\epsilon_3=\epsilon_3(\hat{\alpha})=2n+n(1-p)\frac{<\xi_2(\hat{\alpha}),\xi_1(\hat{\alpha})>_p}{|\xi_1(\hat{\alpha})|_p^p}+\frac{<\xi_1(\hat{\alpha}),\xi_2(\hat{\alpha})>_p}{|\xi_2(\hat{\alpha})|_p^p}$$

where

$$<\eta,\nu>_p=\frac{\sum\limits_1^n|\nu_i|^{p-2}\eta_i^2}{|\nu|_p^p},\quad \eta,\nu \text{ from } R^n.$$

Obviously, $\Delta_1<0$ and $\Delta_2>0$ where

$$\Delta_1=(-\frac{np}{\hat{a}^2(\hat{\alpha})})\text{and } \Delta_2=\det\begin{pmatrix} -\frac{np}{\hat{a}^2(\hat{\alpha})} & 0 \\ 0 & -\frac{np}{\hat{b}^2(\hat{\alpha})} \end{pmatrix}.$$

Let $\Delta_3=\det(HM)$. If $\Delta_3<0$ then $L(a,b,\alpha)$ attains a local maximum at the point $=(a,b,\alpha)=(\hat{a},\hat{b},\hat{\alpha})$, see, e.g., Arens et al. (2013), Section 24.6. Under this assumption, $(\hat{a},\hat{b},\hat{\alpha})=\text{mle}(a,b,\alpha)$. Note that $\Delta_3<0$ if and only if

$$np\epsilon_3(\hat{\alpha})+a^2(\hat{\alpha})\epsilon_1^2(\hat{\alpha})+b^2(\hat{\alpha})\epsilon_2^2(\hat{\alpha})<0. \tag{3}$$

Thus, for finding mle(a,b,α), one has to solve (2) under the constraint (3).
If $p=2$ then the semi-inner product $[.,.]_p$ is symmetric, thus $\hat{\alpha}$ satisfies either the equation $[\xi_1(\hat{\alpha}),\xi_2(\hat{\alpha})]_2=0$, or $|\xi_1(\hat{\alpha})|_2^2=|\xi_2(\hat{\alpha})|^2$. The first and second equations mean that

$$\hat{\alpha}=\frac{1}{2}\arctan\frac{2\cos\angle(x_{(n)},y_{(n)})}{\frac{|y_{(n)}|_2}{|x_{(n)}|_2}-\frac{|x_{(n)}|_2}{|y_{(n)}|_2}}\text{and } \hat{\alpha}=\frac{1}{2}\arctan\frac{\frac{|y_{(n)}|_2}{|x_{(n)}|_2}-\frac{|x_{(n)}|_2}{|y_{(n)}|_2}}{2\cos\angle(x_{(n)},y_{(n)})},$$

respectively, where $\angle(\xi,\eta)$ denotes the angle between the vectors ξ and η, and $\arctan(+(-)\infty)=+(-)\pi/2$.
We consider now the the H_0-restricted likelihood function

$$\tilde{L}(a,\alpha)=L|_{H_0}=L(a,a,\alpha)$$

and put

$$\tilde{\alpha}=\begin{cases}\alpha & \text{if } \alpha \text{ is known}\\ \text{mle}(\alpha)|_{H_0} & \text{if } \alpha \text{ is unknown.}\end{cases}$$

The partial derivative of $\tilde{L}$ with respect to a attains the value zero if $a=\tilde{a}(\alpha)$ where

$$\tilde{a}(\alpha)^p=\frac{1}{2}\left(\hat{a}(\alpha)^p+\hat{b}(\alpha)^p\right)$$

what ever the value of α is. Thus, for suitable choice of $\tilde{\alpha}$, the maximum value of the restricted likelihood function $\tilde{L}$ can be represented as

$$\tilde{L}(\tilde{a},\tilde{\alpha}) = \frac{C_p^{2n}e^{-2n/p}}{(\tilde{a}^2(\tilde{\alpha}))^n}.$$

An angle $\tilde{\alpha}$ solves the second restricted likelihood equation iff it satisfies the equation

$$\left[\xi_1(\tilde{\alpha}),\xi_2(\tilde{\alpha})\right]_p |\xi_2(\tilde{\alpha})|_p^{p-2} = \left[\xi_2(\tilde{\alpha}),\xi_1(\tilde{\alpha})\right]_p |\xi_1(\tilde{\alpha})|_p^{p-2}$$

which can be reformulated as

$$\left[(\cos\tilde{\alpha})y_{(n)} - (\sin\tilde{\alpha})x_{(n)}, (\cos\tilde{\alpha})x_{(n)} + (\sin\tilde{\alpha})y_{(n)}\right]_p |(\cos\tilde{\alpha})x_{(n)} + (\sin\tilde{\alpha})y_{(n)}|_p^{p-2}$$

$$= \left[(\cos\tilde{\alpha})x_{(n)} + (\sin\tilde{\alpha})y_{(n)}, (\cos\tilde{\alpha})y_{(n)} - (\sin\tilde{\alpha})x_{(n)}\right]_p |\cos\tilde{\alpha})y_{(n)} - (\sin\tilde{\alpha})x_{(n)}|_p^{p-2}.$$

If $p = 2$ then every $\tilde{\alpha} \in [0, \frac{\pi}{2})$ solves this equation. Our test statistic

$$Q = \frac{\tilde{L}(\tilde{a},\tilde{\alpha})}{L(\hat{a},\hat{b},\hat{\alpha})} = \frac{\left(\hat{a}(\hat{\alpha})\hat{b}(\hat{\alpha})\right)^n}{(\tilde{a}(\tilde{\alpha}))^{2n}}$$

satisfies the representation

$$\frac{Q^{p/n}}{4} = \frac{\sum\limits_{i=1}^{n} |\mathfrak{x}_i^T\theta_1(\hat{\alpha})|^p \sum\limits_{i=1}^{n} |\mathfrak{x}_i^T\theta_2(\hat{\alpha})|^p}{\left(\sum\limits_{i=1}^{n} |\mathfrak{x}_i^T\theta_1(\tilde{\alpha})|^p + \sum\limits_{i=1}^{n} |\mathfrak{x}_i^T\theta_2(\tilde{\alpha})|^p\right)^2}.$$

The likelihood ratio decision rule means to reject H_0 if for some suitably chosen $t \in (0,1)$ there holds $Q < t$.

We remark that the present statistic becomes the same as that in the axes-aligned p-generalized elliptically contoured case in Section 3 if $\hat{\alpha} = \tilde{\alpha} \in \{0, \frac{\pi}{2}\}$ and $\mu_1 = \mu_2 = 0$. Moreover, the present decision rule can be equivalently reformulated then as to reject H_0 if $T_p = \frac{|Y_{[1]}|_p^p}{|Y_{[2]}|_p^p}$ attains sufficiently small or large values where

$$Y_{[i]}^T = \theta_i(\alpha)^T \begin{pmatrix} X_1 & X_2 & \dots & X_n \\ Y_1 & Y_2 & \dots & Y_n \end{pmatrix}, \; i = 1,2.$$

Example 1 • *If* $\alpha = 0$ *then* $\theta_1(\alpha) = (1,0)^T$ *and* $\theta_2(\alpha) = (0,1)$.

• *If* α *is known (then* $\hat{\alpha} = \tilde{\alpha} = \alpha$ *) then the considered decision rule means in other words to reject* H_0 *for large values of* $\sqrt{R_p} + 1/\sqrt{R_p}$ *where*

$$R_p = \frac{\sum\limits_{i=1}^{n} \left|\mathfrak{x}_i^T\theta_1(\alpha)\right|^p}{\sum\limits_{i=1}^{n} \left|\mathfrak{x}_i^T\theta_2(\alpha)\right|^p}.$$

Note that, for $i = 1, \dots, n$,

$$D(\alpha)\mathfrak{x}_i \sim \Phi_{(a,b),p,(0,0),I_2}$$

where I_2 *denotes the* 2×2*-unit matrix. The statistic* R_p *has therefore independently of the actual value of the angle of rotation* α *the same p-generalized Fisher distribution as the likelihood ratio statistic in Section* 3. *Thus, in this case, rotational dependence is without influence onto the null distribution of the likelihood ratio statistic for proving* $H_0 : \sigma_1^2 = \sigma_2^2$, *or not.*

Remark Since the purpose is to test whether H_0 or not, when there is a correlation between two groups, one might like to consider testing the significance of correlation structure prior to testing H_0 or not. In the present situation where is no rotational correlation, this would mean to test whether the shape-scale parameter satisfies $\tilde{H}_0 : p = 2$ or not. Searching the literature the author was not aware of a significance test for this hypothesis, see for example in González-Farías et al. (2009), Yu et al. (2012), Purczynski and Bednarz-Okrzynska (2014) and Pascal et al. (2017).

5 The semi-inner product $[\,.,.]_p$ appears also in estimating location

Many authors were dealing with estimating parameters of the p-power exponential distribution. Without aiming completeness, and without going into any details, we refer to Stacy and Mihram (1965), Harter (1967), Rahman and Gokhale (1996), Varanasi and Aazhang (1989), Do and Vetterli (1988), Mineo and Ruggieri (2005), González-Farías et al. (2009), Saatci and Akan (2010).

It is well known from the Gauss-Markov theorem that orthogonal projections play a fundamental role in estimating parameters in the theory of linear models. The notion of an orthogonal projection is closely connected with that of a scalar product. If the standard Gaussian distribution is the sample distribution in R^n then it is natural to use the Euclidean norm for several statistical calculations. This norm is generated by the Euclidean scalar product in $R^n \times R^n$. If the density of the sample vector $X_{(n)}$ is, for some $p \geq 1$,

$$f_{X_{(n)}}(x) = \prod_{i=1}^{n} \frac{C_p}{\sigma} \exp\left\{-\frac{|x_i - \mu|^p}{p\sigma^p}\right\}, \; x = (x_1, \ldots, x_n)^T \in R^n$$

then it is natural to work with the norm $|.|_p$ which is not generated by an inner product if $p \neq 2$,

$$f_{X_{(n)}}(x) = \left(\frac{C_p}{\sigma}\right)^n \exp\left\{-\frac{|x - \mu 1_n|_p^p}{p\sigma^p}\right\}, \; x \in R^n, 1_n = (1, \ldots, 1)^T \in R^n.$$

It is known, however, that this norm is generated by the semi-inner product $[\,.,.]_p$ considered in Section 4, $|x|_p \; = [x, x]_p^{1/2}$. The present section is aimed to verify that this semi-inner product plays also a role in estimating the location parameter of a p-power exponential distribution. Let $L(\mu) = f_{X_{(n)}}(x)$. Maximizing L with respect to μ is equivalent to minimizing the function

$$f(\mu) = |x - \mu 1_n|_p^p, \; \mu \in R.$$

Let $x_{(1)} \leq \ldots \leq x_{(n)}$ be the ordered values of the concrete sample vector $x = (x_1, \ldots, x_n)^T$. Given μ, there exists a natural number n_1 such that

$$f(\mu) = \sum_{i=1}^{n_1} (\mu - x_{(i)})^p + \sum_{i=n_1+1}^{n} (x_{(i)} - \mu)^p.$$

Thus, $f'(\mu) = p(f_1(\mu) - f_2(\mu))$ where

$$f_1(\mu) = \sum_{x_{(i)} < \mu} (\mu - x_{(i)})^{p-1} \text{ and } f_2(\mu) = \sum_{x_{(i)} > \mu} (x_{(i)} - \mu)^{p-1}.$$

If $p \in (1, \infty)$, the functions $f_i, i = 1, 2$ are monotonously increasing/decreasing if $i = 1/i = 2$, respectively. Moreover, these functions are continuous, and satisfy $f_1(x_{(1)}) = 0, f_1(x_{(n)}) > 0$ and $f_2(x_{(n)}) = 0, f_2(x_{(1)}) > 0$. Thus there exists a uniquely determined $\hat{\mu}$ such that $(\hat{\mu}, f_1(\hat{\mu})) = (\hat{\mu}, f_2(\hat{\mu}))$ is the intersection point of the curves $\{(\mu, f_1(\mu)) : \mu \in [x_{(1)}, x_{(n)}]\}$ and $\{(\mu, f_2(\mu)) : \mu \in [x_{(1)}, x_{(n)}]\}$. Based upon a bisection algorithm, $\hat{\mu}$ can be numerically calculated and is the solution of the equation $f'(\mu) = 0|_{\mu=\hat{\mu}}$, i.e.

$$\sum_{i=1}^{n} |x_i - \hat{\mu}|^{p-1} \text{sign}(\hat{\mu} - x_i) = 0. \tag{4}$$

Example 2 *If $p = 2$ then (4) reads as*

$$-\sum_{i=1}^{n_1} \left(x_{(i)} - \hat{\mu}\right) + \sum_{i=n_1+1}^{n} \left(x_{(i)} - \hat{\mu}\right)(-1) = 0,$$

thus $\hat{\mu} = \bar{x}_n$.

We consider now two cases excluded so far.

Example 3 *In the case $p = 1$,*

$$f'(\mu) = \sum_{i=1}^{n} \text{sign}\ (\mu - x_i) = \left(\sum_{x_{(i)}<\mu} 1 - \sum_{x_{(i)}>\mu} 1\right),$$

thus $f'(\hat{\mu}) = 0$ iff $\natural\{x_i < \hat{\mu}\} = \natural\{x_i > \hat{\mu}\}$ where $\natural\{\ldots\}$ means the number the event written between the brackets occurs. For odd n, $\hat{\mu} = x_{[n/2]+1}$, and for even n, every $\hat{\mu}$ from $[x_{[n/2]}, x_{[n/2]+1}]$ satisfies this condition, thus $\hat{\mu}$ is the sample median.

Example 4 *In the case $p = \infty$, we first define the notion $|.|_\infty$.*
(a) Let $f(\mu) = \max\{|x_1 - \mu|, \ldots, |x_n - \mu|\}$ and i^* such that $f(\mu) = |x_{i^*} - \mu|$. Then

$$|x_{(n)} - \mu 1_n|_p = |x_{i^*} - \mu| \left(1 + \frac{\sum_{i \neq i^*} |x_i - \mu|^p}{|x_{i^*} - \mu|^p}\right)^{1/p}$$

where

$$\left(1 + \frac{\sum_{i \neq i^*} |x_i - \mu|^p}{|x_{i^*} - \mu|^p}\right)^{1/p} \geq 1 \text{ and } \left(1 + \frac{\sum_{i \neq i^*} |x_i - \mu|^p}{|x_{i^*} - \mu|^p}\right)^{1/p} \leq n^{1/p} \to 1, p \to \infty.$$

By definition,

$$|x_{(n)} - \mu 1_n|_\infty = \lim_{p \to \infty} |x_{(n)} - \mu 1_n|_p.$$

(b) If for some i^* there holds $|x_{i^*} - \hat{\mu}| = \max\{|x_1 - \hat{\mu}|, \ldots, |x_n - \hat{\mu}|\}$ then $\hat{\mu}$ is maximum likelihood estimator of μ. The number $\hat{\sigma} = \max|x_i - \hat{\mu}|$ is the smallest number satisfying $-\hat{\sigma} \leq x_i - \hat{\mu} \leq \hat{\sigma}, \forall i$, thus $x_{(n)} - \hat{\sigma} \leq \hat{\mu} \leq x_{(1)} + \hat{\sigma}$. The smallest possible $\hat{\sigma}$ satisfies $2\hat{\sigma} \geq x_{(n)} - x_{(1)}$, thus $\hat{\sigma} = (x_{(n)} - x_{(1)})/2$. It follows that $\hat{\mu} = (x_{(n)} + x_{(1)})/2 =$ midrange $\{x_1, \ldots, x_n\}$.

With

$$f''(\hat{\mu}) = p(p-1) \sum_{i=1}^{n} |x_i - \hat{\mu}|^{p-2} > 0 \text{ if } p > 1,$$

it follows in the general setting that the uniquely determined solution $\hat{\mu}$ of the equation (4) is a relative maximum point of the likelihood function L, thus $\hat{\mu} = mle(\mu)$. This means that

$$\sum_{x_i<\hat{\mu}} |x_i - \hat{\mu}|^{p-2}(\hat{\mu} - x_i) = \sum_{x_i>\hat{\mu}} |x_i - \hat{\mu}|^{p-2}(x_i - \hat{\mu})$$

or, equivalently,

$$\hat{\mu} \sum_{i=1}^{n} |x_i - \hat{\mu}|^{p-2} = \sum_{i=1}^{n} x_i |x_i - \hat{\mu}|^{p-2}.$$

Thus, on the one hand, $\hat{\mu}$ solves the oscillating fixed point equation

$$\hat{\mu} = \frac{\sum_{i=1}^{n} x_i |x_i - \hat{\mu}|^{p-2}}{\sum_{i=1}^{n} |x_i - \hat{\mu}|^{p-2}}.$$

On the other hand, it follows that $\hat{\mu} = mle(\mu)$ satisfies the equation

$$0 = [1_n, x - \hat{\mu} 1_n]_p \tag{5}$$

which means that $\hat{\mu} = \bar{x}_n$ if $p = 2$. Under suitable assumptions upon the convergence of $\hat{\mu}$ and the limit $\mu^* = \lim\limits_{n\to\infty} \hat{\mu}$, it follows

$$\lim_{n\to\infty} n^{-2/p}[1_n, x - \hat{\mu} 1_n]_p = \frac{E(X - \mu^*)|X - \mu^*|^{p-2}}{(E|X - \mu^*|^p)^{(p-2)/p}},$$

thus $\lim\limits_{n\to\infty} n^{-2/p}[1_n, x - \hat{\mu} 1_n]_p = 0$ can be reformulated as

$$EX|X - \mu^*|^{p-2} - \mu^* E|X - \mu^*|^{p-2} = 0. \tag{6}$$

6 Simulation of star-shaped distributed random vectors

6.1 Preliminary remarks

It may be of interest to determine exact distributions of the statistics dealt with in Sections 2-5. To this end, one might use various analytical tools like, e.g., a geometric measure representation as a starting point of explicit analytical derivations.

As an alternative to such derivations, we present here simulation methods which allow to generate stochastic approximations of statistical distributions.

Let $(X_{1,j}, Y_{1,j})^T, \dots, (X_{n,j}, Y_{n,j})^T$, $j = 1, \dots, N$ be independent samples of independent random vectors following the rotational dependent p-generalized elliptically contoured density $f_{(X,Y)}$ defined in Section 4.1., and let further

$$T_j = T\left((X_{1,j}, Y_{1,j})^T, \dots, (X_{n,j}, Y_{n,j})^T\right), j = 1, \dots, N$$

be a sample of i.i.d. copies of a real valued statistic T. For sufficiently large N, the probability $P(T < t)$ can be stochastically approximated by the relative frequency $\frac{1}{N}\sum_{i=1}^{N} I_{(-\infty,t)}(T_i)$. To this end, we present an acceptance-rejection method for simulating random vectors $(X_{i,j}, Y_{i,j})^T$ in Section 6.2, and a generalized polar method in Section 6.3. This will be done even under the much more general assumption that $(X, Y)^T$ follows an arbitrary star-shaped distribution. This class includes that of p-generalized elliptically contoured distributions. For approaches to general distribution classes see

Fernández et al. (1995), Arnold et al. (2008), Kamiya et al. (2008), Sarabia and Gómez-Déniz (2008), Balkema and Nolde (2010). A geometric representation of star-shaped distributions is given in Richter (2014).

We refer to the letter paper for main notions and recall that $(X, Y)^T$ allows the stochastic representation $X \stackrel{d}{=} R \cdot U$ where R and U are stochastically independent, R is a non-negative random variable, and the singular random vector U follows the star-generalized uniform distribution ω_S on the Borel σ-field $\mathfrak{B}(S)$ of the star-sphere S being the topological boundary of a suitably defined star body K,

$$\omega_S(A) = \frac{O_S(A)}{O_S(S)}, \; A \in \mathfrak{B}(S).$$

For the distribution considered in Section 4.1, K can be chosen as a rotated through the origin axes-aligned p-generalized ellipsoid, $K = D^T(\alpha)B_{(a_1,a_2),p}$, and O_S means the corresponding star-generalized surface content measure.

6.2 Dependent *p*-generalized acceptance-rejection method

General aspects of acceptance-rejection or simply rejection methods are studied in Kalke and Richter (2013) and applied there to the p-generalized rejecting polar method. Platonically generalized uniformly distributed and polyhedral star-shaped distributed random vectors are generated this way in Richter and Schicker (2014, 2016a) as well as Richter and Schicker 2016b, respectively. If the star-spheres are represented in a certain analytical way, Nolan (2016) aims to exploit the geometric measure representation approximatively in a sense, not yet explicitly defined. Here, we demonstrate how to generate in four steps star-shaped distributed vectors.

Step 1. To start with, let $C_i, i = 1, \ldots, d$ be positive constants, $C_{(d)} = (C_1, \ldots, C_d)^T$ and $0_{(d)} = (0, \ldots, 0)^T \in \mathcal{R}^d$. We denote by

$$(0_{(d)}, C_{(d)}) = (0, C_1) \times \ldots \times (0, C_d)$$

an axes-aligned d-dimensional rectangle and by $O \in \mathcal{R}^{d \times d}$ an orthogonal matrix. Using the further notation

$$O(0_{(d)}, C_{(d)}) = \left\{Ox : x \in (0_{(d)}, C_{(d)})\right\},$$

we assume that the random vectors $\xi_n, n = 1, 2, \ldots$ follow the uniform distribution on $O(0_{(d)}, C_{(d)})$, and are independent. Because the vector $O^{-1}\xi_n$ follows the product measure of uniform distributions on univariate intervals, $U_{(0,C_1)} \times \ldots \times U_{(0,C_d)}$, it can immediately be simulated.

Step 2. Let the acceptance region

$$A \in \mathcal{B}^{(d)} \cap [\, O(0_{(d)}, C_{(d)})]$$

be a star body having the origin as an interior point. According to Remark A.1 in Kalke and Richter (2013), the stopping time

$$\tau^A = \inf\{n \in \mathcal{N} : \xi_n \in A\}$$

is almost surely finite if $P(\xi_1 \in A) > 0$. The following lemma says that the stopping element

$$\xi_{\tau^A} = \sum_{n=1}^{\infty} I_{\{\tau^A = n\}} \xi_n$$

is uniformly distributed in the acceptance region A, $\xi_{\tau_A} \sim U_A$.

Lemma 1 The stopping element ξ_{τ^A} satisfies the equation

$$P(\xi_{\tau^A} \in M) = U_A(M) \text{ for all } M \in \mathcal{B}^{(d)} \cap \left[O(0_{(d)}, C_{(d)})\right].$$

Proof It follows from

$$P(\xi_{\tau^A} \in M) = P\left(\bigcup_{n=1}^{\infty} \left\{\tau^A = n, \xi_n \in M\right\}\right)$$

$$= P(\xi_1 \in M)[1 + P(\xi_1 \notin A) + P^2(\xi_1 \notin A) + \ldots]$$

that

$$P(\xi_{\tau^A} \in M) = P(\xi_1 \in M)[1 + \frac{P(\xi_1 \notin A)}{1 - P(\xi_1 \notin A)}]$$

□

Example 5 *For simulating a random vector following a p-generalized elliptically contoured distribution law, put* $A = OB_{a,p}$ *with* $B_{a,p} = \left\{x \in \mathcal{R}^d : \sum_{i=1}^{d} |\frac{x_i}{a_i}|^p \le 1\right\}$, *and* $C_i \ge a_i > 0, i = 1, \ldots, d, a = (a_1, \ldots, a_d)^T$.

Example 6 *For simulating norm or antinorm contoured distributed vectors one can chose the acceptance region* $A = \left\{x \in \mathcal{R}^d : ||x|| \le 1\right\}$, *where* $||.||$ *is an arbitrary norm or antinorm and the constants* $C_i > 0$ *are chosen such that* $A \subset O[0_{(d)}, C_{(d)}]$.

Example 7 *If* $A = P$ *is a star-shaped polyhedron having the origin as an interior point, one can check whether a point belongs to A using the various representations of the Minkowski functional of A given in Richter and Schicker* (2016b).

Example 8 *If A is as described in Nolan (2016), check the condition given there.*

Step 3. It is well known that if A is a star-shaped subset of $\mathcal{R}^d$ *having the origin as an interior point then the Minkowski functional* $h_A(x) = \inf\{\lambda > 0 : x \in \lambda A\}, x \in \mathcal{R}^d$ *is well defined. A normalization of the stopping element based upon this functional is used in the following lemma.*

Step 4. Lemma 2 The random element $X_{\partial A} = \xi_{\tau^A}/h_A(\xi_{\tau^A})$ *follows the star-generalized uniform distribution on* $S = \partial A = \left\{x \in \mathcal{R}^d : h_A(x) = 1\right\}$, $X_{\partial A} \sim \omega_S$ *for short.*

Proof Let $\tilde{M} \in \mathcal{B}^d \cap \partial A$ and *sector*$(\tilde{M}) = \left\{\lambda x : x \in \tilde{M}, 0 \le \lambda \le 1\right\}$. Then

$$P(X_{\partial A} \in \tilde{M}) = P(\xi_{\tau^A} \in sector(\tilde{M})) = U_A(sector(\tilde{M}))$$

□

Example 5, ***continued.*** *The Minkowski functional of the set* $B_{a,p} = O^{-1}A$ *is* $h_{O^{-1}A}(x) = |x|_{a,p} = \left(\sum_{i=1}^{d} |x_i|^p\right)^{1/p}$ *and, with* $S = \partial B_{a,p}$,

$$O_S(\tilde{M}) = \int\limits_{\{(x_1,\dots,x_{d-1}):(x_1,\dots,x_d)^T \in \tilde{M}\}} \frac{d(x_1 \dots x_{d-1})}{(1 - \sum_{1}^{d-1} |\frac{x_i}{a_i}|)^{1/p}}, \tilde{M} \in \mathcal{B}^d \cap E_{a,p}.$$

Example 6, ***continued.*** *For arbitrary norm or antinorm* $||.||$, $h_A(x) = ||x||$, *and for all* $\tilde{M} \in \mathcal{B}^d \cap S$, $S = \{x : ||x|| = 1\}$,

$$O_S(\tilde{M}) = \int\limits_{\{\vartheta=(x_1,\dots,x_{d-1}):(\vartheta,x_d(\vartheta))^T \in \tilde{M}\}} h_{K^*}(N(\vartheta))d\vartheta$$

where K^* *is the unit ball of the norm* $||.||^*$ *being dual to the norm* $||.||$ *in the first case, and the antipolar set of* $K = \{x : ||x|| \leq 1\}$ *in the second one. Moreover,* $N(\vartheta)$ *is the outer/inner normal vector to S at the point* $(\vartheta, x_d(\vartheta)), N(\vartheta) = (grad\, x_d(\vartheta), -1)^T$.

Remark 1 *Let NonNegSim denote the set of all non-negative random variables for which there is known a simulation method. Extensive overviews of simulation algorithms for non-uniform random variables are given in Rubinstein* (1981) *and Devroye* (1986). *If* $R \in$ *NonNegSim is independent of* $X_{\partial A}$ *where* $X_{\partial A}$ *is a star-generalized uniformly on the star sphere* ∂A *distributed random vector then the random vector* $RX_{\partial A}$ *follows a star-shaped distribution centered at the origin,* Φ_A *say.*

As to summarize, Steps *1-4* together constitute an acceptance-rejection algorithm for simulating random vectors following a star-shaped distribution law.

Example 9 *In case of a distribution having a density generating function, g say, the cumulative distribution function of* $R = R(g)$ *is*

$$F_{R(g)}(r) = \frac{1}{I(g)O_A(\partial A)} \int_0^r \rho^{d-1} g(\rho)d\rho \text{ where } I(g) = \int_0^\infty r^{d-1} g(r)dr,$$

thus $R(g)$ *can be simulated accordingly. To this end, let U be uniformly distributed on* $(0, 1)$, *then* $F^{-1}_{R(g)}(U) \stackrel{d}{=} R(g)$.

Remark 2 *If a density generating function g satisfies the equation*

$$O_A(\partial A)I(g) = 1$$

then it is called a density generator. Methods of estimating a density generator are described in Liebscher and Richter (2017).

6.3 Dependent *p*-generalized polar method

The classical polar method is due to Box and Muller (1958). If the acceptance rate of the algorithm described in the previous section is not large enough, or for some other reason, one might seek for a direct star-generalization of the polar method. We just mention here that there are different particular methods for directly generating the star-generalized uniform distribution on a star sphere. For the p-generalized polar method, e.g., such

method has been established in Kalke and Richter (2013) and applied in Richter (2015a). The independent coordinate representation of general two-dimensional norm contoured distributions which is the basis for a norm-generalization of the polar method is proved in Richter (2015b).
Below, we describe an algorithm for an (a, p, O)-generalization of the polar method of Box and Muller where $a = (a_1, \dots, a_d)^T, a_i > 0, i = 1, \dots, d; p > 0$ and O is an orthogonal $d \times d$-matrix. To this end, we assume that a random vector $X = (X_1, \dots, X_d)^T$ follows an axes-aligned p-generalized elliptically contoured distribution having density generating function g and vector of scaling parameters a, $X \sim \Phi_{g,B_{a,p}}$. Let further $Y = OX + \nu$ denote a transformation vector, with an orthogonal $d \times d$-matrix O and $\nu \in R^d$, and put

$$h_0(r) = C_0 r^{d-1} g(r) I_{(0,\infty)}(r)$$

and, for $\phi_i \in [0, \pi)$, $i = 1, \dots, d-2, \phi_{d-1} \in [0, 2\pi)$,

$$h_i(\phi_i) = C_i \frac{(\sin_{(a_i,a_{i+1};p)}(\phi_i))^{d-i-1}}{N^2_{(a_i,a_{i+1};p)}(\phi_i)}$$

where the generalized trigonometric functions $\sin_{(a_i,a_{i+1};p)}(\phi_i), \cos_{(a_i,a_{i+1};p)}(\phi_i)$ and the normalizing functions $N_{(a_i,a_{i+1};p)}(\phi_i)$ are defined in Richter (2014). With suitably chosen constants C_i, the functions h_i are the densities of independent random variables $R, \Phi_1, \dots, \Phi_{n-1}$ jointly satisfying the stochastic representation $X = RU$ where

$$\begin{aligned} U_1 &= a_1 \cos_{(a_1,a_2;p)}(\Phi_1), U_2 = a_2 \sin_{(a_1,a_2;p)}(\Phi_1) \cos_{(a_2,a_3;p)}(\Phi_2), \dots, \\ U_d &= a_d \sin_{(a_1,a_2;p)}(\Phi_1) \dots \sin_{(a_{d-2},a_{d-1};p)}(\Phi_{d-2}) \sin_{(a_{d-2},a_{d-1};p)}(\Phi_{d-1}), \end{aligned} \tag{7}$$

cf. Definition 4 in the same paper.

Step 1 Start the algorithm by generating a non-negative random number R according to the density h_0.
Step 2 Generate random numbers $\Phi_1, \dots, \Phi_{d-2}$ from $[0, \pi)$ and Φ_{d-1} from $[0, 2\pi)$ following the densities $h_1, \dots, h_{d-2}$ and h_{d-1}, respectively.
Step 3 Carry out transformation (7).
Step 4 Return $Y = OR(U_1, \dots, U_d)^T + \nu$.

This algorithm generates a random vector Y following the p-generalized elliptically contoured distribution law $\Phi_{g,a,p,\nu,O}$, see Theorem 4 and Remark 11 in Richter (2014). The particular case $d = 2, a_1 = a_2 = 1$ has been dealt with in Kalke and Richter (2013). Finally, we notice that $X \sim \Phi_{g,a,p,0_d,I_d} = \Phi_{g,B_{a,p},0_d}$.

7 Discussion

Comparing mvv-c with pcv-r models led to some new aspects in testing equality of variances or scaling parameters. Effects of rotational dependence are outlined. A new geometric interpretation of certain likelihood equations is given in terms of a semi-inner product. Based upon the present results for the more specific models dealt with in Sections 2-5, it could be of some interest to re-consider in the future the more general model in Wilcox (2015) and to possibly draw some new conclusions for this model. Our results might further stimulate a comparison of simulation methods, e.g. for particular cases being in the intersection of the work in Nolan (2016) and in Richter and Schicker

(2016a, b). To this end, one would particularly have to determine the Minkowski functionals of the sets considered in Nolan (2016) and then to compare the approximative simulation method there with the exact method presented in Richter and Schicker (2016a,b). Challenging questions are opened for deriving new exact statistical distributions, e.g. of Σ_X/Σ_Y, from dependent sample distributions, and to compare these results with corresponding simulation results. As another open problem it remains to combine rotational and l_p-dependence. Consequences the latter notion has for the derivation of exact distributions of certain statistics have been studied in Müller and Richter (2015, 2016a,b). There, the effects caused by the deviation of a density generating function from that of the p-power exponential law are studied in various situations.

Acknowledgements
The author is grateful to the Associate Editor for his valuable comments leading to an improvement of the paper.

Competing interests
The author declares that he has no competing interest.

References

Arens, T, Hettlich, F, Karpfinger, C, Kockelhorn, U, Lichtenegger, K, Stacehel, H: Mathematik, Spektrum, Heidelberg (2013)
Arnold, BC, Castillo, E, Sarabia, JM: Multivariate distributions defined in terms of contours. J. Stat. Plann. Inference. **138**(12), 4158–4171 (2008). doi:10.1016/j.jspi.2008.03.033
Balkema, G, Nolde, N: Asymptotic independence for unimodal densities. Adv. Appl. Probab. **42**(2), 411–432 (2010). doi:10.1239/aap/1275055236
Box, GEP, Muller, ME: A note on the generation of random normal deviates. Ann. Math. Statist. **29**(2), 610–611 (1958)
Dang, UJ, Browne, RP, McNicholas, PD: Mixtures of multivariate power exponential distributions. Biometrics. **71**(4), 1081–1089 (2015)
Devroye, L: Non-Uniform Random Variate Generation. Springer, Berlin (1986)
Dietrich, T, Kalke, S, Richter, W-D: Stochastic representations and a geometric parametrization of the two-dimensional Gaussian law. Chil J Stat. **4**(2), 27–59 (2013)
Do, MN, Vetterli, M: Wavelet-based texture retrieval using generalized Gaussian density and Kullback-Leibler distance. IEEE Trans. Image. Process. **11**(2), 146–158 (1988). doi:10.1017/S0266466600013384
Dragomir, S: Semi-inner products and applications. 6th edn. Nova Science Publishers, Hauppauge, New York (2004)
Fang, KT, Kotz, S, Ng, KW: Symmetric Multivariate And Related Distributions. Chapman and Hall, London etc (1990)
Fernández, C, Osiewalski, J, Steel, MF: Modeling and inference with ν-spherical distributions. J. Am. Stat. Assoc. **90**(432), 1331–1340 (1995). doi:10.2307/2291523
Games, PA, Winkler, HB, Probert, DA: Robust tests for homogeneity of variance. Educ. Psychol. Meas. **32**(4), 887–909 (1972)
Giles, J: Classes of semi-inner product spaces. Trans. Amer. Math. Soc. **129**(6), 436–446 (1967). doi:10.1080/03610920600672120
Gómez, E, Gómez-Villegas, MA, Marín, JM: A multivariate generalization of the power exponential family of distributions. Commun. Stat. Theory Methods. **27**, 589–600 (1998). doi:10.1080/03610929808832115
Gómez-Villegas, MA, Gómez-Sánchez-Manzano, E, Man, P, Navarro, H: The effect of non-normality in the power exponential distributions. In: Pardo, L, Balakrihnan, N, Gil, MA (eds.), pp. 119–129. Berlin: Springer-Verlag, 972-985, Taylor & Francis, Philadelphia, PA, (2011). doi:10.1080/03610920701762754
González-Farías, G, Domínguez-Molina, A, Rodr+ges-Dagnino, RM: Efficiency of the approximated shape parameter estimator in the generalized Gaussian distribution. IEEE Trans. vehic. techn. **58**(8), 4214–4223 (2009). doi:10.1016/j.jspi.2003.09.008
Harter, L: Maximum-likelihood estimation of the parameters of a four-parameter generalized gamma population from complete and censored samples. Technometrics. **9**(1), 159–165 (1967). doi:10.1016/j.jspi.2003.09.008
Horváth, AG, Lángi, Z, Spirova, M: Semi-inner products and the concept of semi-polarity. **126**(2), 16 (2015). arXiv:13080974v4. doi:10.1016/j.jspi.2003.09.008
Kalke, S, Richter, W-D: Simulation of the p-generalized Gaussian distribution. J. Statist. Comput. Simulation. **83**(4), 639–665 (2013). doi:10.1080/00949655.2011.631187
Kamiya, H, Takemura, A, Kuriki, S: Star-shaped distributions and their generalizations. J. Stat. Plann. Inference. **138**(11), 3429–3447 (2008). doi:10.1016/j.jspi.2006.03.016
Kuwana, Y, Kariya, T: LBI tests for multivariate normality in exponential power distributions. JMultivariate. Anal. **39**, 117–134 (1991)
Levy, KJ: A procedure for testing the equality of p correlated variances. Br. J. Math. Stat. Psychol. **29** (1976). doi:10.1111/j.2044-8317.1976.tb00705.x, http://dx.doi.org/10.1111/j.2044-8317.1976.tb00705.x
Liebscher, E, Richter, W-D: Estimation of star-shaped distributions. Risks xx(x) (2017). doi:10.1111/j.2044-8317.1976.tb00705.x, http://dx.doi.org/10.1111/j.2044-8317.1976.tb00705.x
Lord, F, Novick, M: Statistical theories of mental test scores. Addison-Wesley, Reading, MA (1968)
Lumer, G: Semi-inner product spaces. Trans. Amer. Math. Soc. **100**(1), 29–43 (1961). doi:10.1103/PhysRevE.65.036127
Marsaglia, G, Bray, T: A convenient mathod for generating normal variables. SIAM Rev. **6**, 260–264 (1964). doi:10.1109/TAC.1973.1100374
McCulloch, CE: Tests for equality of variance for paired data. Commun. Stat. Theory Methods. **16** (1987). doi:10.1080/03610928708829445, http://dx.doi.org/10.1080/03610928708829445

Mineo, AM, Ruggieri, M: A software tool for the exponential power distribution: The normalp package. J. Stat. Softw. **12**(4), 24 (2005). doi:10.1017/S0266466600013384
Morgan, WA: A test for the significance of the difference between two variances in a sample from a normal bivariate population. Biometrika. **31** (1939)
Moszyńska, M, Richter, W-D: Reverse triangle inequality. Antinorms and semi-antinorms. Stud. Sci. Math. Hung. **49**(1), 120–138 (2012). doi:10.1556/SScMath.49.2012.1.1192
Mudholkar, GS, Wilding, GE, Mietlowski, WL: Robustness properties of the pitman-morgan test. Commun. Stat. Theory Methods. **32** (2003). doi:10.1081/STA-120022710, http://dx.doi.org/10.1081/STA-120022710
Müller, K, Richter, W-D: Exact extreme value, product, and ratio distributions under non-standard assumptions. AStA Adv. stat. Anal. **99**(1), 1–30 (2015). doi:10.1007/s10182-014-0228-2
Müller, K, Richter, W-D: Exact distributions of order statistics of dependent random variables from $l_{n,p}$-symmetric sample distributions, $n \in \{3, 4\}$. Depend Model. **4**, 1–29 (2016a). doi:10.1515/demo-2016-0001
Müller, K, Richter, W-D: Extreme value distributions for dependent jointly $l_{n,p}$-symmetrically distributed random variables. Depend Model. **4**, 30–62 (2016b). doi:10.1515/demo-2016-0002
Nadarajah, S: The Kotz-type distribution with applications. Statistics. **37**(4), 341–358 (2003). doi:10.1080/0233188031000078060
Nolan, JP: An R package for modeling and simulating generalized spherical and related distributions. J. Stat. Distrib. Appl. **3**(14) (2016)
Pascal, F, Bombrun, L, Tourneret, JY, Berthoumieu, Y: Parameter estimation for multivariate generalized gaussian distribution. IEEE Trans. Signal Process. **61**(23), 5960-5971 (2017). arXiv 1302.6498v2
Pitman, EJG: A note on normal correlation. Biometrika. **31** (1939). doi:10.1093/biomet/31.1-2.9, http://dx.doi.org/10.1093/biomet/31.1-2.9
Purczynski, J, Bednarz-Okrzynska, K: Estimation of the shape parameter of GED distribution for small sample size. Folia Oeconomica Stetinensia. **80**(2), 1–46 (2014)
Rahman, M, Gokhale, D: On estimation of parameters of the exponential power family of distributions. Commun. Stat. Simul. Comput. **25**(2), 291–299 (1996). doi:10.1016/j.jspi.2003.09.008
Richter, W-D: Generalized spherical and simplicial coordinates. J. Math. Anal. Appl. **336**, 1187–1202 (2007). doi:10.1016/j.jmaa.2007.03.047
Richter, W-D: Continuous $l_{n,p}$-symmetric distributions. Lith. Math. J. **49**(1), 93–108 (2009). doi:10.1007/s10986-009-9030-3
Richter, W-D: Geometric disintegration and star-shaped distributions. J. Stat. Distrib. Appl. **1**(20), 1–24 (2014). doi:10.1186/s40488-014-0020-6
Richter, W-D: Convex and radially concave contoured distributions. J. Probab. Stat. **2015**, 1–12 (2015a). doi:10.1155/2015/165468, article ID 165468
Richter, W-D: Norm contoured distributions in R^2. In: Lecture notes of Seminario Interdisciplinare di Matematica. Vol. XII, vol 12, pp. 179–199. Potenza: Seminario Interdisciplinare di Matematica (S.I.M.), University of Basilicata, Italy, (2015b)
Richter, W-D: Exact inference on scaling parameters in norm and antinorm contoured sample distributions. J. Stat. Distrib. Appl. **3**(8), 1–24 (2016). doi:10.1186/s40488-016-0046-z
Richter, WD, Schicker, K: Ball numbers of platonic bodies. J. Math. Anal. Appl. **416**(2), 783–799 (2014). doi:10.1016/j.jmaa.2014.03.007
Richter W-D, Schicker, K: Circle numbers of regular convex polygons. Result Math. **69**(3-4), 521–538 (2016a). doi:10.1007/s00025-016-0534-y
Richter, W-D, Schicker, K: Polyhedral star-shaped distributions. J. Probab. Stat (2016b). Volume 2017, https://doi.org/10.1155/2017/7176897
Rothstein, SM, Bell, WD, Patrick, JA, Miller, H: A jackknife test of homogeneity of variance with paired replicates of data. Psychometrika. **46** (1981). doi:10.1007/BF02293916, http://dx.doi.org/10.1007/BF02293916
Rubinstein, R: Simulation and the Monte Carlo Method. 1st ed. Wiley, New York (1981)
Saatci, E, Akan, A: Respiratory parameter estimation in non-invasive ventilation based on generalized Gaussian noise models. Signal Process. **90**(2), 480–489 (2010). doi:10.1016/j.sigpro.2009.07.015
Sarabia, JM, Gómez-Déniz, E: Construction of multivariate distributions: a review of some recent results. SORT. **32**(1), 3–36 (2008). with discussion by M.C. Pardo and J. Navarro
Snedecor, GW, Cochran, W: Statistical Methods. University Press, Ames, IA (1967)
Stacy, E, Mihram, G: Parameter estimation for a generalized Gamma distribution. Technometrics. **7**(3), 349–358 (1965). doi:10.1016/j.jspi.2003.09.008
Taguchi, T: On a generalization of Gaussian distribution. Ann. Inst. Stat. Math. **30**, 211–242 (1978). doi:10.1007/BF02480215
Tiku, ML, Balakrishnan, N: A robust test for testing the correlation coefficient. Commun. Stat. Simul. Comput. **15** (1986). doi:10.1080/03610918608812554, http://dx.doi.org/10.1080/03610918608812554
Varanasi, M, Aazhang, B: Parametric generalized Gaussian density estimation. J. Acoust. Soc. Amer. **86**(4), 1404–1415 (1989). doi:10.1016/j.jspi.2003.09.008
Wilcox, R: Comparing the variances of two dependent groups. J. Educ Stat. **15**, 237–247 (1990). doi:10.1186/s40488-015-0030-z, http://dx.doi.org/10.2307/1165033
Wilcox, R: Comparing the variances of two dependent variables. J Stat Distrib Appl. **2**(7), 8 (2015). doi:10.1186/s40488-015-0030-z
Yu, S, Zhang, A, Li, H: A review of estimating the shape parameter of generalized Gaussian distribution. J. Comput. Inform. Syst. **8**(21), 9055–9064 (2012)

4

Mean and variance of ratios of proportions from categories of a multinomial distribution

Frantisek Duris[1,2*], Juraj Gazdarica[3], Iveta Gazdaricova[3], Lucia Strieskova[3], Jaroslav Budis[4], Jan Turna[5] and Tomas Szemes[1,3,5]

*Correspondence:
fduris@dcs.fmph.uniba.sk
[1]Geneton s.r.o., Galvaniho 7, 82104 Bratislava, Slovakia
[2]Slovak Centre of Scientific and Technical Information, Lamacska cesta 7315/8A, 81104 Bratislava, Slovakia
Full list of author information is available at the end of the article

Abstract

Ratio distribution is a probability distribution representing the ratio of two random variables, each usually having a known distribution. Currently, there are results when the random variables in the ratio follow (not necessarily the same) Gaussian, Cauchy, binomial or uniform distributions. In this paper we consider a case, where the random variables in the ratio are joint binomial components of a multinomial distribution. We derived formulae for mean and variance of this ratio distribution using a simple Taylor-series approach and also a more complex approach which uses a slight modification of the original ratio. We showed that the more complex approach yields better results with simulated data. The presented results can be directly applied in the computation of confidence intervals for ratios of multinomial proportions.

AMS Subject Classification: 62E20

Keywords: Multinomial distribution, Ratio distribution, Mean, Variance

1 Introduction

Combinations of random variables (e.g., sums, products, ratios) regularly occur in many scientific areas. Particularly useful is the ratio of two random variables. For example, plant scientists use the ratio of leaf area to total plant weight (leaf area ratio) in the plant growth analysis (Poorter and Garnier 1996), and geneticists use the ratio of total genetic diversity distributed among populations to total genetic diversity in the pooled populations as a measure of population differentiation (Culley et al. 2002). The ratio of two fluorescent signals has several applications in fluorescence microscopy, e.g., estimating the DNA sequence copy number as a function of chromosomal location (Piper et al. 1995), and there are many (dimensionless) ratios employed in engineering (Mekic et al. 2012). In case of categorical data (i.e., from a binomial or multinomial distribution), there are numerous applications of ratios as well in consumer preference studies, election poll results, quality control, epidemiology, and so on.

Formally, a ratio distribution is a probability distribution constructed as the distribution of the ratio of two random variables, each having another (known) distribution. More particularly, given two random variables Y_1 and Y_2, the distribution of the random variable Z that is formed as the ratio $Z = Y_1/Y_2$ is a ratio distribution. When using ratio distributions for theoretical and practical purposes, it is helpful to know its mean and variance, preferably in a computationally efficient form. In the case that Y_1 and Y_2 follow normal distributions, and $\mu_{Y_2} = 0$, Z is known as Cauchy distribution (Geary 1930; Fieller 1932; Hinkley 1969; Korhonen and Narula 1989; Marsaglia 2006). Other authors have addressed ratios of binomial proportions (also known as relative risk) (Koopman 1984; Bonett and Price 2006; Price and Bonett 2008), ratios of uniform distributions (Sakamoto 1943), Student's t distributions (Press 1969), Weibull and gamma distributions (Basu and Lochner 1971; Provost 1989; Nadarajah and Kotz 2006), beta distributions (Pham-Gia 2000), Laplace and Bessel distributions (Nadarajah 2005; Nadarajah and Kotz 2005) and others. General notes on the product and ratio of two (not necessarily normal) random variables can also be found in (Frishman 1971; Van Kempen and Van Vliet 2000).

In our paper, we consider a ratio involving two or more random variables that jointly have a multinomial distribution. This situation is similar to relative risk or risk ratio which is the ratio of the probability of an event occurring (for example, developing a disease or being injured) in an exposed group to the probability of the event occurring in a comparison, non-exposed group. However, while the probabilities in the risk ratio are independent (in the sense that they describe two independent events in two independent groups), in our case, the probabilities are tied together through the covariance between multinomial categories. These ratios serve as a common framework for opinion polls, statistical quality control, and consumer preference studies. Confidence intervals for the odds ratio, which can be easily calculated, if the standard deviation is known, are especially important for applications. Nelson (1972) presented estimates, confidence intervals, and hypothesis tests for the odds ratio in trinomial distributions. Piegorsch and Richwine (2001) examined some types of confidence intervals in the context of analysis of genetic mutant spectra. Quesenberry and Hurst (1964) and Goodman (1965) explored methods for obtaining a set of simultaneous confidence intervals for the probabilities of a multinomial distribution. A comparison of performance of various confidence intervals also appeared in Alghamdi (2015); Aho and Bowyer (2015). To the best of our knowledge, however, there has been no analytical treatment of the ratio of multinomial proportions including derivations for formulae for the mean and variance of such a ratio.

A ratio between two or more random variables that jointly have a multinomial distribution also arises in the trending field of the non-invasive prenatal testing of common fetal aneuploidies such as trisomy of the 13th, 18th or 21st chromosome (Chiu et al. 2008; Sehnert et al. 2011; Lau et al. 2012; Minarik et al. 2015). We are currently working on implementation of this model into laboratory practice, and this paper represents a mathematical background of our work. In this paper, we discuss two solutions to the problem of mean and variance of the said ratio. More particularly, we derive asymptotic formulae for the mean and variance of the random variable $Z = Y_1/Y_2$, where $Y_1 = \sum_{k \in I} X_k$ and $Y_2 = \sum_{k \in J} X_k$, $I, J \subset \{1, ..., r\}$ and $I \cap J = \emptyset$, are sums of random variables $X_1, ..., X_r$ which together have a joint multinomial distribution.

2 Solution by Taylor series

There is a simple solution to the mean and variance of the ratio of multinomial proportions that can be derived by using the Taylor series. Formally, let a set of random variables $X_1, ..., X_r$ have a probability function

$$pr\left(X_1 = x_1, ..., X_r = x_r\right) = \frac{n!}{\prod_{i=1}^{r} x_i!} \prod_{i=1}^{r} p_i^{x_i},$$

where x_i are non-negative integers such that $\sum x_i = n$ and p_i are constants with $p_i > 0$ and $\sum p_i = 1$. The joint distribution of $X_1, ..., X_r$ is known as multinomial distribution. Let $u, v \in \{0, 1\}^r$ be two binary vectors such that $\sum u_i > 0, \sum v_i > 0$ and $u_i v_i = 0$ for all i. We define

$$Z_0 = \frac{X \cdot u}{X \cdot v},$$

where $\cdot$ represents a scalar product and $X = (X_1, ..., X_r)$. Without loss of generality, we will restrict our explorations to $r = 3$ and $Z_0 = X_1/X_2$. This holds because the choice vectors u, v have no common X_i; thus, the X_is can be grouped to three disjoint sets: 1) X_is selected by u, 2) X_is selected by v, and 3) all others.

Also, the reader will note that the ratio $Z_0 = X_1/X_2$ can be viewed as a ratio of absolute quantities as well as a ratio of fractions or probabilities because $Z_0 = (X_1/n)/(X_2/n)$.

Before we proceed any further, observe that because of the possible zero in the denominator of Z_0, there is no analytical solution to the mean and variance of the ratio Z_0. A workaround for this problem is to rewrite this ratio using a function that does not have a singularity. Let $Z_0 = f(X_1, X_2) = X_1/X_2$ be a function of two random variables. Then, with $\mu = (\mu_{X_1}, \mu_{X_2})$, we can use the Taylor series to approximate the function f as

$$\begin{aligned} Z_0 = f(X_1, X_2) \approx f(\mu) &+ (X_1 - \mu_{X_1}) \frac{\partial f}{\partial X_1}(\mu) + (X_2 - \mu_{X_2}) \frac{\partial f}{\partial X_2}(\mu) \\ &+ \frac{1}{2}(X_1 - \mu_{X_1})^2 \frac{\partial^2 f}{\partial X_1^2}(\mu) + \frac{1}{2}(X_2 - \mu_{X_2})^2 \frac{\partial^2 f}{\partial X_2^2}(\mu) \\ &+ (X_1 - \mu_{X_1})(X_2 - \mu_{X_2}) \frac{\partial^2 f}{\partial X_1 \partial X_2}(\mu), \end{aligned}$$

from which we have

$$E(Z_0) \approx f(\mu) + \frac{1}{2}\frac{\partial^2 f}{\partial X_1^2}(\mu)\sigma_{X_1}^2 + \frac{1}{2}\frac{\partial^2 f}{\partial X_2^2}(\mu)\sigma_{X_2}^2 + \frac{\partial^2 f}{\partial X_1 \partial X_2}(\mu)\sigma_{X_1, X_2}. \tag{1}$$

Since X_1 and X_2 are terms of a random vector $X = (X_1, X_2, X_3)$ drawn from the multinomial distribution given by (n, p_1, p_2, p_3), we have $\mu_{X_i} = np_i$ and $\sigma_{X_i}^2 = np_i(1 - p_i)$ for $i = 1, 2$, and $\sigma_{X_1, X_2} = -np_1 p_2$. It follows easily that

$$E(Z_0) \approx \frac{p_1}{p_2} + \frac{1}{n}\left(\frac{p_1(1 - p_2)}{p_2^2} + \frac{p_1}{p_2}\right) = \frac{p_1}{p_2}\left(1 + \frac{1}{np_2}\right). \tag{2}$$

For variance, we use a simpler approximation of f

$$f(X_1, X_2) \approx f(\mu) + (X_1 - \mu_{X_1}) \frac{\partial f}{\partial X_1}(\mu) + (X_2 - \mu_{X_2}) \frac{\partial f}{\partial X_2}(\mu),$$

from which we have

$$var(Z_0) \approx \frac{\partial f}{\partial X_1}(\mu)^2 \sigma_{X_1}^2 + \frac{\partial f}{\partial X_2}(\mu)^2 \sigma_{X_2}^2 + 2\frac{\partial f}{\partial X_1}(\mu)\frac{\partial f}{\partial X_2}(\mu)\sigma_{X_1, X_2}, \tag{3}$$

and finally

$$var(Z_0) \approx \frac{1}{n}\left(\frac{p_1(1 - p_1)}{p_2^2} + \frac{p_1^2(1 - p_2)}{p_2^3} + 2\frac{p_1^2}{p_2^2}\right) = \frac{1}{n}\left(\frac{p_1}{p_2}\right)^2\left(\frac{1}{p_1} + \frac{1}{p_2}\right). \tag{4}$$

3 Solution by a modified ratio

3.1 Definition

Let the symbols X, u, and v have the same meaning as in Section 2. We define a new random variable Z_1 as

$$Z_1 = \frac{X \cdot u}{X \cdot v + 1}. \tag{5}$$

The $+1$ in the above definition serves to avoid zero in the denominator, and thus solves the problem with the singularity of Z_0. For the same reasons as in Section 2, we will restrict our explorations to $k = 3$ and $Z_1 = X_1/(X_2 + 1)$.

3.2 Sample space

The sample space $S_{Z_1} \subseteq \mathbb{Q}$ of the random variable Z_1 is limited by the sample space S_X of the multinomially distributed random vector $X = (X_1, X_2, X_3)$. Therefore, if X assumes values from the multinomial distribution given by (n, p_1, p_2, p_3), then Z_1 cannot assume all rational values $a/(b+1)$ for some $a, b \in \mathbb{N}$, but only those that satisfy $a + b \leq n$ and $a, b \geq 0$. Furthermore, values $2/2$ and $4/4$ are considered identical; therefore, different outcomes of random vector X may correspond with the same outcome of Z_1. In other words, each instance (a, b, c) of X corresponds with exactly one instance $a/(b+1)$ of Z_1, while an instance of Z_1 may correspond with multiple instances of X.

Naturally, the probability of a particular value of Z_1 can be determined by summing the probabilities of all (multinomial) vectors that are associated with this value. From this, it follows that if the initial multinomial probability distribution function of random vector X is

$$pr\,(X_1 = a, X_2 = b, X_3 = c) = \binom{n}{a, b, c} p_1^a\, p_2^b\, p_3^c,$$

then the probability distribution function of random variable Z_1 is

$$pr\,(Z_1 = d) = \sum_{\substack{a,b,c \in \{0,\ldots,n\} \\ a+b+c=n \\ a/(b+1)=d}} \binom{n}{a, b, c} p_1^a\, p_2^b\, p_3^c,$$

which can be rewritten as

$$pr\,(Z_1 = d) = \sum_{b=0}^{n} \sum_{\substack{a=0 \\ a/(b+1)=d}}^{n-b} \binom{n}{b} \binom{n-b}{a} p_1^a\, p_2^b\, (1 - p_1 - p_2)^{n-a-b}.$$

3.3 Mean and variance

Now we can state the mean and variance of Z_1. The proofs of the statements can be found in the Appendix.

Theorem 1 *Let $X = (X_1, X_2, X_3)$ be a random vector from the multinomial distribution given by (n, p_1, p_2, p_3). The expected value of the random variable Z_1, given by* (5), *is*

$$E(Z_1) = \frac{p_1}{p_2} \left(1 - (1 - p_2)^n\right).$$

Theorem 2 *Let $X = (X_1, X_2, X_3)$ be a random vector from the multinomial distribution given by (n, p_1, p_2, p_3), where*

$$n > \frac{1-p_2}{p_2}N + \frac{1-2p_2}{p_2}$$

for some natural non-zero N. The variance of the random variable Z_1, given by (5), is

$$var(Z_1) = \left[\frac{p_1}{p_2(1-p_2)}\right]^2 \frac{\frac{1-p_2}{p_1} - 2}{n+2} + \frac{p_1}{p_2(1-p_2)} \frac{\frac{p_1}{1-p_2} - 1}{n+1}$$
$$+ \sum_{k=1}^{N} \frac{\left[\frac{p_1}{p_2(1-p_2)}\right]^2}{\binom{n+k+1}{k} p_2^k} \left[1 - \frac{k+2-\frac{1-p_2}{p_1}}{n+k+2}\right] + O\left(\frac{1}{n^{N+1}}\right).$$

Corollary 1 *For $N = 1$ we have for the variance from Theorem 2*

$$var(Z_1) = \frac{1}{n}\left(\frac{p_1}{p_2}\right)^2 \left(\frac{1}{p_1} + \frac{1}{p_2}\right) + O\left(\frac{1}{n^2}\right).$$

Observe that the formula for the variance is asymptotic in nature, and thus it may not work well for small n and certain configurations of p_1, p_2 and p_3. See Section 5 for more details.

4 Approximate error of solution by a modified ratio

Let

$$Err = g(X_1, X_2) = \frac{X_1}{X_2} - \frac{X_1}{X_2+1} = \frac{X_1}{X_2(X_2+1)}$$

be a function of two random variables expressing the difference between Z_0 and Z_1. Analogous to the Eqs. (1)–(4) from Section 2 and with $f(X_1, X_2) = X_1/\left[X_2(X_2+1)\right]$, we have for the mean and variance of *Err*

$$E(Err) \approx \frac{p_1}{p_2(1+np_2)} + \frac{p_1(1-p_2)\left(1+3np_2+3n^2p_2^2\right)}{np_2^2(1+np_2)^3} + \frac{p_1(1+2np_2)}{np_2(1+np_2)^2}$$
$$= \frac{p_1\left[1+4np_2+(5-p_2)n^2p_2^2+n^3p_2^3\right]}{np_2^2(1+np_2)^3}, \tag{6}$$

$$var(Err) \approx \frac{(1-p_1)p_1}{np_2^2(1+np_2)^2} + \frac{(1-p_2)(p_1+2np_1p_2)^2}{np_2^3(1+np_2)^4} + \frac{2p_1^2(1+2np_2)}{np_2^2(1+np_2)^3}$$
$$= \frac{p_1\left[p_2(1+np_2)^2 + p_1\left\{1+4np_2+(4-p_2)n^2p_2^2\right\}\right]}{np_2^3(1+np_2)^4}. \tag{7}$$

It follows from the Eqs. (6) and (7) that Z_1 is an asymptotically ($n \to \infty$) unbiased estimator of the ratio of multinomial proportions Z_0. Moreover, the Eqs. (6) and (7) can be used to correct the mean and variance of the modified ratio Z_1 to better reflect the mean and variance of the original ratio Z_0. Let $Z_1^{cor} = Z_1 + Err$ be a new random variable. Since the expected value is linear, we have directly

$$E\left(Z_1^{cor}\right) = E(Z_1) + E(Err) \approx$$
$$\approx \frac{p_1}{p_2}\left(1-(1-p_2)^n\right) + \frac{p_1\left[1+4np_2+(5-p_2)n^2p_2^2+n^3p_2^3\right]}{np_2^2(1+np_2)^3}.$$

For the variance, we have

$$var(Z_1^{cor}) = var(Z_1) + var(Err) + 2cov(Z_1, Err),$$

where

$$cov(Z_1, Err) = E(Z_1 \cdot Err) - E(Z_1) \cdot E(Err).$$

To approximate the value of $E(Z_1 \cdot Err)$, we use the Taylor series again, particularly Eq. (1). After some rearrangement, we get

$$E\left(\frac{X_1^2}{X_2(X_2+1)^2}\right) \approx \frac{np_1^2}{p_2(1+np_2)^2} + \frac{(1-p_1)p_1}{p_2(1+np_2)^2} + \frac{p_1^2(1-p_2)\left(1+4np_2+6n^2p_2^2\right)}{p_2^2(1+np_2)^4} + \frac{2p_1^2(1+3np_2)}{p_2(1+np_2)^3} = \frac{p_1\left[p_2(1+np_2)^2 + p_1\left\{1+(5+2p_2)np_2+(8-p_2)n^2p_2^2+n^3p_2^3\right\}\right]}{p_2^2(1+np_2)^4}$$

Thus, we can now easily calculate the value of $var\left(Z_1^{cor}\right)$ (equation omitted due to its length). In the next section, we shall discuss numerical simulations and performance of the presented formulae.

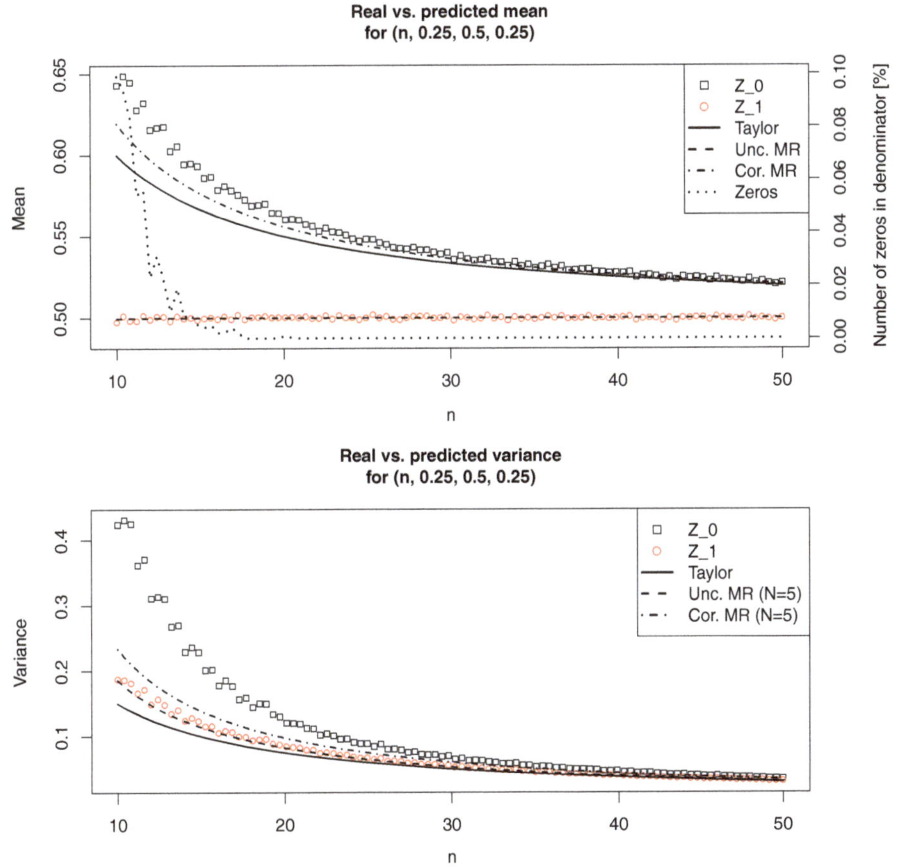

Fig. 1 The simulation results based on the multinomial distribution given by $(n, 0.25, 0.5, 0.25)$, where n ranges from 10 to 50. The mean and variance of the original ratios Z_0 (squares) as well as modified ratios Z_1 (red circles) are compared with models: the Taylor-series model (solid line), the modified ratio model (dashed line), and the corrected modified ratio model (dash-dot line). Additionally, in the upper plot, there is also information about the variable *zeros* on the secondary right axis (dots). In this case, the modified ratio model outperforms the Taylor series model for Z_0 data. Additionally, the uncorrected modified ratio model describes the Z_1 data very well

5 Numerical simulations

Numerical simulations were performed in the following way. We selected several multinomial distributions given by (n, p_1, p_2, p_3) and for each such distribution, we sampled 10^5 random vectors (X_1, X_2, X_3). Vectors with $X_2 = 0$ were counted (variable *zeros*) and omitted from further calculations; that is, they were not replaced by new random vectors. For the vectors with $X_2 \neq 0$, we calculated the ratios $Z_0 = X_1/X_2$, while the ratios $Z_1 = X_1/(X_2 + 1)$ were calculated from all 10^5 sampled vectors. Thus, we obtained $10^5 - zeros$ values of Z_0 and 10^5 values of Z_1. From both sets we calculated the mean and variance of the sampled data. We compared these values with the predictions as follows below.

For the mean, we compared the means of the two data sets with the Taylor-series solution given by Eq. (2), and with the modified ratio (MR) solution given by Theorem 1 with and without the correction given by the Eq. (6).

For the variance, we compared the variances of the two data sets with the Taylor-series solution given by Eq. (4), and with the modified ratio solution given by Theorem 2 with and without the correction (the final formula for corrected variance of the modified ratio was omitted due to its length, but see Section 4 for calculation details). Note that for

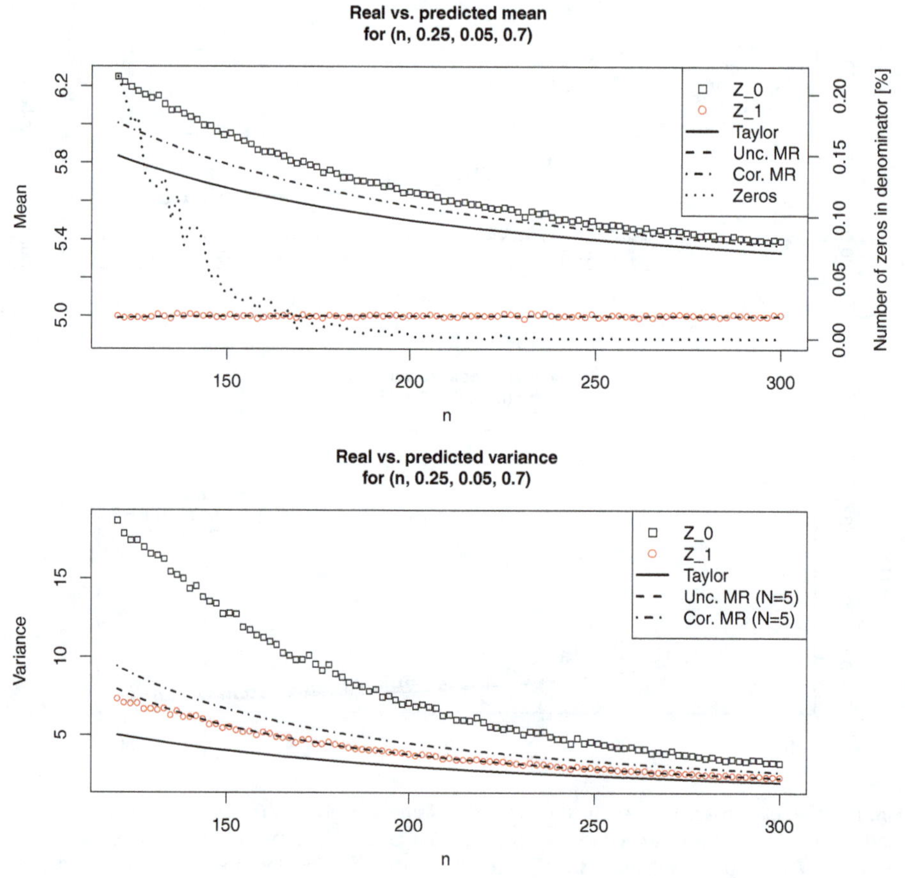

Fig. 2 The simulation results based on the multinomial distribution given by $(n, 0.25, 0.05, 0.7)$, where n ranges from 120 to 300. For the use of symbols see Fig. 1. Again, the modified ratio model outperforms the Taylor-series model for Z_0 data in this case, although the fit is not so close as in Fig. 1. The uncorrected modified ratio model still describes the Z_1 data very well

variance given by Theorem 2, we considered the case $N = 5$ so that its error $O(1/n^6)$ would not interfere with the correction.

Figure 1 shows the simulation results for the multinomial distribution given by $(n = 10, \ldots, 50, p_1 = 0.25, p_2 = 0.5, p_3 = 0.25)$. The corrected modified ratio gives the best model of the mean and variance of Z_0. Observe also that the uncorrected modified ratio is a very precise model of Z_1.

In Fig. 2, when p_2 and n are small, the discrepancy between the models and the data gets larger, although the corrected modified ratio still outperforms the Taylor-series approach. The uncorrected modified ratio is also a very good model of Z_1.

Figures 3 and 4 further explore the limits of the presented models. In Fig. 3, we compared the performance of the variance models in three multinomial distributions (with decreasing value of p_2) for various values of N from Theorem 2. Note that with growing N, there also grows the minimal value of n for which the Theorem 2 holds; therefore, the variance models start from a different n. It will be observed that all models have difficulty describing the initial part of the variance curve of the simulated data. However, one should keep in mind that the formula in Theorem 2 is only asymptotic.

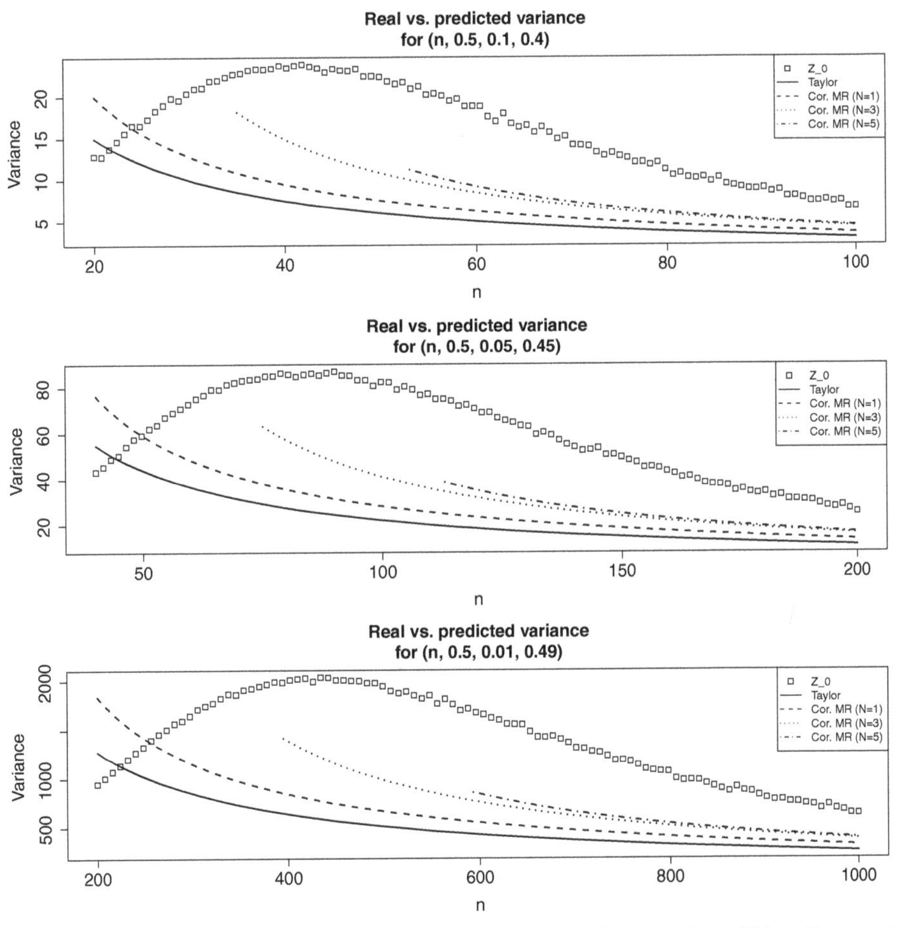

Fig. 3 The simulation results based on three multinomial distributions and various values of N from Theorem 2. Displayed are the results for variance. The simulation data for original ratios Z_0 (squares) are compared with models: the Taylor-series model (solid line) and the corrected modified ratio models with $N = 1$ (dashed line), $N = 3$ (dots), N=5 (dash-dot line). Observe that because of the condition on n in Theorem 2, the modified ratio models do not start at the same value of n for different N

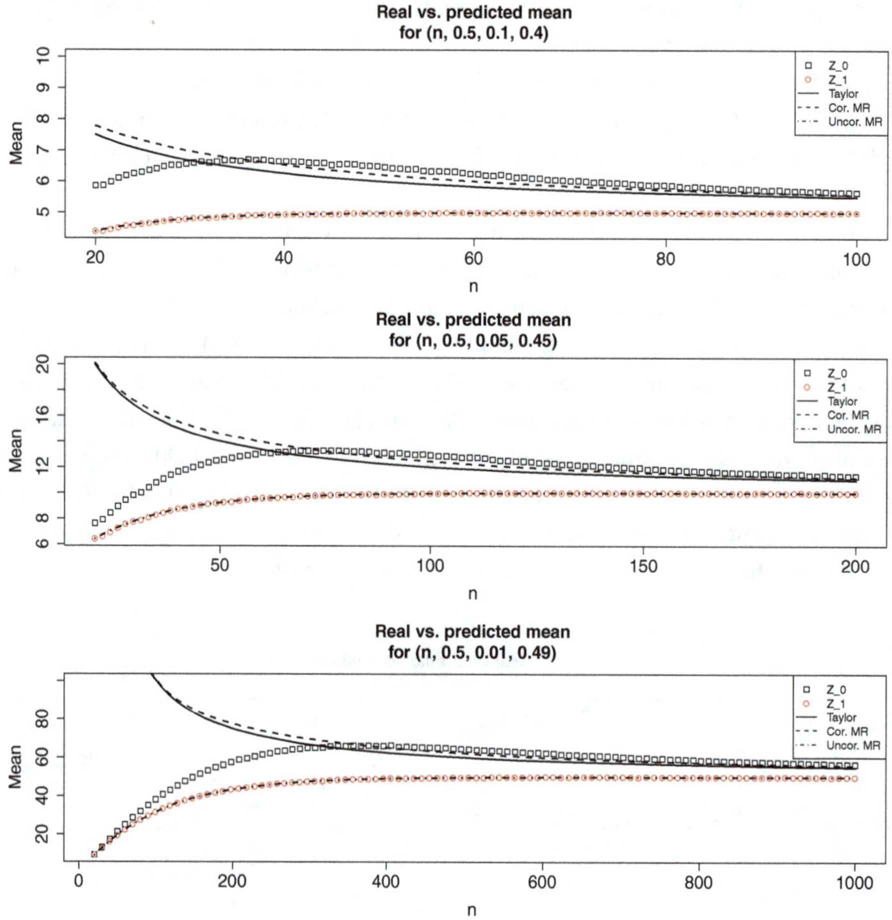

Fig. 4 The results for mean for the data from Fig. 3. The simulation data for original ratios Z_0 (squares) are compared with models: the Taylor-series model (solid line) and corrected modified ratio model (dashed line). Observe the uncorrected modified ratio model (dash-dot line) which exactly models the modified ratios Z_1 (red circles) in all cases

In Fig. 4, we compared the models for mean on the same data as in Fig. 3. Again, for small values of n, the models fail to capture the real trend of the data. On a side note, the data for Z_1 are very well described by the uncorrected modified ratio model from Theorem 1.

The supplemental material contains a script (Additional file 1) to generate similar plots for the user-specified multinomial distribution (n, p_1, p_2, p_3) and a range of n. Given the results from the simulation data, we encourage the reader to use this script and check whether the formulae presented in the paper will provide for a good approximation of Z_0 for his/hers particular multinomial distribution.

Appendix

Proof of Theorem 1

Lemma 1 *Let $n \in \mathbb{N}$ and $R \in \mathbb{R}$. Then it holds*

$$\sum_{k=0}^{n} \binom{n}{k} R^k k = nR\,(1+R)^{n-1}.$$

Proof From $\binom{n}{k} = \frac{n}{k}\binom{n-1}{k-1}$ it directly follows that

$$\sum_{k=0}^{n} \binom{n}{k} R^k k = nR \sum_{k=0}^{n-1} \binom{n-1}{k} R^k = nR(1+R)^{n-1}.$$

□

Proof of Theorem 1 From the definition of the expected value we have

$$E(Z_1) = \sum_{d \in S_{Z_1}} pr(Z_1 = d) \cdot d,$$

where S_{Z_1} is a sample space of Z_1. By using

$$pr(Z_1 = d) = \sum_{b=0}^{n} \sum_{\substack{a=0 \\ a/(b+1)=d}}^{n-b} \binom{n}{b}\binom{n-b}{a} p_1^a p_2^b (1-p_1-p_2)^{n-a-b}$$

from Section 3.2, we can write

$$E(Z_1) = \sum_{d \in S_{Z_1}} \left(\sum_{b=0}^{n} \sum_{\substack{a=0 \\ a/(b+1)=d}}^{n-b} \binom{n}{b}\binom{n-b}{a} p_1^a p_2^b (1-p_1-p_2)^{n-a-b} \right) d.$$

Furthermore, because $\sum_{b=0}^{n}\sum_{a=0}^{n-b}$ enumerates all possible values of a random vector $(X_1, X_2, X_3) = (a, b, n-a-b)$ for the given n, it also enumerates all values of Z_1 including their multiplicities (see Section 3.2). Thus, we can simplify the expression of $E(Z_1)$ into

$$E(Z_1) = \sum_{b=0}^{n}\sum_{a=0}^{n-b} \binom{n}{b}\binom{n-b}{a} p_1^a p_2^b (1-p_1-p_2)^{n-a-b} \frac{a}{b+1}.$$

We rewrite this expression to separate the sums, thus obtaining

$$E(Z_1) = (1-p_1-p_2)^n \sum_{b=0}^{n} \binom{n}{b} \left(\frac{p_2}{1-p_1-p_2}\right)^b \frac{1}{b+1} \cdot \sum_{a=0}^{n-b} \binom{n-b}{a} \left(\frac{p_1}{1-p_1-p_2}\right)^a a. \tag{8}$$

Using Lemma 1, we have for (8)

$$\sum_{a=0}^{n-b} \binom{n-b}{a} \left(\frac{p_1}{1-p_1-p_2}\right)^a a = (n-b)\frac{p_1}{1-p_1-p_2}\left(\frac{1-p_2}{1-p_1-p_2}\right)^{n-b-1}.$$

By putting this back to $E(Z_1)$ and after some rearrangement of the terms, we get

$$E(Z_1) = (1-p_2)^n \left(\frac{p_1}{1-p_2}\right) \sum_{b=0}^{n} \binom{n}{b} \left(\frac{p_2}{1-p_2}\right)^b \frac{n-b}{b+1}. \tag{9}$$

We continue by splitting the following fraction into two terms

$$\frac{n-b}{b+1} = \frac{n+1}{b+1} - 1.$$

By this, the sum in (9) splits into two parts

$$E(Z_1) = A + B,$$

where

$$A = (1-p_2)^n \left(\frac{p_1}{1-p_2}\right) \sum_{b=0}^{n} \binom{n}{b} \left(\frac{p_2}{1-p_2}\right)^b \frac{n+1}{b+1},$$

$$B = (1-p_2)^n \left(\frac{p_1}{1-p_2}\right) \sum_{b=0}^{n} \binom{n}{b} \left(\frac{p_2}{1-p_2}\right)^b (-1).$$

With $\binom{n}{b}\frac{n+1}{b+1} = \binom{n+1}{b+1}$ and some rearrangement of the terms, we obtain

$$A = \frac{p_1}{p_2}\left(\frac{1}{1-p_2} - (1-p_2)^n\right),$$

and a straightforward calculation of B yields

$$B = -\frac{p_1}{1-p_2}.$$

Finally, after putting A and B together, we get

$$E(Z_1) = A + B = \frac{p_1}{p_2} - \frac{p_1}{p_2}(1-p_2)^n = \frac{p_1}{p_2}\left(1 - (1-p_2)^n\right).$$

□

Proof of Theorem 2

The proof of Theorem 2 relies on a series of lemmas and corollaries. For a better navigation through the proof, see Fig. 5 for the proof scheme.

Lemma 2 *Let $n \in \mathbb{N}$ and $R \in \mathbb{R}$. Then it holds*

$$\sum_{k=0}^{n} \binom{n}{k} R^k k^2 = n(n-1)R^2(1+R)^{n-2} + nR(1+R)^{n-1}.$$

Proof From $\binom{n}{k} = \frac{n}{k}\binom{n-1}{k-1}$ and Lemma 1 it follows that

$$\begin{aligned}\sum_{k=0}^{n} \binom{n}{k} R^k k^2 &= nR \sum_{k=0}^{n-1} \binom{n-1}{k} R^k (k+1) \\ &= nR \sum_{k=0}^{n-1} \binom{n-1}{k} R^k k + nR \sum_{k=0}^{n-1} \binom{n-1}{k} R^k \\ &= n(n-1)R^2(1+R)^{n-2} + nR(1+R)^{n-1}.\end{aligned}$$

□

Lemma 3 *Let $n \in \mathbb{N}$ and $R \in \mathbb{R}\backslash\{0\}$. Then, for any $N \in \mathbb{N}$ it holds*

$$\sum_{b=1}^{n} \binom{n}{b} \frac{R^b}{b} = \sum_{k=0}^{N} (A_{2k} - B_{2k}) + A_{2N+1},$$

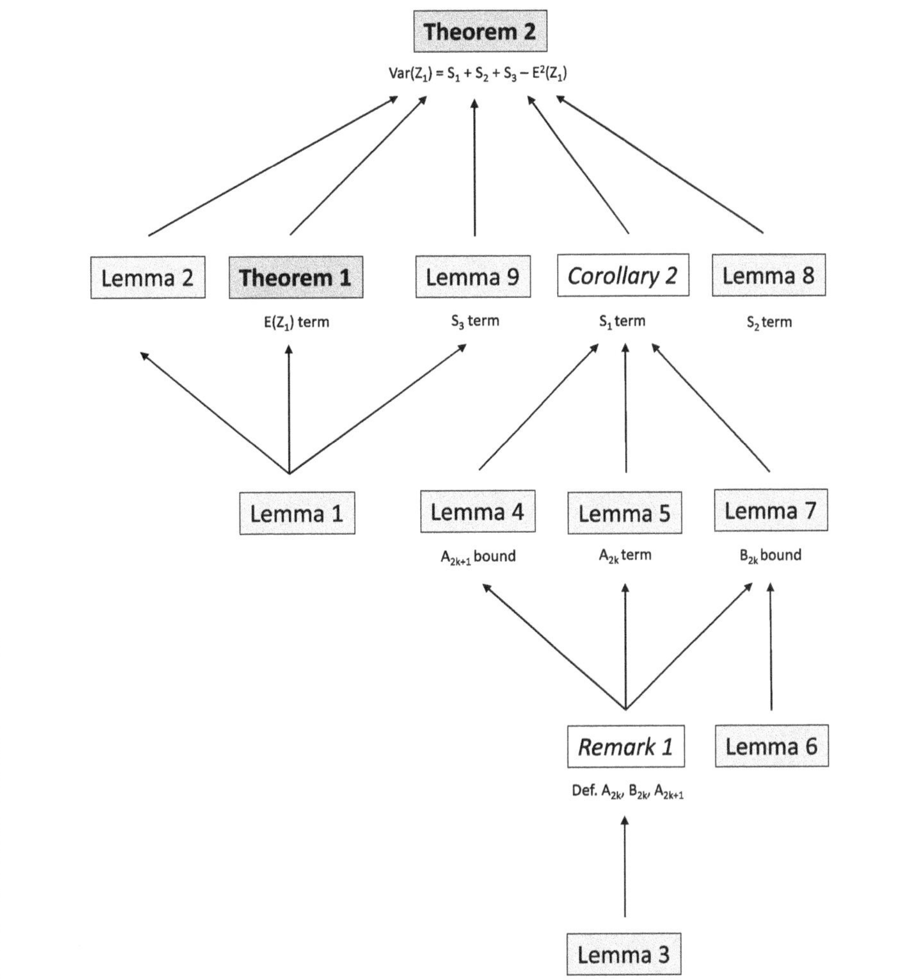

Fig. 5 Scheme of the proof of Theorem 2

where

$$A_{2k} = \left(\prod_{i=1}^{k+1} \frac{1}{n+i}\right) \frac{k!}{R^{k+1}} (1+R)^{n+k+1},$$

$$B_{2k} = \left(\prod_{i=1}^{k+1} \frac{1}{n+i}\right) \frac{k!}{R^{k+1}} \sum_{b=0}^{k+1} \binom{n+k+1}{b} R^b,$$

$$A_{2k+1} = \left(\prod_{i=1}^{k+1} \frac{1}{n+i}\right) \frac{(k+1)!}{R^{k+1}} \sum_{b=k+2}^{n+k+1} \binom{n+k+1}{b} \frac{R^b}{b-(k+1)}.$$

Proof By induction on N. Let $N = 0$. Then, it follows

$$\sum_{b=1}^{n} \binom{n}{b} \frac{R^b}{b} = \sum_{b=1}^{n} \binom{n}{b} \frac{R^b}{b+1} \left(1+\frac{1}{b}\right) = \sum_{b=1}^{n} \binom{n}{b} \frac{R^b}{b+1} + \sum_{b=1}^{n} \binom{n}{b} \frac{R^b}{b(b+1)}.$$

By using $\frac{n+1}{k+1}\binom{n}{k}=\binom{n+1}{k+1}$ and the binomial theorem, we can write

$$\begin{aligned}\sum_{b=1}^{n}\binom{n}{b}\frac{R^b}{b} &= \frac{1}{n+1}\frac{1}{R}\sum_{b=1}^{n}\binom{n+1}{b+1}R^{b+1}+\frac{1}{n+1}\frac{1}{R}\sum_{b=1}^{n}\binom{n+1}{b+1}\frac{R^{b+1}}{(b+1)-1}\\ &= \frac{1}{n+1}\frac{1}{R}\sum_{b=2}^{n+1}\binom{n+1}{b}R^{b}+\frac{1}{n+1}\frac{1}{R}\sum_{b=2}^{n+1}\binom{n+1}{b}\frac{R^{b}}{b-1}\\ &= A_0-B_0+A_1.\end{aligned}$$

The base of the induction holds. Assume that the lemma holds up to some natural N. We prove that it holds for $N+1$ as well. Consider the term A_{2N+1}. We have

$$\begin{aligned}A_{2N+1} &= \left(\prod_{i=1}^{N+1}\frac{1}{n+i}\right)\frac{(N+1)!}{R^{N+1}}\sum_{b=N+2}^{n+N+1}\binom{n+N+1}{b}\frac{R^b}{b+1}\left(1+\frac{N+2}{b-(N+1)}\right)\\ &= X_1+X_2,\end{aligned}$$

where

$$\begin{aligned}X_1 &= \left(\prod_{i=1}^{N+1}\frac{1}{n+i}\right)\frac{(N+1)!}{R^{N+1}}\sum_{b=N+2}^{n+N+1}\binom{n+N+1}{b}\frac{R^b}{b+1},\\ X_2 &= \left(\prod_{i=1}^{N+1}\frac{1}{n+i}\right)\frac{(N+1)!}{R^{N+1}}\sum_{b=N+2}^{n+N+1}\binom{n+N+1}{b}\frac{R^b}{b+1}\frac{N+2}{b-(N+1)}.\end{aligned}$$

Furthermore, by the same trick with the binomial coefficient as above, we rewrite the terms X_1 and X_2 as

$$\begin{aligned}X_1 &= \left(\prod_{i=1}^{N+1}\frac{1}{n+i}\right)\frac{(N+1)!}{R^{N+1}}\frac{1}{n+N+2}\frac{1}{R}\sum_{b=N+2}^{n+N+1}\binom{n+N+2}{b+1}R^{b+1},\\ X_2 &= \left(\prod_{i=1}^{N+1}\frac{1}{n+i}\right)\frac{(N+1)!}{R^{N+1}}\frac{1}{n+N+2}\frac{1}{R}\sum_{b=N+2}^{n+N+1}\binom{n+N+2}{b+1}\frac{R^{b+1}(N+2)}{(b+1)-1-(N+1)}.\end{aligned}$$

After some rearrangement, we finally get (again using the binomial theorem)

$$\begin{aligned}X_1 &= \left(\prod_{i=1}^{N+2}\frac{1}{n+i}\right)\frac{(N+1)!}{R^{N+2}}\sum_{b=N+3}^{n+N+2}\binom{n+N+2}{b}R^{b}=A_{2(N+1)}-B_{2(N+1)},\\ X_2 &= \left(\prod_{i=1}^{N+2}\frac{1}{n+i}\right)\frac{(N+2)!}{R^{N+2}}\sum_{b=N+3}^{n+N+2}\binom{n+N+2}{b}\frac{R^{b}}{b-(N+2)}=A_{2(N+1)+1}.\end{aligned}$$

□

Remark 1 *We will often use Lemma* 3 *with $n+1$ instead of n. Therefore, we restate the Lemma* 3 *with this change. Let $n\in\mathbb{N}$ and $R\in\mathbb{R}\backslash\{0\}$. Then, for any $N\in\mathbb{N}$ it holds*

$$\sum_{b=1}^{n+1}\binom{n+1}{b}\frac{R^b}{b}=\sum_{k=0}^{N}(A_{2k}-B_{2k})+A_{2N+1},$$

where

$$A_{2k} = \left(\prod_{i=2}^{k+2} \frac{1}{n+i}\right) \frac{k!}{R^{k+1}} (1+R)^{n+k+2},$$

$$B_{2k} = \left(\prod_{i=2}^{k+2} \frac{1}{n+i}\right) \frac{k!}{R^{k+1}} \sum_{b=0}^{k+1} \binom{n+k+2}{b} R^b,$$

$$A_{2k+1} = \left(\prod_{i=2}^{k+2} \frac{1}{n+i}\right) \frac{(k+1)!}{R^{k+1}} \sum_{b=k+2}^{n+k+2} \binom{n+k+2}{b} \frac{R^b}{b-(k+1)}.$$

Lemma 4 *Let $p_1, p_2 \in (0,1)$ be some real constants. Let k, n be some non-zero natural numbers. Let A_{2k+1} be the term from Remark 1. Furthermore, let $R = p_2/(1-p_2)$, and let*

$$A = (n+1)n\left(\frac{p_1}{1-p_2}\right)^2 + (n+1)\frac{p_1}{1-p_2},$$

$$D = \frac{(1-p_2)^n}{n+1}\frac{1-p_2}{p_2}.$$

Then, for $\alpha \in [1, k+2]$, it holds

$$ADA_{2k+1} \leq \alpha \frac{n}{(k+2)\binom{n+k+3}{k+2}} \frac{p_1}{p_2^{k+3}(1-p_2)} \left(\frac{p_1}{1-p_2} + \frac{1}{n}\right) = O\left(\frac{1}{n^{k+1}}\right).$$

Proof First of all, for $\alpha \in [1, k+2]$ we have

$$A_{2k+1} = \alpha \left(\prod_{i=2}^{k+3} \frac{1}{n+i}\right) \frac{(k+1)!}{R^{k+2}} \sum_{b=k+3}^{n+k+3} \binom{n+k+3}{b} R^b.$$

This follows easily by applying the inequality

$$\frac{k+2}{b+1} \geq \frac{1}{b-(k+1)} \geq \frac{1}{b+1}$$

to the term A_{2k+1} from Remark 1, which holds for any natural b, k except for pairs $b = k+1$ (in our case $b > k+1$). We can see this by solving the inequality

$$\frac{1+x}{b+1} \geq \frac{1}{b-(k+1)}$$

for x. By this, we get an upper and lower bound on the term A_{2k+1}, which differ by a multiplicative constant $k+2$. Finally, the lemma follows by extending the summation through index b in the term A_{2k+1} to a full range from 0 to $n+k+3$, by applying the binomial theorem and some simple rearrangement of the terms. The O bound follows from the fact that $\binom{n}{k} \geq \left(\frac{n}{k}\right)^k$. □

Lemma 5 *Let $p_1, p_2 \in (0,1)$ be some real constants. Let k, n be some non-zero natural numbers. Let A_{2k} be the term from Remark 1. Furthermore, let $R = p_2/(1-p_2)$, and let*

$$A = (n+1)n\left(\frac{p_1}{1-p_2}\right)^2 + (n+1)\frac{p_1}{1-p_2},$$

$$D = \frac{(1-p_2)^n}{n+1}\frac{1-p_2}{p_2}.$$

Then, it holds

$$ADA_{2k} = \frac{\left(\frac{p_1}{p_2(1-p_2)}\right)^2}{\binom{n+k+1}{k} p_2^k}\left(1 - \frac{k+2-\frac{1-p_2}{p_1}}{n+k+2}\right).$$

Proof The lemma follows easily by a straightforward multiplication of the terms A, D and A_{2k}, and some rearrangement of the terms. □

The following lemma is an extension of one borrowed from Graham et al. (1994).

Lemma 6 *Let* $0 < \alpha < R/(1+R)$ *for some real* $R > 0$. *Then, it holds*

$$\sum_{k \le \alpha n} \binom{n}{k} R^k = R^m 2^{nH(\alpha) - \frac{1}{2}\lg n + O(1)},$$

where $m = \lfloor \alpha n \rfloor$ *and*

$$H(\alpha) = \alpha \lg \frac{1}{\alpha} + (1-\alpha) \lg \frac{1}{1-\alpha}.$$

Proof First of all, we have

$$\frac{\binom{n}{k-1} R^{k-1}}{\binom{n}{k} R^k} = \frac{k}{n-k+1}\frac{1}{R} \le \frac{\alpha n}{n - \alpha n + 1}\frac{1}{R} < \frac{\alpha}{1-\alpha}\frac{1}{R}.$$

Let $m = \lfloor \alpha n \rfloor = \alpha n - \epsilon$. It holds

$$\binom{n}{m} R^m < \sum_{k \le m} \binom{n}{k} R^k < \binom{n}{m} R^m \left(1 + \frac{\alpha}{1+\alpha}\frac{1}{R} + \left(\frac{\alpha}{1-\alpha}\frac{1}{R}\right)^2 + \ldots\right)$$
$$= \binom{n}{m} R^m \frac{(1-\alpha)R}{(1-\alpha)R - \alpha}$$

because

$$\frac{\alpha}{1-\alpha}\frac{1}{R} < 1,$$

which follows from $\alpha < R/(1+R)$. Thus,

$$\sum_{k \le m} \binom{n}{k} R^k = \binom{n}{m} R^m O(1).$$

By Stirling's approximation, we have

$$\log \binom{n}{m} = -\frac{1}{2}\log n - (\alpha n - \epsilon)\log\left(\alpha - \frac{\epsilon}{n}\right) - ((1-\alpha)n + \epsilon)\log\left(1 - \alpha + \frac{\epsilon}{n}\right) + O(1)$$
$$= -\frac{1}{2}\log n - n\alpha \log \alpha - n(1-\alpha)\log(1-\alpha) + O(1),$$

and the lemma follows. □

Lemma 7 *Let* $p_1, p_2 \in (0,1)$ *be some real constants. Let* k, n *be some non-zero natural numbers such that*

$$n > \frac{1-p_2}{p_2}k + \frac{1-2p_2}{p_2}.$$

Let B_{2k} be the term from Remark 1. Furthermore, let $R = p_2/(1-p_2)$, and let

$$A = (n+1)n\left(\frac{p_1}{1-p_2}\right)^2 + (n+1)\frac{p_1}{1-p_2},$$
$$D = \frac{(1-p_2)^n}{n+1}\frac{1-p_2}{p_2}.$$

Then, it holds

$$ADB_{2k} = n(1-p_2)^n \frac{p_1}{p_2(1-p_2)}\left(p_1 + \frac{1-p_2}{n}\right)\frac{2^{k+1}O(1)}{(k+1)(n+k+2)^{\frac{1}{2}}} = O\left(n^{\frac{1}{2}}(1-p_2)^n\right).$$

Proof Let $\alpha = (k+1)/(n+k+2)$. One can easily verify that $\alpha < R/(1+R) = p_2$ because of the choice of n. Thus, we can apply Lemma 6 to the sum from the term B_{2k}. From this, it follows that

$$\sum_{b=0}^{k+1}\binom{n+k+2}{b}\left(\frac{p_2}{1-p_2}\right)^b = \left(\frac{p_2}{1-p_2}\right)^{k+1} 2^{(n+k+2)H(\alpha)-\frac{1}{2}\lg(n+k+2)+O(1)}, \tag{10}$$

where

$$H(\alpha) = \alpha \lg\frac{1}{\alpha} + (1-\alpha)\lg\frac{1}{1-\alpha}.$$

Moreover, for $H(\alpha)$ we have

$$H(\alpha) = \frac{k+1}{n+k+2}\lg\left(\frac{2(n+k+2)}{k+1}\right) - O\left(\frac{1}{n^2}\right),$$

which follows from

$$\lg(1-\alpha) = -\sum_{i=1}^{\infty}\frac{\alpha^i}{i}.$$

Plunging this into (10), we get

$$\sum_{b=0}^{k+1}\binom{n+k+2}{b}\left(\frac{p_2}{1-p_2}\right)^b = \left(\frac{p_2}{1-p_2}\right)^{k+1}\frac{\left(\frac{2(n+k+2)}{k+1}\right)^{k+1}O(1)}{(n+k+2)^{\frac{1}{2}}}.$$

With this, we can write for the whole B_{2k} term from Remark 1

$$B_{2k} = \frac{\left(\frac{2(n+k+2)}{k+1}\right)^{k+1}O(1)}{\binom{n+k+1}{k}(n+k+2)^{\frac{3}{2}}} \le \frac{2^{k+1}\binom{n+k+2}{k+1}O(1)}{\binom{n+k+1}{k}(n+k+2)^{\frac{3}{2}}} = \frac{2^{k+1}O(1)}{(k+1)(n+k+2)^{\frac{1}{2}}} \tag{11}$$

because $\left(\frac{n}{k}\right)^k \le \binom{n}{k}$. Similarly, with $\binom{n}{k} < \left(\frac{ne}{k}\right)^k$, we have for B_{2k}

$$B_{2k} \ge \frac{\left(\frac{2}{e}\right)^{k+1}O(1)}{(k+1)(n+k+2)^{\frac{1}{2}}},$$

if we use

$$\binom{n+k+1}{k}(n+k+2)^{\frac{3}{2}} = \binom{n+k+2}{k+1}(k+1)(n+k+1)^{\frac{1}{2}}.$$

Thus, we have

$$B_{2k} = \frac{2^{k+1}O(1)}{(k+1)(n+k+2)^{\frac{1}{2}}},$$

and the lemma easily follows by multiplying B_{2k} with the term AD. □

Corollary 2 *Let $p_1, p_2 \in (0,1)$ be some real constants. Let n, N be some non-zero natural numbers such that*

$$n > \frac{1-p_2}{p_2}N + \frac{1-2p_2}{p_2}.$$

Let A_{2k}, B_{2k}, $k = 0, ..., N$, and A_{2N+1} be terms from Remark 1. Furthermore, let $R = p_2/(1-p_2)$, and let

$$A = (n+1)n\left(\frac{p_1}{1-p_2}\right)^2 + (n+1)\frac{p_1}{1-p_2},$$

$$D = \frac{(1-p_2)^n}{n+1}\frac{1-p_2}{p_2}.$$

Then, it holds

$$AD\sum_{b=1}^{n+1}\binom{n+1}{b}\left(\frac{p_2}{1-p_2}\right)^b\frac{1}{b} = \left(\frac{p_1}{p_2(1-p_2)}\right)^2\sum_{k=0}^{N}\frac{1-\frac{k+2-\frac{1-p_2}{p_1}}{n+k+2}}{\binom{n+k+1}{k}p_2^k} + O\left(\frac{1}{n^{N+1}}\right).$$

Proof Follows from Lemmas 4, 5 and 7. □

Lemma 8 *Let $p_1, p_2 \in (0,1)$ be some real constants and n some non-zero natural number. Let*

$$B = (2n+1)\left(\frac{p_1}{1-p_2}\right)^2 + \frac{p_1}{1-p_2},$$

$$D = \frac{(1-p_2)^n}{n+1}\frac{1-p_2}{p_2}.$$

Then, it holds

$$BD\sum_{b=1}^{n+1}\binom{n+1}{b}\left(\frac{p_2}{1-p_2}\right)^b = 2\left(\frac{p_1}{1-p_2}\right)^2\frac{1}{p_2} + \frac{1}{n+1}\frac{p_1}{p_2(1-p_2)}\left(1-\frac{p_1}{1-p_2}\right) + O\left((1-p_2)^n\right).$$

Proof Straightforward by binomial theorem. □

Lemma 9 *Let $p_1, p_2 \in (0,1)$ be some real constants and n some non-zero natural number. Let*

$$C = \left(\frac{p_1}{1-p_2}\right)^2,$$

$$D = \frac{(1-p_2)^n}{n+1}\frac{1-p_2}{p_2}.$$

Then, it holds

$$CD\sum_{b=1}^{n+1}\binom{n+1}{b}\left(\frac{p_2}{1-p_2}\right)^b b = \left(\frac{p_1}{1-p_2}\right)^2.$$

Proof Straightforward by Lemma 1 and binomial theorem. □

Proof of Theorem 2 The variance of the random variable Z_1 can be calculated as

$$var(Z_1) = E(Z_1^2) - E^2(Z_1).$$

By Theorem 1, we have

$$E(Z_1) = \frac{p_1}{p_2}\left(1 - (1-p_2)^n\right).$$

So, we only need to determine the value of $E\left(Z_1^2\right)$. From the definition of the expected value, we have

$$E\left(Z_1^2\right) = \sum_{b=0}^{n}\sum_{a=0}^{n-b}\binom{n}{b}\binom{n-b}{a}p_1^a p_2^b(1-p_1-p_2)^{n-a-b}\left(\frac{a}{b+1}\right)^2 = (1-p_1-p_2)^n V_1 V_2,$$

where

$$V_1 = \sum_{b=0}^{n}\binom{n}{b}\left(\frac{p_2}{1-p_1-p_2}\right)^b\left(\frac{1}{b+1}\right)^2,$$

$$V_2 = \sum_{a=0}^{n-b}\binom{n-b}{a}\left(\frac{p_1}{1-p_1-p_2}\right)^a a^2.$$

By application of Lemma 2 to V_2, we obtain

$$E(Z_1^2) = (1-p_2)^n\sum_{b=0}^{n}\binom{n}{b}\left(\frac{p_2}{1-p_2}\right)^b\left(\frac{1}{b+1}\right)^2 W,$$

$$W = (n-b)(n-b-1)\left(\frac{p_1}{1-p_2}\right)^2 + (n-b)\frac{p_1}{1-p_2}.$$

By using the equality

$$\binom{n}{b}\left(\frac{1}{b+1}\right)^2 = \binom{n+1}{b+1}\frac{1}{n+1}\frac{1}{b+1}$$

and adjustment of the summation borders, we get

$$E\left(Z_1^2\right) = \frac{(1-p_2)^n}{n+1}\cdot\frac{1-p_2}{p_2}\cdot\sum_{b=1}^{n+1}\binom{n+1}{b}\left(\frac{p_2}{1-p_2}\right)^b\frac{1}{b}W,$$

$$W = (n-b+1)(n-b)\left(\frac{p_1}{1-p_2}\right)^2 + (n-b+1)\frac{p_1}{1-p_2}.$$

Next, we split the term W according to powers of b, thus obtaining

$$W = A - Bb + Cb^2,$$

where

$$A = (n+1)n\left(\frac{p_1}{1-p_2}\right)^2 + (n+1)\frac{p_1}{1-p_2},$$

$$B = (2n+1)\left(\frac{p_1}{1-p_2}\right)^2 + \frac{p_1}{1-p_2},$$

$$C = \left(\frac{p_1}{1-p_2}\right)^2.$$

If we set

$$D = \frac{(1-p_2)^n}{n+1}\cdot\frac{1-p_2}{p_2},$$

then we can write

$$E\left(Z_1^2\right) = D\sum_{b=1}^{n+1}\binom{n+1}{b}\left(\frac{p_2}{1-p_2}\right)^b\left(\frac{A}{b} - B + Cb\right) = S_1 + S_2 + S_3,$$

where

$$S_1 = AD\sum_{b=1}^{n+1}\binom{n+1}{b}\left(\frac{p_2}{1-p_2}\right)^b\frac{1}{b},$$

$$S_2 = -BD\sum_{b=1}^{n+1}\binom{n+1}{b}\left(\frac{p_2}{1-p_2}\right)^b,$$

$$S_3 = CD\sum_{b=1}^{n+1}\binom{n+1}{b}\left(\frac{p_2}{1-p_2}\right)^b b,$$

and by Corollary 2 (S_1) and Lemmas 8 (S_2) and 9 (S_3) we get

$$\begin{aligned} E(Z_1^2) = &\sum_{k=0}^{N}\left(\frac{p_1}{p_2(1-p_2)}\right)^2\frac{1}{\binom{n+k+1}{k}p_2^k}\left(1-\frac{k+2-\frac{1-p_2}{p_1}}{n+k+2}\right) - \\ &-2\left(\frac{p_1}{1-p_2}\right)^2\frac{1}{p_2} - \frac{1}{n+1}\frac{p_1}{p_2(1-p_2)}\left(1-\frac{p_1}{1-p_2}\right) \\ &+\left(\frac{p_1}{1-p_2}\right)^2 + O\left(\frac{1}{n^{N+1}}\right). \end{aligned}$$

The rest of the proof follows from adding the term $-E^2(Z_1)$ to the derived expression for $E\left(Z_1^2\right)$, separating the term for $k = 0$ from the rest of the sum, and simple rearrangement of the resulting terms. □

Acknowledgements
This contribution is the result of implementation of the project *REVOGENE – Research centre for molecular genetics* (ITMS 26240220067) supported by the Research & Developmental Operational Programme funded by the European Regional Development Fund.

Authors' contributions
All authors contributed equally to the research. FD wrote the manuscript. JG prepared the figures. All authors read and approved the final manuscript.

Competing interests
The authors declare that they have no competing interests.

Author details
[1]Geneton s.r.o., Galvaniho 7, 82104 Bratislava, Slovakia. [2]Slovak Centre of Scientific and Technical Information, Lamacska cesta 7315/8A, 81104 Bratislava, Slovakia. [3]Comenius University, Faculty of Natural Sciences, Ilkovicova 3278/6, 84104 Bratislava, Slovakia. [4]Comenius University Faculty of Mathematics, Physics and Informatics, Mlynska dolina, 84248 Bratislava, Slovakia. [5]Comenius University, Science Park, Ilkovicova 8, 84104 Bratislava, Slovakia.

References
Aho, K, Bowyer, RT: Confidence intervals for ratios of proportions: implications for selection ratios. Methods Ecol. Evol. **6**(2), 121–132 (2015)
Alghamdi, N: Confidence intervals for ratios of multinomial proportions (2015). Master's thesis, University if Nebraska at Omaha
Basu, A, Lochner, RH: On the distribution of the ratio of two random variables having generalized life distributions. Technometrics. **13**(2), 281–287 (1971)
Bonett, DG, Price, RM: Confidence intervals for a ratio of binomial proportions based on paired data. Stat. Med. **25**(17), 3039–3047 (2006)

Chiu, RW, Chan, KA, Gao, Y, Lau, VY, Zheng, W, Leung, TY, Foo, CH, Xie, B, Tsui, NB, Lun, FM, et al: Noninvasive prenatal diagnosis of fetal chromosomal aneuploidy by massively parallel genomic sequencing of dna in maternal plasma. Proc. Natl. Acad. Sci. **105**(51), 20458–20463 (2008)

Culley, TM, Wallace, LE, Gengler-Nowak, KM, Crawford, DJ: A comparison of two methods of calculating gst, a genetic measure of population differentiation. Am. J. Bot. **89**(3), 460–465 (2002)

Fieller, E: The distribution of the index in a normal bivariate population. Biometrika. **24**, 428–440 (1932)

Frishman, F: On the arithmetic means and variances of products and ratios of random variables (1971). Technical report, DTIC Document

Geary, R: The frequency distribution of the quotient of two normal variates. J. R. Stat. Soc. **93**(3), 442–446 (1930)

Goodman, LA: On simultaneous confidence intervals for multinomial proportions. Technometrics. **7**(2), 247–254 (1965)

Graham, RL, Knuth, DE, Patashnik, O: Concrete Mathematics: A Foundation for Computer Science, 2nd edn, p. 492. Addison-Wesley Longman Publishing Co., Inc., Boston (1994). exercise 42

Hinkley, DV: On the ratio of two correlated normal random variables. Biometrika. **56**(3), 635–639 (1969)

Koopman, P: Confidence intervals for the ratio of two binomial proportions. Biometrics. **40**, 513–517 (1984)

Korhonen, PJ, Narula, SC: The probability distribution of the ratio of the absolute values of two normal variables. J. Stat. Comput. Simul. **33**(3), 173–182 (1989)

Lau, TK, Chen, F, Pan, X, Pooh, RK, Jiang, F, Li, Y, Jiang, H, Li, X, Chen, S, Zhang, X: Noninvasive prenatal diagnosis of common fetal chromosomal aneuploidies by maternal plasma dna sequencing. J. Matern. Fetal Neonatal Med. **25**(8), 1370–1374 (2012)

Marsaglia, G: Ratios of normal variables. J. Stat. Softw. **16**(4), 1–10 (2006)

Mekic, E, Sekulovic, N, Bandjur, M, Stefanovic, M, Spalevic, P: The distribution of ratio of random variable and product of two random variables and its application in performance analysis of multi-hop relaying communications over fading channels. Przegl. Elektrotechniczny. **88**(7A), 133–137 (2012)

Minarik, G, Repiska, G, Hyblova, M, Nagyova, E, Soltys, K, Budis, J, Duris, F, Sysak, R, Bujalkova, MG, Vlkova-Izrael, B, *et al*: Utilization of benchtop next generation sequencing platforms ion torrent pgm and miseq in noninvasive prenatal testing for chromosome 21 trisomy and testing of impact of in silico and physical size selection on its analytical performance. PloS ONE. **10**(12), 0144811 (2015)

Nadarajah, S: On the product and ratio of laplace and bessel random variables. J. Appl. Math. **2005**(4), 393–402 (2005)

Nadarajah, S, Kotz, S: On the ratio of pearson type vii and bessel random variables. Adv. Decis. Sci. **2005**(4), 191–199 (2005)

Nadarajah, S, Kotz, S: On the product and ratio of gamma and weibull random variables. Econ. Theory. **22**(2), 338–344 (2006)

Nelson, W: Statistical methods for the ratio of two multinomial proportions. Am. Stat. **26**(3), 22–27 (1972)

Pham-Gia, T: Distributions of the ratios of independent beta variables and applications. Commun. Stat. Theory Methods. **29**(12), 2693–2715 (2000)

Piegorsch, WW, Richwine, KA: Large-sample pairwise comparisons among multinomial proportions with an application to analysis of mutant spectra. J. Agric. Biol. Environ. Stat. **6**(3), 305–325 (2001)

Piper, J, Rutovitz, D, Sudar, D, Kallioniemi, A, Kallioniemi, O-P, Waldman, FM, Gray, JW, Pinkel, D: Computer image analysis of comparative genomic hybridization. Cytometry. **19**(1), 10–26 (1995)

Poorter, H, Garnier, E: Plant growth analysis: an evaluation of experimental design and computational methods. J. Exp. Bot. **47**(9), 1343–1351 (1996)

Press, SJ: The t-ratio distribution. J. Am. Stat. Assoc. **64**(325), 242–252 (1969)

Price, RM, Bonett, DG: Confidence intervals for a ratio of two independent binomial proportions. Stat. Med. **27**(26), 5497–5508 (2008)

Provost, S: On the distribution of the ratio of powers of sums of gamma random variables. Pak. J. Stat. **5**, 157–174 (1989)

Quesenberry, CP, Hurst, D: Large sample simultaneous confidence intervals for multinomial proportions. Technometrics. **6**(2), 191–195 (1964)

Sakamoto, H: On the distributions of the product and the quotient of the independent and uniformly distributed random variables. Tohoku Math. J. First Ser. **49**, 243–260 (1943)

Sehnert, AJ, Rhees, B, Comstock, D, de Feo, E, Heilek, G, Burke, J, Rava, RP: Optimal detection of fetal chromosomal abnormalities by massively parallel dna sequencing of cell-free fetal dna from maternal blood. Clin. Chem. **57**(7), 1042–1049 (2011)

Van Kempen, G, Van Vliet, L: Mean and variance of ratio estimators used in fluorescence ratio imaging. Cytometry. **39**(4), 300–305 (2000)

5

Parameters of stochastic models for electroencephalogram data as biomarkers for child's neurodevelopment after cerebral malaria

Maria A. Veretennikova[1*], Alla Sikorskii[2] and Michael J. Boivin[3]

*Correspondence: mveretennikova@hse.ru
[1]Department of Statistics and Data Analysis, Faculty of Economic Science, National Research University, Higher School of Economics, Shabolovka 28/11, 9, 19049 Moscow, Russia
Full list of author information is available at the end of the article

Abstract
The objective of this study was to test statistical features from the electroencephalogram (EEG) recordings as predictors of neurodevelopment and cognition of Ugandan children after coma due to cerebral malaria. The increments of the frequency bands of EEG time series were modeled as Student processes; the parameters of these Student processes were estimated and used along with clinical and demographic data in a machine-learning algorithm for the prediction of children's neurodevelopmental and cognitive scores 6 months after cerebral malaria illness. The key innovation of this work is in the identification of stochastic EEG features that can serve as language-independent markers of the impact of cerebral malaria on the developing brain. The results can enhance prognostic determination of which children are in most need of rehabilitative interventions, which is especially important in resource-constrained settings such as sub-Saharan Africa.

Mathematics Subject Classification: 60E05, 62M10, 62P10

Keywords: Student processes coma EEG wavelets regression regularization

1 Introduction

Cerebral malaria (CM) affects over half a million people annually and has high prevalence in sub-Saharan Africa. Different sources indicate distinct mortality rates, but for children it is above 40 percent (Idro et al. 2010). For those who survive, the sequelae could include neurodevelopmental impairments and metabolic disturbances.

During CM the red blood cells are parasitized, most often by *Plasmodium Falciparum*. *P. Vivax* and *P. Knowlesi* are also known to cause severe malaria, but unlike *P. Falciparum* they do not lead to coma (WHO 2014). Coma is the principal diagnostic difference between severe malaria and CM, with lumbar puncture ruling out bacterial central nervous system (CNS) infection or other cause of coma beyond severe malaria. CM is specific to *P. Falciparum* and is distinctive from other forms of malaria because of the sequestration of infected red blood cells in the microvasculature of the brain and compromise of the blood/brain barrier contributing to an immunopathogenic inflammatory cascade. This cascade together with ischemic and metabolic effects cause coma and often seizures during illness and contribute to the neuropathogenic basis of subsequent neurological and neurocognitive sequelae (John et al. 2008). Since coma is defined as prolonged unconsciousness and unresponsiveness (usually between 1 h and 4 weeks), clinical language-independent data provide the only opportunity to gauge the extent to which brain injury may impact subsequent neurodevelopment and cognitive function.

Today there is a number of different techniques for the statistical analysis of the brain. Despite some limitations, EEG is still widely used as a noninvasive way to monitor patients, predict seizure onsets and to determine the amount of activity in near-death states. In Awal et al. (2016) the authors use EEG features to predict neurodevelopmental outcomes for term infants with hypoxic ischaemic encephalopathy (HIE). The association between brain activity during coma and trauma outcomes was investigated in Malagurski et al. (2017) and Juan et al. (2015). Statistical analysis of EEG has been used to identify quasi-brain-death from coma (Li et al. 2014) and to confirm brain death (Chen et al. 2008).

Much of the work on the analysis of EEG data has focused on classification (Kirch et al. 2015; Piryatinska et al. 2009; Temko et al. 2011) or prediction of seizures (Duncan et al. 2013) without specification of the model for the underlying stochastic process. The use of spectral methods for the EEG time series, while popular, is problematic due to the evidence of non-stationarity of the process (Ignaccolo et al. 2009). In this paper we propose a stochastic model for the EEG time series, where for each frequency band the increment process is assumed to be a Student process, realized as Lévy driven Ornstein-Uhlenbeck-type process. The parameters of the Student marginal distribution are estimated and entered into a machine learning algorithm to test their association with children's neurodevelopmental and cognitive scores 6 months after cerebral malaria illness. The use of the Student distribution parameters markedly improves explained variation of neurodevelopment and cognition compared to using only demographic and clinical characteristics including plasma and cerebrospinal fluid biomarkers, or prediction based upon frequently used traditional EEG features. Identification of biomarkers such as the parameters of stochastic models for the EEG data has the potential to enhance diagnostic and prognostic determination by complementing the very limited clinical expertise in neurologists able to read and interpret EEG in resource constrained settings. Further, language-independent markers of neurodevelopment and cognition based on stochastic features of EEG data can complement the limited expertise available in sub-Saharan Africa for developmental and cognitive evaluations.

2 Dataset and preprocessing

Data used in this analysis were collected during the observational study of the pathogenesis of severe malaria (cerebral malaria (CM) and severe malarial anemia (SMA)) in surviving children, along community control children from their households who did not have a history of severe malaria (Bangirana et al. 2016). The study was performed at Mulago National Referral and Teaching Hospital in Kampala, Uganda in 2008 – 2015. Children with cerebral malaria, severe malarial anemia, or community control children were enrolled if they were between 18 months and 12 years of age. Cerebral malaria was defined as: 1) coma (Blantyre Coma Score [BCS] $\leq$ 2; 2) Plasmodium falciparum on blood smear; and 3) no other known cause of coma (e.g., hypoglycemia-associated coma reversed by glucose infusion, meningitis, or a prolonged postictal state). Children were enrolled after obtaining written informed consent in the local language from their parent(s) and signed assent from children 7 years of age and older.

The observational study was approved by the Institutional Review Boards of the Makerere University School of Medicine and the University of Minnesota. Data from CM malaria children were included in this study as EEGs were done only for them and not SMA or community control children. Data from community controls were used to create the z-scores of neurodevelopmental and cognitive outcomes of the CM children as described below.

2.1 EEG data

MATLAB software was used for the processing of the EEG data set that comprised the standard 10–20 EEG recordings for 78 children with the sampling rate of 500 Hz and the average record duration of 30 min. Persyst software (Persyst, Prescott, USA) was used to remove artifacts due to breath, muscle movement and heartbeat from the raw EEG data. For most children there were 19 channels, which means the electrodes were not located very densely, so there was no necessity to use the average reference. Hence we've chosen CZ to be the reference electrode to avoid the laterality bias. CZ is one of the predominant choices for a reference (Teplan 2002). Then we used EEGLAB (Makeig and Delorme 2004) to identify problematic channels based on properties of their voltage measurements, leaving 16 channels for analysis and excluding three (PZ-CZ, C4-CZ, O2-CZ). Data for several of the included 16 channels had substantial numbers of zero observations, which could be due to poor connection between the electrode and the skin. For these channels we extracted features described below with and without zeros, with the rationale that if a feature resulted from an artifact and was not important, it would not be selected by the machine learning algorithm.

We used Daubechies wavelets for splitting the clean signal into frequency bands. Daubechies orthogonal wavelets have a number of vanishing moments, which is used as an index for referencing, e.g. the standard notation Db4 means Daubechies wavelet with 4 vanishing moments. Research indicates particular suitability of Db4 for statistical analysis of EEG. We examined the relative average mean squared error (MSE) between the wavelet signal approximation and the actual signal for different Daubechies wavelets. For these data Db4 yielded the reconstruction error or order 10^{-9}, which is sufficiently low. Also, Db4 frequency band separation resulted in frequency intervals which are very close to the traditional frequency ranges: delta, theta, beta, alpha and gamma bands (Daubechies 1992), see Table 1.

Due to occasional spikes and irregular patterns in the original time series, the idea was not to split them into epochs as it is often done (Fraschini et al. 2016). We hypothesized that some of the statistical features, such as the frequency of flat line measurements

Table 1 Frequency band correspondence

Traditional	Db4 band's central frequency
Delta 0 – 3.5 Hz	2.7 Hz
Theta 3.5 – 7.5 Hz	5.57 Hz
Alpha 7.5 – 13 Hz	11 Hz
Beta 13 – 30 Hz	22.3 Hz
Gamma > 30 Hz	four subsequent bands

relative to the whole EEG record, could be useful as explanatory variables for neurodevelopment and cognition, whilst dividing the record into epochs would complicate extraction of useful information.

Empirical studies show that a low activity level in the gamma frequencies is closely related to the coma state (Chen et al. 2008), and generally, gamma band oscillations are thought to be related to higher cortical functioning, such as consciousness, memory, perception and learning (Uhlhaas et al. 2009). In Deng et al. (2015) it was shown that EEG gamma band activity characteristics are associated with the outcomes of targeted temperature management for brain recovery after cardiac arrest. On the other hand, it is also advised to exclude the highest EEG gamma frequency bands from the analysis, because it is most likely to be noise, rather than the real deterministic signal. In view of these recommendations and the goal of this research to investigate stochastic features of EEG, we excluded only the highest gamma band (D1), keeping the rest. Our rationale was that if the other gamma bands (D2, D3, and D4 in Daubechies' frequency band notation) were indeed useless for the prediction of post-comatose neurodevelopmental and cognitive scores, then this would be empirically determined in statistical algorithms for the extraction of important features.

2.2 Measures of neurodevelopment and cognition

Children had neurodevelopmental assessments (appropriate for those 5 years of age or younger) or cognitive assessments (appropriate for children over 5 years of age) a week after discharge from the hospital (or at enrollment for community control children) and then at 6 and 12 months after enrollment. Data from the assessment at 6 month post-enrollment were used for this analysis.

Neurodevelopmental assessment for children 5 years of age or younger. The Mullen Scales of Early Learning (MSEL) (Mullen 1995) were used to quantify neurodevelopment. MSEL is based on a comprehensive test assessing specific developmental domains: visual reception, gross motor skills, fine motor skills, receptive language, and expressive language. A composite score derived from standardized t-scores of the four domains (excluding gross motor) provides a measure of *g*, the general measure of fluid intelligence.

Cognitive assessment for children over 5 years of age. The Kaufman Assessment Battery for Children, second edition (KABC-II) (Kaufman and Kaufman 2004) evaluates sequential and simultaneous processing, learning, reasoning, and crystallized intellectual ability (knowledge). The knowledge subscale was not administered because it was not suitable in this setting (Bangirana et al. 2009). Summation of scores for the domains of sequential processing, simultaneous processing, learning, and planning yielded the Mental Processing Index (MPI) which was the measure of overall cognitive ability in this age group.

The United States of America (USA) norms were used to arrive at the MSEL composite *g* score and the KABC-II MPI score, because using such norms to adjust for the child's age was necessary to compute these global measures. To obtain a single measure of neurodevelopment or cognition for all children regardless of age, we computed the means and standard deviations of the age appropriate measures, the MSEL composite or the KABC-II MPI, among the community control children. Then for CM children the z-scores in each age group were computed by subtracting the means and dividing by the standard deviation of the community control children.

2.3 Other measures

Home Observation for the Measurement of the Environment (HOME) (Bradley and Caldwell 1979) is a composite measure designed to assess the quality and quantity of stimulation that the child is exposed to in their home environment. A total HOME score was generated by summing the number of *"yes"* responses to a checklist of items; higher HOME scores indicate higher quality of home environment.

Demographic and anthropometric data included age, sex, height-for-age and weight-for-age z-scores computed using the World Health Organization reference norms (Wold Health Organization Growth Standards 2009). Socioeconomic status (SES) was assessed using a checklist of material possessions, housing quality, cooking resources and water accessibility. Clinical variables and biomarker panels from plasma and cerebrospinal fluid were collected during each child's hospitalization for CM.

3 Creating the feature matrices

3.1 Non-EEG features

The non-EEG features included demographic and anthropometric data, SES and HOME scores, the Blantyre Coma Score, and plasma and cerebrospinal fluid biomarker panels, for a total of 54 potential explanatory variables.

3.2 Commonly used EEG features

We have evaluated 362 EEG features that have been commonly used in the past analyses of EEG data. Presence of seizures was reflected by binary variable that was defined using Persyst software indicators (Persyst, Prescott, USA). Frequencies of peaks in the original cleaned time series differing from the nearest measurements from both sides by 1/3, 1, 2 and 3 standard deviations were calculated and denoted by fp1/3, fp1, fp2 and fp3, respectively. Proportion of flat line EEG for each of the 16 channels was evaluated for the original cleaned time series.

For delta (a7), theta (D7), alpha (D6), beta (D5) and gamma (D4, D3, D2) frequency bands for each channel, we calculated amplitude variances and Shannon time entropy using wavelets (De Oliveira 2015; Smolentsev 2014). This version of entropy is defined in MATLAB as

$$S(x_i) = -\sum_{i=1}^{N-1} x_i^2 \log\left(x_i^2\right)$$

where x_i is the i-th measurement in the time series for the signal (MATLAB wavelet packet). Relative frequency band energy was defined as the sum of wavelet coefficients divided by the total sum of the coefficients for all the frequency bands (Smolentsev 2014, Chapter 3.10).

Hjorth complexity and mobility parameters (Hjorth 1970) were calculated for the entire time series based on the second moment as well as the first and second order differences.

3.3 New stochastic features

After splitting the EEG time series into frequency bands using Daubechies wavelets, for each frequency band we constructed the histograms for the increment process at different time blocks, and evaluated sample means and variances. The histograms were consistently approximately bell-shaped but with peaks higher and tails heavier than normal (Fig. 1).

The means were consistently close to zero, and the variances ratios for different time blocks fell within a rule of thumb range of [0.25, 4] (Montgomery 2012). Based on these empirical features, we selected a modeling approach that uses a stationary stochastic process for the increments of time series. To reflect the leptokurtic distribution seen in the data, we propose a stationary Student process as a model for the increments of the EEG time series for each frequency band. The symmetric scaled Student marginal distribution has the density

$$f_{\nu,\delta}(x) = \frac{\Gamma\left(\frac{\nu+1}{2}\right)}{\delta\sqrt{\pi}\Gamma\left(\frac{\nu}{2}\right)}\left(1+\left(\frac{x-\mu}{\delta}\right)^2\right)^{-\frac{\nu+1}{2}}, \ x \in \mathbb{R}, \tag{1}$$

$\nu > 0$ is degrees of freedom, $\delta > 0$ is a scale parameter, $\mu \in \mathbb{R}$ is location parameter. We denote this distribution by $T_{\mu,\delta,\nu}$. The expectation exists when $\nu > 1$, the variance exists when $\nu > 2$, and generally the n-th central moment exists when $\nu > n$.

There are several processes with the Student marginal distribution (Heyde and Leonenko 2005), of which we chose the Lèvy driven Ornstein-Uhlenbeck (OU)-type process as the process with the least restrictive potential parameter range. We have considered Student diffusion process (Leonenko and Suvak 2010) as another possible model. However, the mean reversion term needs to be included in the stochastic differential equation (SDE) defining a stationary Student diffusion process. Thus the mean has to exist, restricting the parameter range to $\nu > 1$. The parameter estimates based on these data did not support such restriction across all channels. The Student OU type process provides a flexible model for the increments of the EEG data, and is the solution of SDE

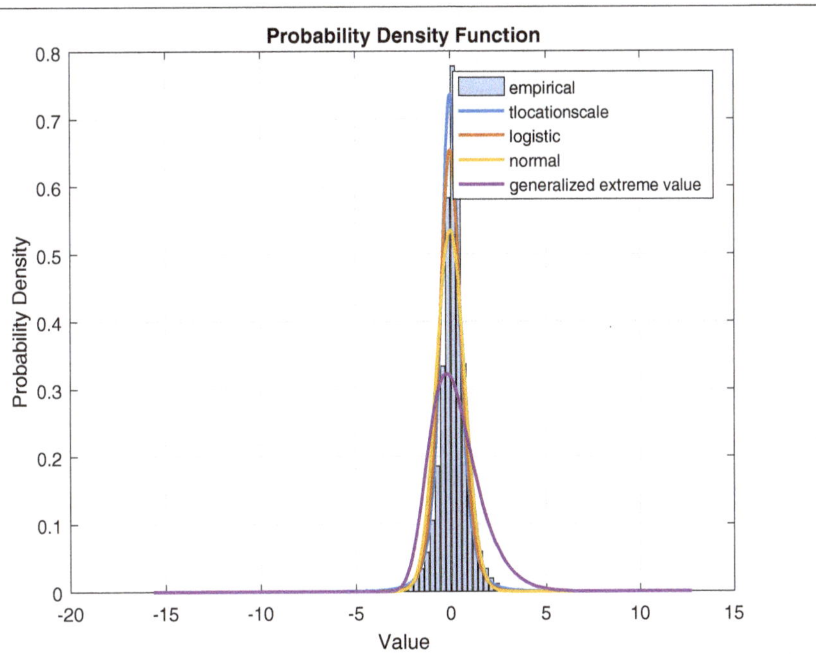

Fig. 1 Alpha frequency band

$$dX(t) = -\lambda X(t)dt + dY(\lambda t), \quad t \geq 0 \tag{2}$$

where $\{Y(t), t \geq 0\}$ is the background driving Lévy process (BDLP). This process is defined by specifying the distribution of $Y(1)$ through the cumulant transform

$$\log e^{i\zeta Y(1)} = i\zeta\mu - \delta|\zeta|\frac{K_{\nu/2-1}(\delta|\zeta|)}{K_{\nu/2}(\delta|\zeta|)}, \quad \zeta \in \mathbb{R}, \ \zeta \neq 0, \tag{3}$$

where K_s is the modified Bessel function of the third kind:

$$K_s(x) = \frac{1}{2}\int_0^{\infty} u^{s-1}\exp\left[\frac{1}{2}x\left(u + \frac{1}{u}\right)\right]du, \ x > 0, \ s \in \mathbb{R}.$$

Since the Student distribution is self-decomposable, the distribution of $Y(1)$ is infinitely divisible, and the Student OU-type process exists, as formally stated in the following Theorem (Heyde and Leonenko 2005).

Theorem 1 *There exists a strictly stationary stochastic process $\{X(t), t \geq 0\}$ that has the marginal $T_{\mu,\delta,\nu}$ distribution with the density function (1). The process solves SDE (2) for any $\lambda > 0$. The BLDP $\{Y(t), t \geq 0\}$ has the cumulant transform (3). The solution is given by*

$$X_t = e^{-\lambda t}X_0 + e^{-\lambda t}\int_0^t e^{\lambda s}dY(\lambda s).$$

Note that parametrization in (2) is such that the marginal distribution of the Student OU-type process $X(t)$ does not depend on λ. If $\nu > 1$, then the first moment of the marginal distribution exits, and $\mathbb{E}X(t) = \mu$. If $\nu > 2$, then the correlation function exists and depends only on the parameter λ, namely for $0 < s < t$

$$\text{corr}(X(s), X(t)) = e^{-\lambda(t-s)}.$$

In this work we used the first-order properties of the process $X(t)$ reflected by the parameters of the marginal distribution. We set $\mu = 0$ based on the empirical evidence as all histograms were centered on zero with virtually no variability in this respect among EEG channels. We used quasi-likelihood estimation (Heyde 1997) to evaluate the parameters ν and δ of the stationary Student process. It was possible to use this approach because of the specification of the distribution. A general method for estimation of the tail parameter (ν in this case) without specifying the marginal distribution is discussed in Grahovac et al. (2015) and could be used in other applications. Quasi-likelihood estimation was performed for the lowest four frequency bands for 15 channels, resulting in 120 features. Additional 30 features were derived by repeating the parameter estimation for 15 channels delta frequency band containing excess zeros potentially due to the disconnection of the electrode from the skin. Note that in this set of features there is no CZ-CZ channel at all. The estimated 150 parameters were used as features in the matrix of potential explanatory variables for the neurodevelopment and cognition 6 months post CM illness. We refer to these features as stochastic features since they were derived based on the proposed stochastic process model.

4 Data analysis

Three feature matrices X with 78 rows each, one containing 54 non-EEG features (columns), the second containing 362 traditional EEG features, and the third containing 150 new stochastic EEG features, were prepared as described above for the entry into a machine learning algorithm to predict the neurodevelopmental and cognitive scores of children surviving cerebral malaria. Out of 54 columns for non-EEG features 20 had at least 19 missing values. The maximum number of empty entries in a column of the non-EEG part of the feature matrix was 23. The pattern of missing data could not be assumed to be missing at random (MAR), because some of the data were missing due to clinical reasons. Therefore we used imputation method that is not dependent upon the MAR assumption. Soft Impute is a matrix completion method based on Singular Value Decomposition (SVD) of a matrix (Mazumder et al. 2010). Application of this algorithm requires the following assumption to hold: the rank of the approximating matrix $rank(Z) << min(n,p)$, which is reasonable in our case. This assumption makes sense due to the nature of features, which may be grouped by correlation into a smaller number of clusters due to inherent synchrony between channels. We present the central lemma behind this method for completeness.

Lemma 1 *Suppose the matrix $W_{m\times n}$ has rank r. The solution to the optimization problem*

$$\min_{Z}\left(\frac{1}{2}\|W-Z\|_F^2+\lambda\|Z\|_*\right) \tag{4}$$

is given by $\hat{Z}=S_\lambda(W)$, where $S_\lambda(W)=UD_\lambda V^T$, with $D_\lambda = diag[\,(d_1-\lambda,\cdots,(d_r-\lambda)_+]$, where UDV^T is the SVD of W, $D=diag[\,d_1,\cdots,d_r]$, and $t_+=\max(t,0)$.

Here $\|A\|_F$ is the Frobenius norm of a matrix A, whilst $\|A\|_*$ is the sum of the singular values of the matrix A. In our case, in computing the Frobenius norm in (4) we only look at the pairs of indices (i,j), for which there are no missing values. The algorithm implemented in SoftImpute iteratively updates the matrix Z through the use of this lemma, until convergence is reached in approximating the matrix of interest. R was used for the imputation of missing values.

Machine learning was performed using the Elastic Net technique that solves the following optimization problem:

$$\min_{\bar{\beta}}\left(\sum_{i=1}^{n}\left(y_i-\sum_{j=1}^{p}\beta_j x_{ij}\right)^2+\alpha\times l_1\sum_{j=1}^{p}|\beta_j|+0.5\times\alpha\times(1-l_1)\sum_{j=1}^{p}\beta_j^2\right), \tag{5}$$

Elastic Net was chosen because it is more suitable than the Least Absolute Shrinkage and Selection Operator (LASSO) in face of multicollinearity (Oyeyemi et al. 2015), which we expected among the selected features. There are two hyper-parameters, and leave-one-out cross validation (LOOCV) was used to select the optimal pair. We have also used LOOCV for the estimation of the mean squared error in the prediction models produced by the algorithm. The Elastic Net algorithm was implemented using Python 3 with Anaconda.

The non-parametric missForest technique for matrix completion (Stekhoven and Bühlmann 2012) is based on averaging over a random forest of regression trees and was used for additional validation of the results. Trees were built based on observed and bootstrapped parts of the training data set.

5 Results

5.1 Imputations

We have run over a grid of different parameter values for α, considering powers of 10, starting with $\alpha = 0.00001$ and finishing with $\alpha = 1000$, whilst for the l_1 ratio coefficient we considered 7 possible values starting with 0.0001 and finishing with 1. In all models, setting the regularization parameter $\lambda = 100$ in SoftImpute gave better mean squared error (MSE) than other values of λ and resulted in the least complex model, after applying the Elastic Net following the matrix completion. This value is used in reporting of the results.

5.2 Results from elastic net for three sets of features

Table 2 summarizes the results of applying the regression methods listed in the previous section.

The lowest LOOCV MSE of 0.15 was obtained with $\alpha = 0.001$, l_1 ratio$= 0.5$ in the objective function subject to minimization:

$$\frac{1}{2 * 78} \|y - Xw\|_2^2 + \alpha * \text{l1 ratio} * \|w\|_1 + 0.5 * \alpha * (1 - \text{l1 ratio}) * \|w\|_2^2 \quad (6)$$

with leave-one-out cross-validation. This combination of α and l_1 ratio was the best across combinations described in the Section 4.

Similarly to the result for the matrix with 150 features after SoftImpute, using the random forest technique for matrix completion and again, those features in the top 5 in feature importance by Breiman were for the channels FP2-CZ, O1-CZ, F7-CZ, T6-CZ, whilst F8-CZ and T6-CZ appeared most often in the top 12 non-zero coefficients by absolute value. So this result almost copies the outcome with the Elastic Net after SoftImpute

Table 2 Elastic Net - best results, after Soft Impute with $\lambda = 100$

Feature set	LOOCV MSE	Number of nonzero coefficients	Sample features with non-zero coefficients
54 anthropomentic, socio-economic, and medical non-EEG features	0.3982	15	Weight, hemoglobin level, weight, BCS, the HOME score, white blood cell count, cerebrospinal fluid levels of interleukin (IL)-1 receptor antagonist (RA), IL-6, RANTES (an acronym for Regulated on Activation, Normal T Expressed and Secreted), IL-8, and plasma levels of vascular endothelial growth factor and von Willebrand factor.
362 traditional EEG features in frequency bands	0.5285	62	Features include: fp2 in T6-CZ, FZ-CZ and P4-CZ, fp1 F4-CZ, wave energy for theta in channel T5-CZ, variance in theta for FP2-CZ, variance in alpha frequency band for T5-CZ.
150 stochastic features for increment processes in frequency bands	0.1511	85	12 coefficients have the absolute values over 0.5. Top 5 coefficients are for the channels FP2-CZ, O1-CZ, T6-CZ, FZ-CZ. Channels F8-CZ and T6-CZ appear most often among the 12 top coefficients by absolute value, 4 and 3 times respectively. Four of the top 12 are for fitting a stochastic process model for the a7 frequency band.

with $\lambda = 100$ and yields a marginally different Elastic Net LOOCV MSE value. This was anticipated and confirms validity of the matrix completion method for such data.

6 Conclusions

We conclude that stochastic modeling brings a noticeable improvement in explaining the variation in neurodevelpmental and cognitive outcomes of children 6 months after surviving cerebral malaria. Stochastic features alone do an even better job than the tested sets of medical non-EEG or traditional EEG features which are not based on stochastic models for the underlying time series.

Regarding medical non-EEG biomarkers, only tumor necrosis factor alpha (TNF-alpha) in cerebrospinal fluid but not in plasma was predictive of 6-month later cognitive scores of children older than 5 years, but not of neurodevelopment of younger children (Shabani et al. 2017). So our finding of TNF-alpha not being among top predictors is in line with the developing literature on the role of biomarkers collected at the time of acute illness in predicting later neurodevelopment and cognition in children. This paper extends the results of Shabani et al. (2017) to testing more than one biomarker using modern statistical and probabilistic methods. We have identified biomarkers that can be further considered in future research as potentially important prognostic factors for neurodevelopment and cognition.

Regarding the traditional EEG features, their performance in explaining the variation in neurodevelopmental and cognitive outcomes was inferior to that of anthropometric, socio-economic, and non-EEG medical features. This finding may be due to the fact that these features are not based on underlying stochastic models. For example, the traditional computation of Shannon's entropy build into software assumes that the underlying stochastic process is stationary, which could be reasonable in some populations (Piryatinska et al. 2009), but is at odds with other literature (Ignaccolo et al. 2009). Whether or not the underlying stochastic process is stationary in a given population is an empirical question that needs to be addressed in methodology of analyzing EEG data. For the population of Ugandan children in coma from cerebral malaria, we have found that the assumption of stationarity of the time series was unreasonable, while for the increment process it was. Further, stochastic modeling for the increment process had clear advantages, as seen from our results.

When considering the stochastic features for the increment process, the combination of channels for which the stochastic features proved to be particularly useful is FP2-CZ, 01-CZ, T6-CZ, FZ-CZ. Parameters from channels located on the right side dominated the most important features. Taken as a group, these locations are associated with visual and attention processes that are related to visual-spatial simultaneous processing working memory. It would be interesting to see if these channels also arise as important in relation to child neurodevelopment and cognition in other infectious diseases that could affect the brain.

Abbreviations

BCS: Blantyre coma score; CM: Cerebral malaria; Dbi: Daubechies wavelet with *i* vanishing moments; D1,. . ., D7, a7: Frequency bands from the highest gamma (D1) to delta (a7), see the table; EEG: Electroencephalogram; fp1/3, fp1, fp2 and fp3: Frequencies of peaks in the original cleaned time series differing from the nearest measurements from both sides by 1/3, 1, 2 and 3 standard deviations were calculated; HOME: Home observation for the measurement of the environment; Hz: Hertz; KABC-II: Kaufman assessment battery for children, second edition; LOOCV MSE: Leave-one-out cross-validation mean squared error; MAR: Missing at random; MPI: Mental processing index; MSEL: Mullen scales of early learning; SDE: Stochastic differential equation; SES: Socioeconomic status; SMA: Severe malarial anemia; SVD: Singular value decomposition; TNF-alpha: Tumor necrosis factor alpha

Acknowledgments
The authors thank C. C. John and D. Postels for the acquisition and provision of data, as well as G. Shtekh and A. Hall for fruitful discussions and advice with programming. We also thank the company Persyst for providing a trial copy of their software.

Funding
Data collection was supported by the National Institutes of Health (NIH) grant R01 NS055349 (PI: C. John). For M. Veretennikova the study has been funded by the Russian Science Foundation (project number 17–11-01098). M. J. Boivin and A. Sikorskii were supported by the grant NIH R01 HD064416 (PI: M. Boivin).

Authors' contributions
Conceptualization (MV, AS, MJB), data curation (MV, AS, MJB), data analysis (MV), funding acquisition (MJB), investigation (MV, AS, MJB), methodology (MV, AS, MJB), manuscript writing and editing (MV, AS, MJB). All authors read and approved the final manuscript.

Competing interests
None of the authors have any competing interests.

Author details
[1]Department of Statistics and Data Analysis, Faculty of Economic Science, National Research University, Higher School of Economics, Shabolovka 28/11, 9, 19049 Moscow, Russia. [2]Department of Psychiatry and Department of Statistics and Probability, Michigan State University, 909 Wilson Road, 48824, East Lansing MI, USA. [3]Department of Psychiatry and Department of Neurology and Ophtalmology, Michigan State University, 909 Wilson Road, 48824 East Lansing MI, USA.

References
Awal, M. A., Lai, M. M., Azemi, G., Colditz, P. B.: EEG background features that predict outcome in term neonates with hypoxic ischaemic encephalopathy: A structured review. Clin. Neurophysiol. **125**(1), 285–296 (2016)

Bangirana, P., Opoka, R. O., Boivin, M. J., Idro, R., Hodges, J. S., John, C. C.: Neurocognitive domains affected by cerebral malaria and severe malarial anemia in children. Learn. Individ. Differ. **46**, 38–44 (2016)

Bangirana, P., Seggane, M., Allebeck, P., Giordani, B., John, C. C., Byarugaba, J., Ehnvall, A., Boivin, M. J.: A preliminary investigation of the construct validity of the KABC-II in Ugandan children with prior cerebral insult. Afr. Health Sci. **9**(3), 186–192 (2009)

Bradley, R. H., Caldwell, B. M.: Home observation for measurement of the environment. University of Arkansas Press, Little Rock (1979)

Chen, Z., Cao, J., Cao, Y., Zhang, Y., Gu, F., Zhu, G., Hoong, Z., Wang, B., Cichocki, A.: An empirical EEG analysis in brain death diagnosis for adults. Cogn. Neurodyn. **2**(3), 257–271 (2008)

Daubechies, I.: Ten Lectures on Wavelets. SIAM, Philadelphia (1992)

De Oliveira, H.: Shannon and Renyi Entropy of Wavelets. International Journal of Mathematics and Computer Science. **10**, 13–26 (2015)

Deng, R., Koenig, M. A., Young, L. M., Jia, X.: Early Quantitative Gamma-Band EEG Marker is Associated with Outcomes After Cardiac Arrest and Targeted Temperature Management. Neurocrit. Care. **23 2**, 262–73 (2015)

Duncan, D., Talmon, R., Zaveri, H., Coifman, R.: Identifying preseizure state in intracranial EEG data using diffusion kernels. Math. Biosci. Eng. **10**, 579–90 (2013)

Fraschini, M., Demuru, M., Crobe, A., Marrosu, F., Stam, CJ., Hillebrand, A.: The effect of epoch length on estimated EEG functional connectivity and brain network organisation. J. Neural Eng. **13**, 036015 (2016)

Grahovac, D., Jia, M., Leonenko, N. N., Taufer, E.: Asymptotic properties of the partition function and applications in tail index inference of heavy-tailed data. Statistics. **49**(6), 1221–1242 (2015)

Heyde, C. C.: Quasi-likelihood and its application: A general approach to optimal parameter estimation. Springer, New York (1997)

Heyde, C. C., Leonenko, N. N.: Student processes. Adv. Appl. Probab. **37**, 342–365 (2005)

Hjorth, B.: EEG analysis based on time domain properties. Electroencephalography and Clinical Neurophysiology. **29**(3), 306–310 (1970)

Idro, R., Marsh, K., John, C. C., Newton, C. R.: Cerebral malaria; mechanisms of brain injury and strategies for improved neuro-cognitive outcome. Pediatr. Res. **68**(4), 267–274 (2010)

Ignaccolo, M., Latka, M., Jernajczyk, W., Grigolini, P., West, B.: The dynamics of EEG entropy. J. Biol. Phys. **36**, 185–96 (2009)

John, C. C., Bangirana, P., Byarugaba, J., Opoka, R. O., Idro, R., Jurek, A. M., Wu, B., Boivin, M. J.: Cerebral malaria in children is associated with long-term cognitive impairment. Pediatrics. **122**(1), e92-99 (2008)

Juan, E., Kaplan, P. W., Oddo, M., Rossetti, A. O.: EEG as an indicator of cerebral functioning in postanoxic coma. J. Clin. Neurophysiol. **32**, 465–47 (2015)

Kaufman, N. L., Kaufman, A. S.: Manual for the Kaufman AssessmentBbattery for Children. 2nd. American Guidance Service Publishing/Pearson Products Inc., Circle Pines (2004)

Kirch, C., Muhsa, B., Ombao, H.: Detection of Changes in Multivariate Time Series With Application to EEG Data. J. Am. Stat. Assoc. **110**(511), 1197–1216 (2015). Taylor and Francis, https://doi.org/10.1080/01621459.2014.957545

Leonenko, N. N., Suvak, N.: Statistical inference for Student diffusion process. Stoch. Anal. Appl. **28**(6), 972–1002 (2010)

Li, L., Wilton, A., Marcora, S., Bowman, H., Mandic, D. P.: EEG-based brain connectivity analysis of states of awareness. **1002–5** (2014)

Makeig, S., Delorme, A.: EEGLAB: an open source toolbox for analysis of single-trial EEG dynamics. J. Neurosci. Methods. **134**(1), 9–21 (2004)

Malagurski, B., Péran, P., Sarton, B., Riu, B., Vardon-Bounes, F., Seguin, T., Geeraerts, T., Fourcade, O., de Pasquale, F., Silva, S.: Neural signature of coma revealed by posteromedial cortex connection density analysis. NeuroImage Clinical. **15**, 315–324 (2017)

MATLAB wavelet packet. https://www.mathworks.com/help/wavelet/ref/wentropy.html. Accessed 21 Sept 2018

Mazumder, R., Hastie, T., Tibshirani, R.: Spectral Regularization Algorithms for Learning Large Incomplete Matrices. J. Mach. Learn. Res. **11**, 2287–2322 (2010)

Montgomery, D. C.: Design of Experiments. 8th. Wiley, Hoboken (2012)

Mullen, E. M.: Mullen Scales of Early Learning. American Guidance Service, Circle Pines (1995)

Oyeyemi, G. M., Ogunjobi, E. O., Folorunsho, A. I.: On performance of shrinkage methods - a Monte Carlo study. Int. J. Stat. Appl. **5**(2), 72–76 (2015)

Piryatinska, A., Terdik, G., Woyczynski, W. A., Lopar, K. A., Scher, M. S., Zlotnik, A.: Automated detection of neonate EEG sleep stages. Comput. Methods Prog. Biomed. **95**(1), 31- 46 (2009)

Shabani, E., Ouma, B. J., Idro, R., Bangirana, P., Opoka, R. O., Park, G. S., Conrov, A. L., John, C. C.: Elevated cerebrospinal fluid tumour necrosis factor is associated with acute and long-term neurocognitive impairment in cerebral malaria. Parasite Immunol. **39**(7) (2017). https://doi.org/10.1111/pim.12438. Epub 2017 May 28

Smolentsev, N. K.: Fundamentals of the theory of wavelets. Wavelets in MATLAB, DMK Publishing Press, Moscow (2014)

Stekhoven, D., Bühlmann, P.: MissForest - Non-parametric missing value imputation for mixed-type data. Bioinformatics (Oxf. Engl.) **28**, 112–118 (2012)

Temko, A., Thomas, E., Marnane, W., Lightbody, G., Boylan, G.: EEG-based neonatal seizure detection with Support Vector Machines. Clin. Neurophysiol. **122**(3), 464–473 (2011)

Teplan, M.: Fundamentals of EEG Measurement. Meas. Sci. Rev. **2**(2), 1–11 (2002)

Uhlhaas, P., Pipa, G., Lima, B., Melloni, L., Neuenschwander, S., Nikolić, D., Singer, W.: Neural synchrony in cortical networks: history, concept and current status. Front. Integr. Neurosci. **3**, 17 (2009)

Wold Health Organization Growth Standards (2009). https://www.who.int/growthref/en/. Accessed 20 Jan 2017

WHO: Tropical Medicine and International Health. Wiley, Hoboken (2014). 19 (Suppl. 1)

6

The Kumaraswamy transmuted Pareto distribution

Sher B. Chhetri[1], Alfred A. Akinsete[2*], Gokarna Aryal[3] and Hongwei Long[1]

*Correspondence:
akinsete@marshall.edu
[2]Department of Mathematics, Marshall University, Huntington, WV, USA
Full list of author information is available at the end of the article

Abstract

In this work, a new five-parameter Kumaraswamy transmuted Pareto (*Kw*TP) distribution is introduced and studied. We discuss various mathematical and statistical properties of the distribution including obtaining expressions for the moments, quantiles, mean deviations, skewness, kurtosis, reliability and order statistics. The estimation of the model parameters is performed by the method of maximum likelihood. We compare the distribution with few other distributions to show its versatility in modeling data with heavy tail.

Keywords: Kumaraswamy distribution, Quadratic rank transmutation map (QRTM), Pareto distribution, Hazard function, Maximum likelihood estimation

AMS Subject Classification: $60E05, 62E15, 62H12$

1 Introduction

Furtherance to the work by Eugene et al. (2002), who proposed and defined the beta-generated class of distributions for a continuous random variable, derived from the logit of the beta random variable, many statistical distributions have been proposed and studied by numerous authors. According to Eugene et al. (2002), suppose X is a random variable with cumulative distribution function (CDF) $F(x)$, the CDF for the beta-generated family is obtained by applying the inverse probability transformation to the beta density function. The CDF for the beta-generated family of distributions is given by

$$G(x) = \frac{1}{B(u,v)} \int_0^{F(x)} t^{u-1}(1-t)^{v-1}\, dt, \ 0 < u, v < \infty, \tag{1}$$

where $B(u,v) = \Gamma(u)\Gamma(v)/\Gamma(u+v)$. The corresponding probability density function (PDF) is given by

$$g(x) = \frac{1}{B(u,v)} [F(x)]^{u-1} [1-F(x)]^{v-1} \left[\frac{d}{dx}F(x)\right]. \tag{2}$$

A closely related generalized distribution similar to using the beta random variable as the baseline distribution as defined in Eqs. (1) and (2), is the work due to Cordeiro and de Castro (2011), where the authors combined the works of Eugene et al. (2002) and Jones (2009) by replacing the baseline distribution with the Kumaraswamy (*Kw*) distribution to construct a new class of *Kw-G* distributions. See also Bourguignon et al. (2013), Cordeiro et al. (2010), and Elbatal (2013). The work by Shams (2013) also introduced and studied another type of generalization of the Kumaraswamy distribution.

A new approach of constructing and generalizing statistical distributions was proposed by Shaw and Buckley (2007,2009). Their concept is defined by transmutation maps that are functional composition of the cumulative distribution function of one distribution with the quantile function of another distribution. One of the maps is referred to as the Quadratic Rank Transmutation Map (QRTM). Specifically, given two distributions with a common sample space with CDFs F_1 and F_2, we can define a pair of general rank transmutation map

$$G_{R_{12}}(u) = F_2\left(F_1^{-1}(u)\right), \quad G_{R_{21}}(u) = F_1\left(F_2^{-1}(u)\right),$$

where the pair of maps $\{G_{R_{12}}(u), G_{R_{21}}(u)\}$ takes the unit interval $I_{[0,1]}$ into itself, and under suitable assumptions are mutual inverses, satisfying the conditions $G_{ij}(0) = 0$ and $G_{ij}(1) = 1$. A useful additional condition requires that the rank transmutation maps are monotone and continuously differentiable, otherwise transmuted density may be discontinuous.

In this paper, we use the transmutation map approach outlined in Shaw and Buckley (2007), and also defined in Shaw and Buckley (2009). This approach considered for $|\lambda| \leq 1$,

$$G_{R_{12}}(u) = u + \lambda u(1 - u).$$

The consequence of the above results in the relationship

$$F_2(x) = (1 + \lambda)F_1(x) - \lambda F_1(x)^2,$$

which on differentiation yields,

$$f_2(x) = f_1(x)[(1 + \lambda) - 2\lambda F_1(x)],$$

where $f_1(x)$ and $f_2(x)$ are the corresponding PDFs associated with $F_1(x)$ and $F_2(x)$ respectively.

We will use the above formulation for a pair of distributions $F(x)$ and $G(x)$, where $G(x)$ is a submodel or the baseline distribution of $F(x)$. By a formal definition, a random variable X is said to have a transmuted probability distribution with CDF $F(x)$ if

$$F(x) = (1 + \lambda)G(x) - \lambda G(x)^2, \qquad |\lambda| \leq 1 \tag{3}$$

The corresponding PDF of the transmuted probability distribution is

$$f(x) = g(x)(1 + \lambda - 2\lambda G(x)).$$

Note that when $\lambda = 0$, the above relation reduces to the distribution $(G(x))$ of the baseline random variable.

Following the work by Aryal and Tsokos (2009) on the transmuted extreme value distribution, a number of transmuted family of distributions have been proposed and discussed by many authors in the literature. A few of them are the work by Aryal and Tsokos (2011): transmuted Weibull distribution; Khan and King (2013): transmuted modified Weibull distribution; Ashour and Eltehiwy (2013): transmuted Lomax distribution; Hady and Ebraheim (2014): exponentiated transmuted Weibull distribution; Hussian (2014): transmuted exponentiated gamma distribution; Merovci and Puka (2014): transmuted Pareto distribution; Owokolo et al. (2015): transmuted exponential distribution, and Khan et al. (2016): transmuted Kumaraswamy distribution; Afify et al. (2014): transmuted complementary Weibull geometric distribution; Yousof et al. (2015): the transmuted exponentiated generalized-G family of distributions, and a host of many others. Tahir and Cordeiro (2016) provided a comprehensive list of contributed works on transmuted distributions. This paper is outlined as follows. In Section 2, we introduce the Kumaraswamy transmuted Pareto distribution. Some of its mathematical properties are discussed in Sections 3, 4, 5, 6, 7, 8 and 9. Section 10 discusses the estimation of the parameters of the distribution by the method of maximum likelihood. We provide the application of the distribution to two real life data in Section 11. In Section 12 we have simulation, and concluding remarks in Section 13.

2 The Kumaraswamy transmuted Pareto (KwTP) distribution

Following the transmutation map by Shaw and Buckley (2007) as defined in Eq. (3), we combine the Kumaraswamy distribution and the transmuted Pareto to form what we refer to as the Kumaraswamy transmuted Pareto (*Kw*TP) distribution. Without loss of generality, we give a brief description of these *component* distributions in what follows.

2.1 Component distributions

*2.1.1 The Kumaraswamy-*Kw *distribution*

Kumaraswamy (1980) proposed and discussed a probability distribution for handling double-bounded random processes with varied hydrological applications. Let T be a random variable with the Kumaraswamy's distribution. The PDF and CDF are defined, respectively, as

$$f_T(t) = abt^{a-1}(1-t^a)^{b-1}, \quad 0 < t < 1,$$

and

$$F_T(t) = 1-(1-t^a)^b, \quad 0 < t < 1, \tag{4}$$

where both $a > 0$ and $b > 0$ are shape parameters. A generalized form of this distribution is obtained when t is replaced by the CDF $G(t)$ of another random variable, to have what is referred to as the Kumaraswamy-G (*Kw-G*) distribution. The beta and Kumaraswamy distributions share similar properties. For example, the Kumaraswamy distribution, also referred to as the minimax distribution, is unimodal, uniantimodal, increasing, decreasing or constant depending on the values of its parameters. A more detailed description, background, genesis, and properties of Kumaraswamy distribution are outlined in Jones (2009), where the author highlighted several advantages of the Kumaraswamy distribution over the beta distribution, namely; its simple normalizing constant, simple explicit formulas for the distribution and quantile functions, and simple random variate generation procedure. There have been many contributions to the theory and applications of the Kumaraswamy distribution in the literature. See for example the works by Cordeiro et al. (2010): Kumaraswamy Weibull; Cordeiro et al. (2012): Kumaraswamy Gumbel; Pascoa et al. (2011): Kumaraswamy generalized gamma; Saulo (2012): Kumaraswamy Birnbaum-Saunders; Akinsete et al. (2014): Kumaraswamy geometric, and a host of many others.

2.1.2 The Pareto and the transmuted Pareto (TP) distributions

The Pareto distribution was named after a sociologist Vilfredo Pareto. It is used in modelling the distribution of incomes and other financial variables, and in the description of social and other phenomena. Many distributions have been derived using the Pareto distribution. A few examples are the gamma-Pareto distribution by Alzaatreh et al. (2012), the beta exponentiated Pareto distribution by Zea et al. (2012), and the beta-Pareto distribution by Akinsete et al. (2008).

The PDF and CDF of the Pareto distribution are given respectively by

$$r(x) = \frac{k\theta^k}{x^{k+1}},$$

and

$$R(x) = 1 - (\theta/x)^k, \;\; k > 0, \;\; \theta > 0, \;\; x \geq \theta.$$

Using the quadratic rank transmutation map, Merovci and Puka (2014) generalized the Pareto distribution to obtain what the authors called the transmuted Pareto distribution. They provided a comprehensive description of the mathematical properties of the distribution and its application in modeling real life data. Formally, a random variable X is said to have the transmuted Pareto distribution with scale parameter $\theta > 0$, shape parameter $\alpha > 0$, and the transmuted parameter λ ($|\lambda| \leq 1$), if its PDF is given by

$$g(x;\alpha,\theta,\lambda) = \frac{\alpha\theta^\alpha}{x^{\alpha+1}}\left[1 - \lambda + 2\lambda\left(\frac{\theta}{x}\right)^\alpha\right],$$

where θ is the (necessarily positive) minimum possible value of X. The corresponding CDF of the transmuted Pareto distribution is given by

$$G(x;\alpha,\theta,\lambda) = \left[1 - \left(\frac{\theta}{x}\right)^\alpha\right]\left[1 + \lambda\left(\frac{\theta}{x}\right)^\alpha\right]. \tag{5}$$

The distribution in Eq. (5) becomes the Pareto when $\lambda = 0$.

2.2 The *Kw*TP distribution

By substituting Eq. (5) into Eq. (4), the cumulative distribution function of the five-parameter Kumaraswamy transmuted Pareto (*Kw*TP) distribution is given by

$$\begin{aligned} F(x;\alpha,\theta,\lambda,a,b) &= 1-[1 - G(x;\alpha,\theta,\lambda)^a]^b \\ &= 1 - \left\{1 - \left[1 - \left(\frac{\theta}{x}\right)^\alpha\right]^a\left[1 + \lambda\left(\frac{\theta}{x}\right)^\alpha\right]^a\right\}^b, \end{aligned} \tag{6}$$

where $x > \theta$, $\alpha > 0$, $a > 0$, and $b > 0$ are shape parameters, $\theta > 0$ is the scale parameter, while $|\lambda| \leq 1$ is the transmuted parameter. The corresponding probability density function of the *Kw*TP distribution may be expressed as

$$\begin{aligned} f(x;\alpha,\theta,\lambda,a,b) &= abg(x)G(x)^{a-1}[1-G(x)^a]^{b-1} \\ &= \frac{ab\alpha\theta^\alpha}{x^{\alpha+1}}\left[1-\lambda+2\lambda\left(\frac{\theta}{x}\right)^\alpha\right]\left\{\left[1-\left(\frac{\theta}{x}\right)^\alpha\right]\left[1+\lambda\left(\frac{\theta}{x}\right)^\alpha\right]\right\}^{a-1} \\ &\times\left\{1-\left[1-\left(\frac{\theta}{x}\right)^\alpha\right]^a\left[1+\lambda\left(\frac{\theta}{x}\right)^\alpha\right]^a\right\}^{b-1}. \end{aligned} \tag{7}$$

Figure 1 shows the graphs of the PDF of *Kw*TP distribution for selected values of the parameters. The graphs indicate that *Kw*TP is unimodal, and may be used in modeling different phenomena exhibited by real world data. Figure 2 also shows the graphs of the CDF for various values of the parameters.

3 Special cases of *Kw*TP

Some special submodels may be obtained from *Kw*TP by varying the values of the parameters. Here is a list of few these submodels.

- When $\lambda = 0$, *Kw*TP reduces to *Kw*P distribution by Bourguignon et al. (2013) with PDF

$$f(x;\alpha,\theta,a,b) = \frac{ab\alpha\theta^\alpha}{x^{\alpha+1}} \times \left\{\left[1-\left(\frac{\theta}{x}\right)^\alpha\right]\right\}^{a-1}\left\{1-\left[1-\left(\frac{\theta}{x}\right)^\alpha\right]^a\right\}^{b-1}.$$

- When $a = b = 1$, *Kw*TP reduces to the transmuted Pareto distribution by Merovci and Puka (2014) with PDF

$$f(x;\alpha,\theta,\lambda) = \frac{\alpha\theta^\alpha}{x^{\alpha+1}}\left[1-\lambda+2\lambda\left(\frac{\theta}{x}\right)^\alpha\right].$$

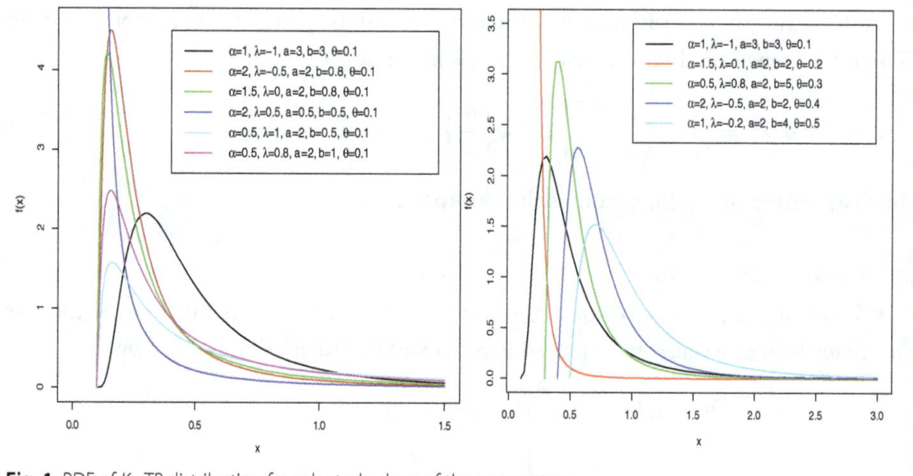

Fig. 1 PDF of *Kw*TP distribution for selected values of the parameters

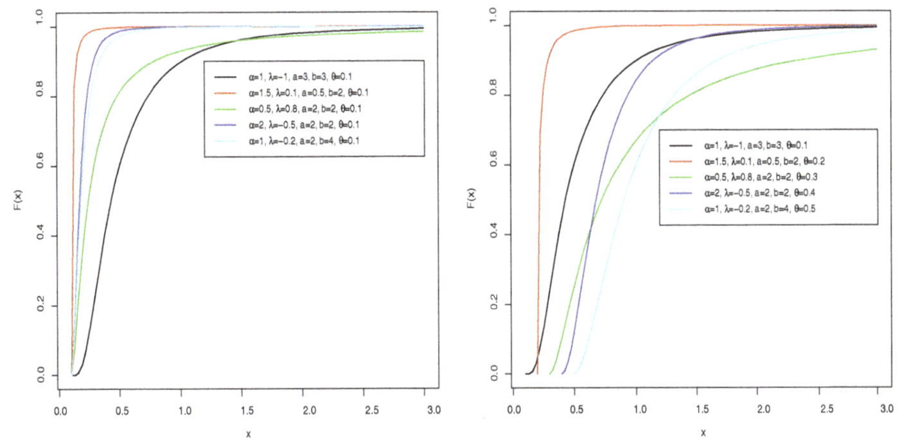

Fig. 2 CDF of *KwTP* distribution for selected values of the parameters

- By setting $b = 1$ and $\lambda = 0$, the *KwTP* reduces to the exponentiated Pareto distribution, defined by Nadarajah (2005), with PDF

$$f(x;\alpha,\theta,a) = \frac{a\alpha\theta^{\alpha}}{x^{\alpha+1}} \times \left\{\left[1-\left(\frac{\theta}{x}\right)^{\alpha}\right]\right\}^{a-1}.$$

- When $a = 1 = b$ and $\lambda = 0$, *KwTP* reduces to the classical Pareto distribution with PDF

$$f(x;\alpha,\theta) = \frac{\alpha\theta^{\alpha}}{x^{\alpha+1}}.$$

4 Mixture representation

In this section, we provide alternative and useful expressions for the PDF and the CDF of *KwTP* distribution. Let X be a random variable having the *KwTP* density (7).

For any $b > 0$ real non-integer and $|z| < 1$, consider the power series

$$(1-z)^{b-1} = \sum_{k=0}^{\infty} \frac{(-1)^k \Gamma(b)}{k!\,\Gamma(b-k)} z^k. \tag{8}$$

As mentioned by Afify et al. (2016), applying the power series (8) to Eq. (7), we can write the PDF as a mixture of exponentiated-transmuted Pareto with the power parameter $(k+1)a$ as

$$f(x) = \sum_{k=0}^{\infty} v_k g(x) G(x)^{(k+1)a-1}, \tag{9}$$

where

$$v_k = \frac{(-1)^k ab\Gamma(b)}{k!\,\Gamma(b-k)} = ab\binom{b-1}{k}(-1)^k,$$

and $g(x)$ and $G(x)$ are the PDF and CDF of the transmuted Pareto distribution respectively.

The PDF (9) can be expressed as a mixture of exp-G densities

$$f(x) = \sum_{k=0}^{\infty} w_k \gamma_k g(x) G(x)^{\gamma_k - 1}, \tag{10}$$

where $w_k = \frac{v_k}{(k+1)a}$, and $\gamma_k = (k+1)a$. By integrating (10), we get a similar mixture representation for the CDF

$$F(x) = \sum_{k=0}^{\infty} w_k G(x)^{(k+1)a}, \tag{11}$$

where $G(x)$ is the CDF of the transmuted Pareto distribution with power parameter $(k+1)a$. Using the binomial expansions of the expressions

$$\left[1-\left(1-(\theta/x)^{\alpha}\right)\left(1+\lambda(\theta/x)^{\alpha}\right)\right]^{b-1}, \quad \left(1-(\theta/x)^{\alpha}\right)^{(k+1)a-1} \text{ and } \left(1+\lambda(\theta/x)^{\alpha}\right)^{(k+1)a-1},$$

the PDF of the KwTP distribution can be expressed as

$$\begin{aligned} f(x) &= \frac{ab\alpha\theta^{\alpha}}{x^{\alpha+1}}\left[1-\lambda+2\lambda\left(\frac{\theta}{x}\right)^{\alpha}\right] \times \left\{\left[1-\left(\frac{\theta}{x}\right)^{\alpha}\right]\left[1+\lambda\left(\frac{\theta}{x}\right)^{\alpha}\right]\right\}^{a-1} \\ &\quad \times \left\{1-\left[1-\left(\frac{\theta}{x}\right)^{\alpha}\right]^{a}\left[1+\lambda\left(\frac{\theta}{x}\right)^{\alpha}\right]^{a}\right\}^{b-1} \\ &= \sum_{k=0}^{\infty} \frac{ab\alpha\theta^{\alpha}}{x^{\alpha+1}}\binom{b-1}{k}(-1)^{k}\left[1-\lambda+2\lambda\left(\frac{\theta}{x}\right)^{\alpha}\right] \\ &\quad \times \left[1-\left(\frac{\theta}{x}\right)^{\alpha}\right]^{(k+1)a-1}\left[1+\lambda\left(\frac{\theta}{x}\right)^{\alpha}\right]^{(k+1)a-1} \\ &= \sum_{k=0}^{\infty} \frac{ab\alpha\theta^{\alpha}}{x^{\alpha+1}}\binom{b-1}{k}(-1)^{k}\left[1-\lambda+2\lambda\left(\frac{\theta}{x}\right)^{\alpha}\right] \\ &\quad \times \sum_{i=0}^{\infty}\sum_{j=0}^{\infty}\binom{(k+1)a-1}{i}\binom{(k+1)a-1}{j}(-1)^{i}\lambda^{j}\left(\frac{\theta}{x}\right)^{\alpha(i+j)}, \end{aligned}$$

where $\alpha > 0, a > 0, b > 0, \theta > 0, |\lambda| \leq 1$ and $x > \theta$. Let

$$w_{ij}^{k} = \binom{(k+1)a-1}{i}\binom{(k+1)a-1}{j}(-1)^{i}\lambda^{j}.$$

By setting $i + j = m$, and $w_{k,m}^{*} = \sum_{i,j:i+j=m} w_{ij}^{k} = \sum_{i=0}^{m}\sum_{j=0}^{m-i} w_{ij}^{k}$, we can further express the PDF of the *KwTP* in the form

$$f(x) = \frac{ab\alpha\theta^{\alpha}}{x^{\alpha+1}}\left[1-\lambda+2\lambda\left(\frac{\theta}{x}\right)^{\alpha}\right]\sum_{m=0}^{\infty} u_m \left(\frac{\theta}{x}\right)^{\alpha m}, \tag{12}$$

where

$$u_m = \sum_{k=0}^{\infty} w_{k,m}^{*}\binom{b-1}{k}(-1)^{k}. \tag{13}$$

In particular,

$$u_0 = \sum_{k=0}^{\infty} w_{k,0}^{*}\binom{b-1}{k}(-1)^{k}, \text{ where } w_{k,0}^{*} = \sum_{i=0}^{0}\sum_{j=0}^{0-i} w_{ij}^{k}.$$

That is, $m = 0 \implies i = 0, j = 0$. From $w_{ij}^k = \binom{(k+1)a-1}{i}\binom{(k+1)a-1}{j}(-1)^i\lambda^j$ and $i = 0, j = 0$, we have $w_{0,0} = 1$. Thus, $w_{k,0}^* = 1$, and hence, $u_0 = \sum_{k=0}^{\infty}\binom{b-1}{k}(-1)^k$.

Similarly, using (11) we can express $F(x)$ as

$$\begin{aligned}
F(x) &= \sum_{k=0}^{\infty} w_k G(x)^{(k+1)a} \\
&= \sum_{k=0}^{\infty} w_k \left[1 - \left(\frac{\theta}{x}\right)^{\alpha}\right]^{(k+1)a} \times \left[1 + \lambda\left(\frac{\theta}{x}\right)^{\alpha}\right]^{(k+1)a} \\
&= \sum_{k=0}^{\infty}\sum_{i=0}^{\infty}\sum_{j=0}^{\infty} w_k(-1)^i\lambda^j \binom{(k+1)a}{i}\binom{(k+1)a}{j}\left(\frac{\theta}{x}\right)^{\alpha(i+j)}.
\end{aligned}$$

Let $v_{ij}^k = \binom{(k+1)a}{i}\binom{(k+1)a}{j}(-1)^i\lambda^j$, where in particular $v_{00}^k = 1$.

Setting $i + j = m$, $v_{k,m}^* = \sum_{i,j:i+j=m} v_{ij}^k = \sum_{i=0}^{m}\sum_{j=0}^{m-i} v_{ij}^k$, $v = \left(\frac{\theta}{x}\right)^{\alpha}$, and

$$\rho_m = \sum_{k=0}^{\infty} v_{k,m}^* w_k, \tag{14}$$

we can express the CDF of the *KwTP* as,

$$\begin{aligned}
F(x) &= \sum_{m=0}^{\infty}\sum_{k=0}^{\infty} v_{k,m}^* w_k \left(\frac{\theta}{x}\right)^{\alpha m} \\
&= \sum_{m=0}^{\infty}\left(\frac{\theta}{x}\right)^{\alpha m}\rho_m \\
&= \sum_{m=0}^{\infty}\rho_m v^m,
\end{aligned} \tag{15}$$

where ρ_0 may be expressed as

$$\begin{aligned}
\rho_0 &= \sum_{k=0}^{\infty} v_{k,0}^* w_k = \sum_{k=0}^{\infty} v_{00}^k w_k = \sum_{k=0}^{\infty} w_k \\
&= \sum_{k=0}^{\infty}\frac{v_k}{a(k+1)} = \sum_{k=0}^{\infty}(-1)^k\frac{ab\binom{b-1}{k}}{a(k+1)} \\
&= b\sum_{k=0}^{\infty}\frac{(-1)^k}{k+1}\binom{b-1}{k}.
\end{aligned} \tag{16}$$

5 Quantile function

The quantile function of a distribution is the real solution of $F(x_q) = q$ for $0 \le q \le 1$. It is defined as the inverse function of the CDF of a random variable X. Using Eq. (6), the quantiles (x_q) of *KwTP* distribution may be expressed in the form

$$x_q = F^{-1}(q) = \theta\left(\frac{\lambda - 1 + \sqrt{(1+\lambda)^2 - 4\lambda\left(\left\{1 - (1-q)^{\frac{1}{b}}\right\}^{\frac{1}{a}}\right)}}{2\lambda}\right)^{-\frac{1}{\alpha}}.$$

For specified values of parameters $\alpha, \theta, \lambda, a$ and b, and a set of random number $u \in (0,1)$, we can generate a random variable X having the KwTP distribution (6), and satisfying

$$X = Q_u = F^{-1}(u) = \theta \left(\frac{\lambda - 1 + \sqrt{(1+\lambda)^2 - 4\lambda \left(\left\{ 1 - (1-u)^{\frac{1}{b}} \right\}^{\frac{1}{a}} \right)}}{2\lambda} \right)^{-\frac{1}{\alpha}}. \tag{17}$$

In general, the quantile function enables one to find the relationship between one random variable and another random variable. For example, if Y is a random variable with CDF $G(y)$, we can show that $X \sim KwTP(x|\alpha, \theta, \lambda, a, b)$ if

$$X = \theta \left(\frac{2\lambda}{\lambda - 1 + \sqrt{(1+\lambda)^2 - 4\lambda \left(\left\{ 1 - (G(y))^{\frac{1}{b}} \right\}^{\frac{1}{a}} \right)}} \right)^{\frac{1}{\alpha}}.$$

6 Moment functions

We define in this section, various moment expressions for the KwTP distribution. These consist of the raw moments and their related functions, e.g., skewness and kurtosis, and the moment generating function. By definition, the rth-moment of the KwTP random variable is defined by

$$\begin{aligned}
E(X^r) &= \mu_r'(\alpha, \theta, \lambda, a, b) = \int_\theta^\infty x^r f(x) dx \\
&= ab\alpha\theta^\alpha \left[(1-\lambda) \sum_{m=0}^\infty u_m \theta^{\alpha m} \int_\theta^\infty x^{r-1-\alpha-\alpha m} dx + 2\lambda \sum_{m=0}^\infty u_m \theta^{\alpha+\alpha m} \right. \\
&\quad \left. \times \int_\theta^\infty x^{r-1-2\alpha-\alpha m} dx \right] \\
&= ab\alpha\theta^\alpha \left\{ (1-\lambda) \sum_{m=0}^\infty u_m \theta^{\alpha m} \left[\frac{x^{r-\alpha-\alpha m}}{r-\alpha-\alpha m} \right] \Bigg|_\theta^\infty \right. \\
&\quad \left. + 2\lambda \sum_{m=0}^\infty u_m \theta^{\alpha+\alpha m} \left[\frac{x^{r-2\alpha-\alpha m}}{r-2\alpha-\alpha m} \right] \Bigg|_\theta^\infty \right\} \\
&= (1-\lambda) ab\alpha\theta^r \sum_{m=0}^\infty \frac{u_m}{\alpha m + \alpha - r} + 2\lambda ab\alpha\theta^r \sum_{m=0}^\infty \frac{u_m}{\alpha m + 2\alpha - r} \\
&= ab\alpha\theta^r \sum_{m=0}^\infty u_m \left[\frac{1-\lambda}{\alpha + \alpha m - r} + \frac{2\lambda}{2\alpha + \alpha m - r} \right], \quad r < \alpha.
\end{aligned} \tag{18}$$

In particular if $\lambda = 0, a = b = 1$, we have the Pareto distribution. In this case, we must have $m = 0, u_0 = 1$, so that the moment expression becomes

$$E(X^r) = \mu_r'(\alpha, \theta, 0, 1, 1) = \frac{\alpha\theta^r}{\alpha - r}, \quad r < \alpha.$$

It is interesting to note that the mean of X is

$$E(X) = \mu_1'(\alpha, \theta, 0, 1, 1) = \frac{\alpha\theta}{\alpha - 1}, \quad \alpha > 1.$$

By similar procedure, the moment generating function is defined as follows:

$$
\begin{aligned}
M_X(t) &= \int_\theta^\infty e^{tx} f(x) dx \\
&= ab\alpha\theta^\alpha \left[(1-\lambda) \sum_{m=0}^\infty u_m \theta^{\alpha m} \int_\theta^\infty e^{tx} x^{-\alpha-\alpha m-1} dx + 2\lambda \sum_{m=0}^\infty u_m \theta^{\alpha+\alpha m} \right. \\
&\quad \left. \times \int_\theta^\infty e^{tx} x^{-\alpha m-2\alpha-1} dx \right] \\
&= ab\alpha \left[(1-\lambda) \sum_{m=0}^\infty u_m (-t\theta)^{\alpha+\alpha m} \Gamma(-\alpha-\alpha m, -t\theta) \right. \\
&\quad \left. + 2\lambda \sum_{m=0}^\infty u_m (-t\theta)^{2\alpha+\alpha m} \Gamma(-2\alpha-\alpha m, -t\theta) \right], \\
&= ab\alpha \sum_{m=0}^\infty u_m (-t\theta)^{\alpha+\alpha m} \big[(1-\lambda)\Gamma(-\alpha-\alpha m, -t\theta) \\
&\quad + 2\lambda(-t\theta)^\alpha \Gamma(-2\alpha-\alpha m, -t\theta) \big], \qquad t < 0,
\end{aligned}
$$

where $\Gamma(.,.)$ denotes the upper incomplete gamma function. That is, $\Gamma(s,\theta) = \int_\theta^\infty e^{-t} t^{s-1} dt$.

The variance, skewness and kurtosis of the *Kw*TP distribution can be calculated from (18) using the relations given below.

$$
\begin{aligned}
Variance(X) &= E(X^2) - [E(X)]^2, \ \alpha > 2 \\
Skewness(X) &= \frac{E(X^3) - 3E(X)E(X^2) + 2E^3(X)}{Var^{3/2}(X)}, \ \alpha > 3 \\
Kurtosis(X) &= \frac{E(X^4) - 4E(X)E(X^3) + 6E(X^2)E^2(X) - 3E^4(X)}{Var^2(X)}, \ \alpha > 4.
\end{aligned}
$$

We provide in Table 1, the numerical measures of the median, mean, variance, skewness and kurtosis of the *Kw*TP distribution for selected values of the parameters, to illustrate the effects of the parameters on these measures. We had arbitrarily fixed the value of parameter θ at 0.5.

The values of the statistics; mean, median, variance, skewness and kurtosis increase or decrease according to the changes in parameter values. For example, these statistics decrease as α or b increases, while other parameters are held constant. We also observe that each statistic in the set consisting of the mean, median and variance increases by increasing the value of parameter a, while each statistic in the set decreases as λ increases. On the other hand, the skewness and kurtosis both decrease for increasing value of a, while the values in this set both increase, and then decrease, for increasing value of λ. The effects of parameters a and b on the mean, variance, skewness and kurtosis for given values of λ, θ and α are displayed in Figs. 3 and 4 respectively.

In addition to the traditional measures of skewness and kurtosis, we also obtain the Bowley's measure of skewness, introduced by Bowley (1920), and the Moors' measure of kurtosis, introduced by Moors (1988). These measures are quartile alternatives to the traditional skewness and kurtosis, and are more robust estimation of these measures, see Kenney and Keeping (1962).

Table 1 Median, mean, variance, skewness, kurtosis for selected values of the parameters

α	λ	a	b	Median	Mean	Variance	Skewness	Kurtosis
5	0.2	1	1	0.5636	0.6111	0.0224	4.9196	81.8250
6	0.2	1	1	0.5525	0.5891	0.0130	4.0458	42.9488
7	0.2	1	1	0.5446	0.5744	0.0084	3.5980	30.9806
8	0.2	1	1	0.5388	0.5638	0.0059	3.3243	25.2955
10	0.2	1	1	0.5309	0.5497	0.0034	3.0056	19.8668
5	-1	1	1	0.6392	0.6944	0.0386	4.1113	60.5271
5	-0.5	1	1	0.6061	0.6997	0.0335	4.2366	63.3030
5	0.1	1	1	0.5688	0.6181	0.0243	4.7743	77.4267
5	0.5	1	1	0.5505	0.5903	0.0162	5.4825	101.8249
5	0.8	1	1	0.5408	0.5694	0.0091	6.0430	135.7943
5	1	1	1	0.5359	0.5556	0.0039	2.8111	17.8286
5	0.5	1	1	0.5505	0.5903	0.0162	5.4825	101.8249
5	0.5	2	1	0.5964	0.6422	0.0249	4.7467	78.6519
5	0.5	3	1	0.6304	0.6802	0.0313	4.4534	70.3107
5	0.5	4	1	0.6578	0.7106	0.0366	4.2881	65.8432
5	0.5	5	1	0.6810	0.7362	0.0411	4.1793	63.0006
5	0.2	1	1	0.5636	0.6111	0.0224	4.9196	81.8250
5	0.2	1	2	0.5303	0.5480	0.0031	3.0122	20.2613
5	0.2	1	3	0.5199	0.5305	0.0011	2.6351	15.2360
5	0.2	1	4	0.5150	0.5224	0.0006	2.4635	13.3142
5	0.2	1	5	0.5118	0.5176	0.0003	2.3646	12.2905

- Bowley skewness (B_{sk}): By definition, the Bowley's measure of skewness is expressed as

$$B_{sk} = \frac{Q_3 + Q_1 - 2Q_2}{Q_3 - Q_1} = \frac{Q_{0.75} - 2Q_{0.5} + Q_{0.25}}{Q_{0.75} - Q_{0.25}}.$$

- Moors kurtosis M_{kur}: This is defined as

$$M_{kur} = \frac{(E_7 - E_5) + (E_3 - E_1)}{E_6 - E_2} = \frac{Q_{0.875} - Q_{0.625} + Q_{0.375} - Q_{0.125}}{Q_{0.75} - Q_{0.25}},$$

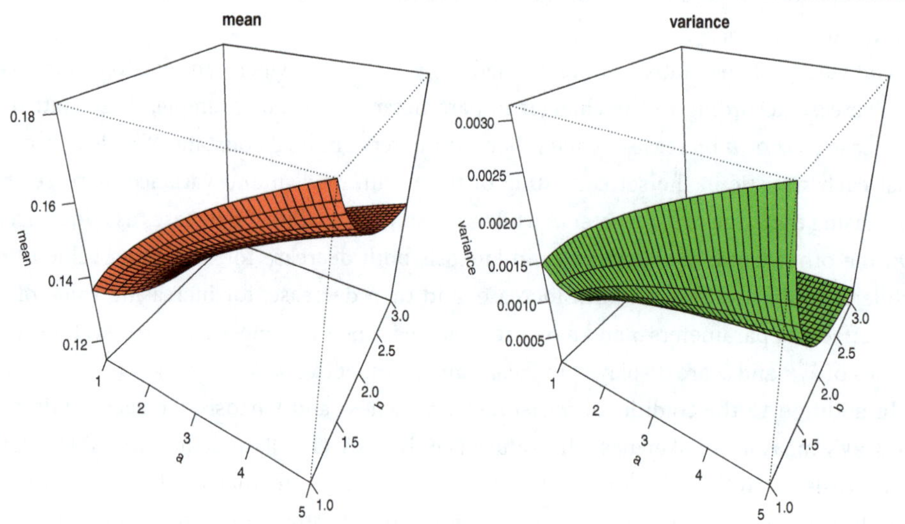

Fig. 3 Plots of mean and variance of *Kw*TP distribution

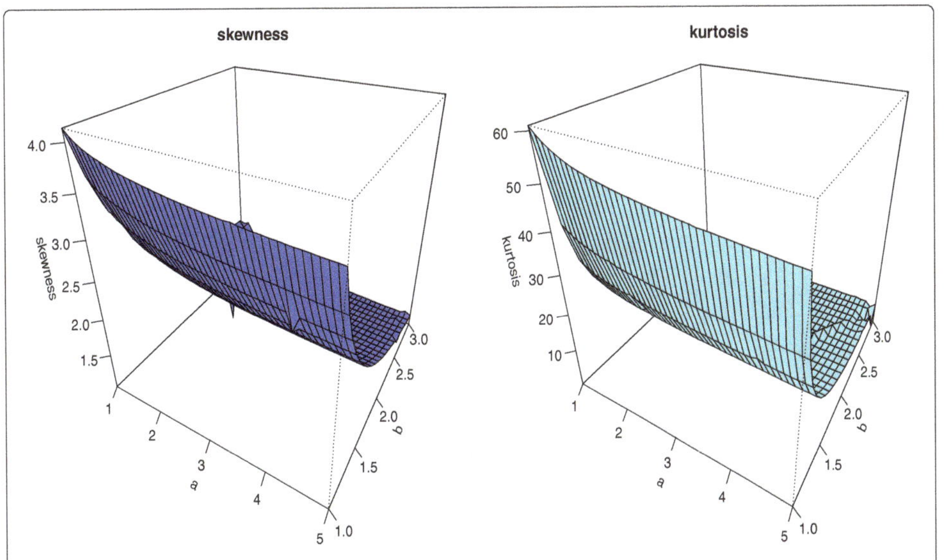

Fig. 4 Plots of skewness and kurtosis of *KwTP* distribution

where Q_i is the ith quartile for $i = 1, 2, 3$, and $E_i = F^{-1}(i/8)$, $i = 1, 2, \cdots, 7$ represents the ith octile.

The graphs of the Bowley skewness and Moors kurtosis for different values of a and b, with $\lambda = 0.2, \alpha = 2$ and $\theta = 0.1$, are shown in Fig. 5. The graphs show that both Bowley skewness and Moors kurtosis depend on the choice of the parameters.

7 Mean deviation

The deviation from the mean, in the case of symmetric distributions, or deviation from the median, in the case of skewed distributions, can be used as a measure of

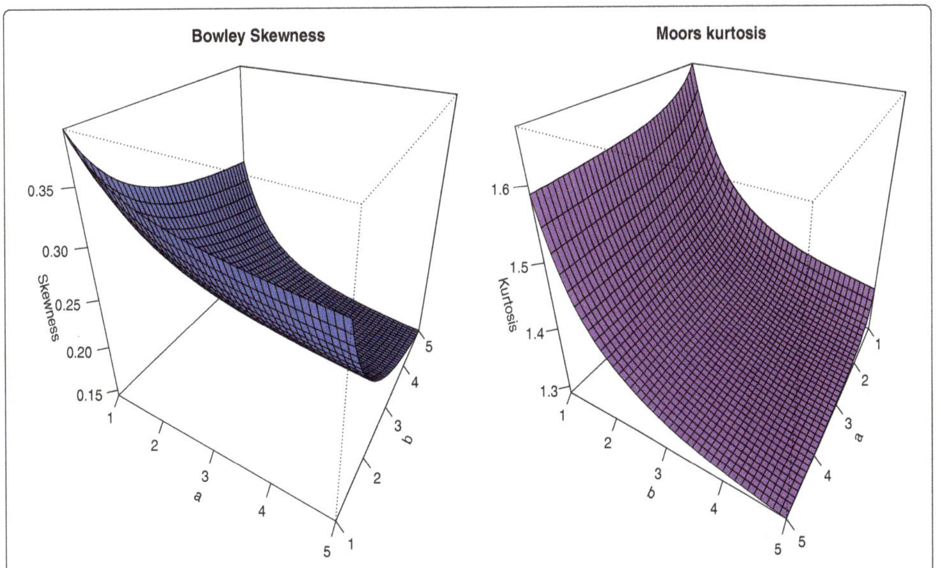

Fig. 5 Bowley skewness (*left*) and Moors kurtosis (*right*) for *KwTP* distribution as a function of the parameters *a* and *b*

spread in a population. Given that the random variable X has the *KwTP* distribution. Let $\mu = E(X)$ be the mean and M be the median of X. The mean deviation about the mean $D(\mu)$ and the mean deviation about the median $D(M)$ can be expressed respectively as

$$
\begin{aligned}
D(\mu) &= \int_{\theta}^{\infty} |x-\mu| f(x)dx = \int_{\theta}^{\mu} (\mu - x) f(x)dx + \int_{\mu}^{\infty} (x-\mu) f(x)dx \\
&= 2\mu F(\mu) - 2J(\mu),
\end{aligned} \tag{19}
$$

and

$$
\begin{aligned}
D(M) &= \int_{\theta}^{\infty} |x-M| f(x)dx = \int_{\theta}^{M} (M - x) f(x)dx + \int_{M}^{\infty} (x-M) f(x)dx \\
&= \mu - 2J(M),
\end{aligned} \tag{20}
$$

where $F(.)$ is the CDF of the *KwTP* distribution, and $J(t) = \int_{\theta}^{t} xf(x)dx$.

We compute $J(t)$ as follows:

$$
\begin{aligned}
J(t) &= \int_{\theta}^{t} xf(x)dx \\
&= ab\alpha\theta^{\alpha}\Bigg[(1-\lambda)\sum_{m=0}^{\infty} u_m \theta^{\alpha m} \int_{\theta}^{t} x^{-\alpha-\alpha m} dx + 2\lambda \sum_{m=0}^{\infty} u_m \theta^{\alpha+\alpha m} \\
&\quad \times \int_{\theta}^{t} x^{-2\alpha-\alpha m} dx\Bigg] \\
&= ab\alpha\Bigg[(1-\lambda)\sum_{m=0}^{\infty} u_m \theta^{\alpha+\alpha m}\left(\frac{t^{-\alpha-\alpha m+1}}{1-\alpha-\alpha m} + \frac{\theta^{-\alpha-\alpha m+1}}{\alpha+\alpha m-1}\right) \\
&\quad + 2\lambda \sum_{m=0}^{\infty} u_m \theta^{2\alpha+\alpha m}\left(\frac{t^{-2\alpha-\alpha m+1}}{1-2\alpha-\alpha m} + \frac{\theta^{-2\alpha-\alpha m+1}}{2\alpha+\alpha m-1}\right)\Bigg] \\
&= ab\alpha \sum_{m=0}^{\infty} u_m \Bigg[\frac{(1-\lambda)t}{1-\alpha-\alpha m}\left(\frac{\theta}{t}\right)^{\alpha} + \frac{(1-\lambda)\theta}{\alpha+\alpha m-1} + \frac{2\lambda t}{1-2\alpha-\alpha m}\left(\frac{\theta}{t}\right)^{2\alpha+\alpha m} \\
&\quad + \frac{2\lambda\theta}{2\alpha+\alpha m-1}\Bigg].
\end{aligned} \tag{21}
$$

Using Eq. (21), we can write appropriate expressions for $J(\mu)$ and $J(M)$. Combining these with the expression for $F(\mu)$, it is easy to obtain the expressions for $D(\mu)$ and $D(M)$ in Eqs. (19) and (20) respectively.

8 Reliability

According to Ashour and Eltehiwy (2013), the reliability function, also called the survival function, is the characteristic of an explanatory variable that maps a set of events, usually associated with mortality or failure of some system onto time. In other words, the reliability function $R(t)$ is the probability of an item not failing prior to some time t, and is defined by $R(t) = 1 - F(t)$. The reliability function of the KwTP distribution is

$$\begin{aligned} R(t) &= 1 - F(t) \\ &= \left\{1 - \left[1 - \left(\frac{\theta}{t}\right)^{\alpha}\right]^{a}\left[1 + \lambda\left(\frac{\theta}{t}\right)^{\alpha}\right]^{a}\right\}^{b}. \end{aligned}$$

The other characteristic of interest of a random variable is the hazard rate function $h(t)$ which is also known as instantaneous failure rate. The hazard function of a random variable X with PDF $f(x)$ and associated CDF $F(x)$ is defined as

$$h(t) = \frac{f(t)}{1 - F(t)}.$$

Substituting the PDF and CDF for the KwTP distribution into the above expression, we have

$$\begin{aligned} h(t) &= \frac{f(t)}{1 - F(t)} \\ &= \frac{ab\alpha\theta^{\alpha}\left[1 - \lambda + 2\lambda\left(\frac{\theta}{t}\right)^{\alpha}\right]\left[1 - \left(\frac{\theta}{t}\right)^{\alpha}\right]^{a-1}\left[1 + \lambda\left(\frac{\theta}{t}\right)^{\alpha}\right]^{a-1}}{t^{\alpha+1}\left\{1 - \left[1 - \left(\frac{\theta}{t}\right)^{\alpha}\right]^{a}\left[1 + \lambda\left(\frac{\theta}{t}\right)^{\alpha}\right]^{a}\right\}}. \end{aligned}$$

The flexibility of KwTP distribution to model reliability data is illustrated by varying shapes of the hazard rate function in Fig. 6.

The following lemma shows the limiting behavior of the hazard function:

Lemma 1 *If $h(t)$ is the hazard function of the Kumaraswamy transmuted Pareto distribution, then*

$$\lim_{t \to \theta} h(t) = \begin{cases} 0, & a > 1 \\ \frac{ab\alpha(1+\lambda)}{\theta}, & a = 1 \\ \infty, & a < 1, \end{cases}$$

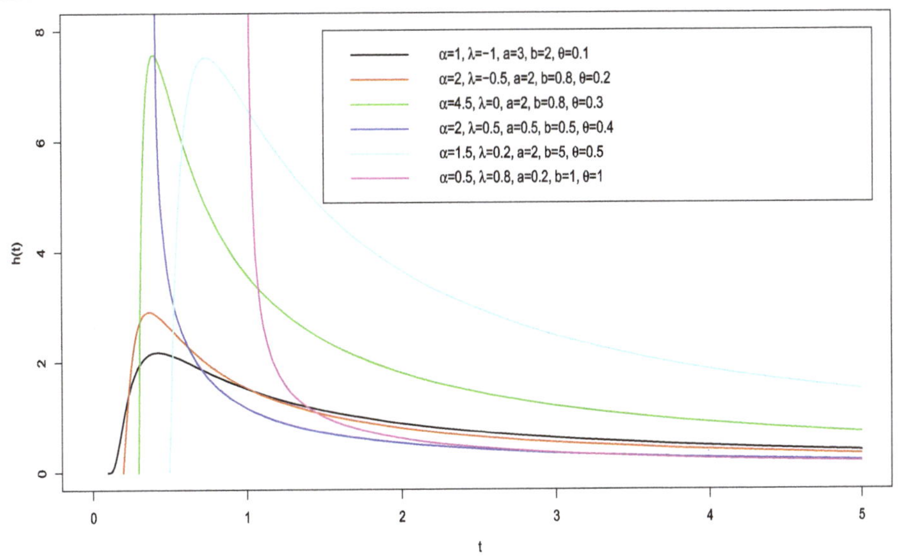

Fig. 6 Hazard rate function of KwTP for selected values of $\alpha, \theta, \lambda, a$ and b

and

$$\lim_{t\to\infty} h(t) = 0.$$

Proof First note that

$$\lim_{t\to\theta} G(t) = \lim_{t\to\theta}\left[1-\left(\frac{\theta}{t}\right)^{\alpha}\right]\left[1+\lambda\left(\frac{\theta}{t}\right)^{\alpha}\right] = 0,$$

and

$$\lim_{t\to\theta} g(t) = \lim_{t\to\theta}\frac{\alpha\theta^{\alpha}}{t^{\alpha+1}}\left[1-\lambda+2\lambda\left(\frac{\theta}{t}\right)^{\alpha}\right] = \frac{\alpha(1+\lambda)}{\theta}.$$

Then we have

$$\begin{aligned}\lim_{t\to\theta} h(t) &= \lim_{t\to\theta}\frac{f(t)}{1-F(t)} = \lim_{t\to\theta}\frac{abg(t)G(t)^{a-1}}{1-G(t)^{a}}\\ &= \begin{cases}0, & \text{if } a>1\\ \frac{ab\alpha(1+\lambda)}{\theta}, & \text{if } a=1\\ \infty, & \text{if } a<1.\end{cases}\end{aligned}$$

For the second part, we already know that $\lim_{t\to\infty} G(t) = 1$, $\lim_{t\to\infty} g(t) = 0$, $\lim_{t\to\infty} F(t) = 1$, and $\lim_{t\to\infty} f(t) = 0$. Then using the L'Hospital's rule, we have

$$\begin{aligned}\lim_{t\to\infty} h(t) &= \lim_{t\to\infty}\frac{abg(t)G(t)^{a-1}}{1-G(t)^{a}}\\ &= \lim_{t\to\infty}\frac{abg(t)}{G(t)^{1-a}-G(t)}\\ &= \lim_{t\to\infty}\frac{abg'(t)}{(1-a)G(t)^{-a}g(t)-g(t)}\\ &= \lim_{t\to\infty}\frac{ab}{[(1-a)G(t)^{-a}-1]}\times\lim_{t\to\infty}\frac{g'(t)}{g(t)}\\ &= -b\lim_{t\to\infty}\frac{g'(t)}{g(t)} = 0.\end{aligned}$$

□

9 Order statistics

Let $X_1, X_2, \ldots\ldots, X_n$ be a simple random sample from $KwTP\,(x;\alpha,\theta,\lambda,a,b)$ with CDF (6) and PDF (7). Also let $X_{(1)}, X_{(2)}, \ldots\ldots, X_{(n)}$ denote the order statistics from this sample. The PDF $f_{(i:n)}(x)$ of the i-th order statistics is given by

$$f_{i:n}(x) = \frac{1}{B(i,n-i+1)}[F(x)]^{i-1}[1-F(x)]^{n-i}f(x), \tag{22}$$

and the CDF is given by

$$F_{i:n}(x) = \sum_{k=i}^{n}\binom{n}{k}[F(x)]^{k}[1-F(x)]^{n-k} = \int_0^{F(x)}\frac{1}{B(i,n-i+1)}t^{i-1}(1-t)^{n-i}dt.$$

Using Eqs. (12), (13), (14), (15) and binomial expansions, Eq. (22) becomes

$$
\begin{aligned}
f_{i:n}(x) &= \frac{1}{B(i,n-i+1)} f(x) \sum_{s=0}^{n-i} (-1)^s \binom{n-i}{s} F(x)^{i+s-1} \\
&= \frac{1}{B(i,n-i+1)} \left(\frac{ab\alpha\theta^\alpha}{x^{\alpha+1}} \left[1-\lambda+2\lambda \left(\frac{\theta}{x}\right)^\alpha \right] \sum_{l=0}^{\infty} u_l \left(\frac{\theta}{x}\right)^{\alpha l} \right) \\
&\quad \times \sum_{s=0}^{n-i} (-1)^s \binom{n-i}{s} \times \left(\sum_{m=0}^{\infty} \rho_m v^m \right)^{i+s-1}. \qquad (23)
\end{aligned}
$$

From Gradshteyn and Ryzhik (2000, sec. 0.314), for any positive integer r,

$$
\left(\sum_{m=0}^{\infty} \rho_m v^m \right)^r = \sum_{m=0}^{\infty} d_{r,m} v^m,
$$

where the coefficients $d_{r,m}$, for $m = 1, 2, 3, ...$, can be determined from the recurrence equation

$$
d_{r,m} = (m\rho_0)^{-1} \sum_{q=1}^{m} \{q(r+1)-m\} \rho_q d_{r,m-q}
$$

and $d_{r,0} = (\rho_0^*)^r$.

We can obtain $d_{r,m}$ from $d_{r,0}, \cdots, d_{r,m-1}$ and, therefore, from $\rho_0, \rho_1, \cdots, \rho_m$. Hence (23) becomes

$$
\begin{aligned}
f_{i:n}(x) &= \frac{ab\alpha\theta^\alpha (1-\lambda+2\lambda v)}{x^{\alpha+1} B(i,n-i+1)} \sum_{l=0}^{\infty} u_l v^l \\
&\quad \times \sum_{s=0}^{n-i} (-1)^s \binom{n-i}{s} \times \left(\sum_{m=0}^{\infty} d_{i+s-1,m} v^m \right) \\
&= \frac{ab\alpha\theta^\alpha (1-\lambda+2\lambda v)}{x^{\alpha+1} B(i,n-i+1)} \sum_{l=0}^{\infty} u_l v^l \sum_{m=0}^{\infty} c_i(m) v^m \\
&= \frac{ab\alpha\theta^\alpha (1-\lambda+2\lambda v)}{x^{\alpha+1} B(i,n-i+1)} \sum_{l=0}^{\infty} \sum_{m=0}^{\infty} u_l c_i(m) v^{l+m},
\end{aligned}
$$

where

$$
v = \left(\frac{\theta}{x}\right)^\alpha, \quad c_i(m) = \sum_{s=0}^{n-i} (-1)^s \binom{n-i}{s} d_{i+s-1,m},
$$

$$
d_{i+s-1,m} = (m\rho_0)^{-1} \sum_{q=1}^{m} \left[q(i+s)-m \right] \rho_q d_{i+s-1,m-q}
$$

and $d_{i+s-1,0} = (\rho_0)^{i+s-1}$, where ρ_0 is defined in Eq. (16).

10 Parameter estimation

In this section we estimate the parameters of the Kumaraswamy transmuted Pareto distribution by the method of maximum likelihood estimation.

Let $X_1, X_2, \ldots\ldots, X_n$ be a random sample from the Kumaraswamy transmuted Pareto distribution with observed values $x_1, x_2, \ldots\ldots, x_n$, and $\boldsymbol{\gamma} = (\alpha, \theta, \lambda, a, b)^T$ be the vector of the model parameters. The log-likelihood function for $\boldsymbol{\gamma}$ can be written as

$$
\begin{aligned}
l(\gamma) &= \log L(x;\alpha,\theta,\lambda,a,b) \\
&= n\log a + n\log b + n\log\alpha + n\alpha\log\theta - (\alpha+1)\sum_{i=1}^{n}\log(x_i) \\
&+ \sum_{i=1}^{n}\log\left(1-\lambda+2\lambda\left(\frac{\theta}{x_i}\right)^{\alpha}\right) \\
&+ (a-1)\left[\sum_{i=1}^{n}\log\left(1-\left(\frac{\theta}{x_i}\right)^{\alpha}\right)+\sum_{i=1}^{n}\log\left(1+\lambda\left(\frac{\theta}{x_i}\right)^{\alpha}\right)\right] \\
&+ (b-1)\sum_{i=1}^{n}\log\left\{1-\left[1-\left(\frac{\theta}{x_i}\right)^{\alpha}\right]^{a}\left[1+\lambda\left(\frac{\theta}{x_i}\right)^{\alpha}\right]^{a}\right\}. \qquad (24)
\end{aligned}
$$

Since $x \in (\theta,\infty)$, the maximum likelihood estimator of θ is the first order statistic $X_{(1)}$. Following the normal routine of parameter estimation for the maximum likelihood estimation of α,λ,a and b, we differentiate Eq. (24) with respect to α,λ,a and b to obtain the score vector $\left(\frac{\partial l}{\partial\alpha},\frac{\partial l}{\partial\lambda},\frac{\partial l}{\partial a},\frac{\partial l}{\partial b}\right)^{T}$. The elements of score vectors are

$$
\begin{aligned}
\frac{\partial l}{\partial\alpha} &= \frac{n}{\alpha}+n\log\theta-\sum_{i=1}^{n}\log(x_i)+\sum_{i=1}^{n}\frac{2\lambda\left(\frac{\theta}{x_i}\right)^{\alpha}.\ln\left(\frac{\theta}{x_i}\right)}{\left(1-\lambda+2\lambda\left(\frac{\theta}{x_i}\right)^{\alpha}\right)} \\
&-(a-1)\left[\sum_{i=1}^{n}\frac{\left(\frac{\theta}{x_i}\right)^{\alpha}.\ln\left(\frac{\theta}{x_i}\right)}{\left(1-\left(\frac{\theta}{x_i}\right)^{\alpha}\right)}-\sum_{i=1}^{n}\frac{\lambda\left(\frac{\theta}{x_i}\right)^{\alpha}\ln\left(\frac{\theta}{x_i}\right)}{\left(1+\lambda\left(\frac{\theta}{x_i}\right)^{\alpha}\right)}\right] \qquad (25) \\
&-a(b-1)\sum_{i=1}^{n}\frac{G^{a-1}(x_i;\alpha,\theta,\lambda)\left(\frac{\theta}{x_i}\right)^{\alpha}\left[1-\lambda+2\lambda\left(\frac{\theta}{x_i}\right)^{\alpha}\right]\ln\left(\frac{\theta}{x_i}\right)}{1-G^{a}(x_i;\alpha,\theta,\lambda)},
\end{aligned}
$$

$$
\begin{aligned}
\frac{\partial l}{\partial\lambda} &= \sum_{i=1}^{n}\frac{2\left(\frac{\theta}{x_i}\right)^{\alpha}-1}{1-\lambda+2\lambda\left(\frac{\theta}{x_i}\right)^{\alpha}}+(a-1)\sum_{i=1}^{n}\frac{\left(\frac{\theta}{x_i}\right)^{\alpha}}{1+\lambda\left(\frac{\theta}{x_i}\right)^{\alpha}} \\
&-a(b-1)\sum_{i=1}^{n}\frac{\left(\frac{\theta}{x_i}\right)^{\alpha}\left[1-\left(\frac{\theta}{x_i}\right)^{\alpha}\right]^{a}\left[1+\lambda\left(\frac{\theta}{x_i}\right)^{\alpha}\right]^{a-1}}{1-G^{a}(x_i;\alpha,\theta,\lambda)}, \qquad (26)
\end{aligned}
$$

$$
\begin{aligned}
\frac{\partial l}{\partial a} &= \frac{n}{a}+\left[\sum_{i=1}^{n}\log\left(1-\left(\frac{\theta}{x_i}\right)^{\alpha}\right)+\sum_{i=1}^{n}\log\left(1+\lambda\left(\frac{\theta}{x_i}\right)^{\alpha}\right)\right] \\
&-(b-1)\sum_{i=1}^{n}\frac{G^{a}(x_i;\alpha,\theta,\lambda)\times\log\{G^{a}(x_i;\alpha,\theta,\lambda)\}}{1-G^{a}(x_i;\alpha,\theta,\lambda)}, \qquad (27)
\end{aligned}
$$

and

$$
\frac{\partial l}{\partial b} = \frac{n}{b}+\sum_{i=1}^{n}\log\left\{1-G^{a}(x_i;\alpha,\theta,\lambda)\right\}, \qquad (28)
$$

where $G(x_i;\alpha,\theta,\lambda)$ is the PDF of the transmuted Pareto distribution.

The maximum likelihood estimates $\hat{\alpha}, \hat{\lambda}, \hat{a}, \hat{b}$ of the unknown parameters α, λ, a, b respectively, can be obtained by setting Eqs. (25) - (28) equal zero and solving for the parameters. We can use numerical methods such as the quasi-Newton algorithm to numerically optimize the log-likelihood function given in Eq. (24), to get the maximum likelihood estimates of the parameters α, λ, a, b. To compute the standard error and the asymptotic confidence interval, we use the usual large sample approximation in which the maximum likelihood estimators for $\boldsymbol{\gamma}$ can be treated as being approximately normal. This will require the computations of the second order derivatives of Eq. (24) with respect to the vector of parameters.

This procedure will result in

$$\begin{pmatrix} \hat{\alpha} \\ \hat{\lambda} \\ \hat{a} \\ \hat{b} \end{pmatrix} \sim Normal \left[\begin{pmatrix} \alpha \\ \lambda \\ a \\ b \end{pmatrix}, \begin{pmatrix} \hat{V}_{\alpha\alpha} & \hat{V}_{\alpha\lambda} & \hat{V}_{\alpha a} & \hat{V}_{\alpha b} \\ \hat{V}_{\lambda\alpha} & \hat{V}_{\lambda\lambda} & \hat{V}_{\lambda a} & \hat{V}_{\lambda b} \\ \hat{V}_{a\alpha} & \hat{V}_{a\lambda} & \hat{V}_{aa} & \hat{V}_{ab} \\ \hat{V}_{b\alpha} & \hat{V}_{b\lambda} & \hat{V}_{ba} & \hat{V}_{bb} \end{pmatrix} \right]$$

with $\hat{V}_{u,v} = V_{(u,v))}\big|_{(u,v)=(\hat{u},\hat{v})}$, and the asymptotic variance-covariance matrix of the MLEs is,

$$\begin{pmatrix} \hat{V}_{\alpha\alpha} & \hat{V}_{\alpha\lambda} & \hat{V}_{\alpha a} & \hat{V}_{\alpha b} \\ \hat{V}_{\lambda\alpha} & \hat{V}_{\lambda\lambda} & \hat{V}_{\lambda a} & \hat{V}_{\lambda b} \\ \hat{V}_{a\alpha} & \hat{V}_{a\lambda} & \hat{V}_{aa} & \hat{V}_{ab} \\ \hat{V}_{b\alpha} & \hat{V}_{b\lambda} & \hat{V}_{ba} & \hat{V}_{bb} \end{pmatrix} = -E \begin{pmatrix} V_{\alpha\alpha} & V_{\alpha\lambda} & V_{\alpha a} & V_{\alpha b} \\ V_{\lambda\alpha} & V_{\lambda\lambda} & V_{\lambda a} & V_{\lambda b} \\ V_{a\alpha} & V_{a\lambda} & V_{aa} & V_{ab} \\ V_{b\alpha} & V_{b\lambda} & V_{ba} & V_{bb} \end{pmatrix}^{-1}$$

where entries are obtained from

$$V_{\alpha\alpha} = \frac{\partial^2 l}{\partial \alpha^2}, \qquad V_{\alpha\lambda} = \frac{\partial^2 l}{\partial \alpha \partial \lambda}, \qquad V_{\alpha a} = \frac{\partial^2 l}{\partial \alpha \partial a}, \qquad V_{\alpha b} = \frac{\partial^2 l}{\partial \alpha \partial b}$$

$$V_{\lambda\alpha} = \frac{\partial^2 l}{\partial \lambda \partial \alpha}, \qquad V_{\lambda\lambda} = \frac{\partial^2 l}{\partial \lambda^2}, \qquad V_{\lambda a} = \frac{\partial^2 l}{\partial \lambda \partial a}, \qquad V_{\lambda b} = \frac{\partial^2 l}{\partial \lambda \partial b}$$

$$V_{a\alpha} = \frac{\partial^2 l}{\partial a \partial \alpha}, \qquad V_{a\lambda} = \frac{\partial^2 l}{\partial a \partial \lambda}, \qquad V_{aa} = \frac{\partial^2 l}{\partial a^2}, \qquad V_{ab} = \frac{\partial^2 l}{\partial a \partial b}$$

$$V_{b\alpha} = \frac{\partial^2 l}{\partial b \partial \alpha}, \qquad V_{b\lambda} = \frac{\partial^2 l}{\partial b \partial \lambda}, \qquad V_{ba} = \frac{\partial^2 l}{\partial b \partial a}, \qquad V_{bb} = \frac{\partial^2 l}{\partial b^2}$$

and l is the log-likelihood function given in (24). Approximate $100(1-\phi)\%$ two sided confidence intervals for α, λ, a and b are, respectively, given by

$$\hat{\alpha} \pm z_{\frac{\phi}{2}} \sqrt{\hat{V}_{\alpha\alpha}}\,, \qquad \hat{\lambda} \pm z_{\frac{\phi}{2}} \sqrt{\hat{V}_{\lambda\lambda}}\,, \qquad \hat{a} \pm z_{\frac{\phi}{2}} \sqrt{\hat{V}_{aa}}\,, \qquad \text{and} \qquad \hat{b} \pm z_{\frac{\phi}{2}} \sqrt{\hat{V}_{bb}}\,,$$

where z_{ϕ} is the upper $\phi-$th percentile of the standard normal distribution.

Table 2 Descriptive statistics for the Wheaton river data

Min.	Q1	Median	Mean	Q3	Max.	Skewness	Kurtosis
0.100	2.125	9.500	12.200	20.12	64.00	1.4725	5.8895

11 Applications of KwTP

In this section we apply the *Kw*TP to two data sets. These are exceedances of flood peaks (in m^3/s) of the Wheaton River near Carcross in Yukon Territory, Canada, and the Norwegian fire insurance claims data.

11.1 The exceedances of flood peaks (in m^3/s) of the Wheaton River near Carcross in Yukon Territory, Canada

The data are the exceedances of flood peaks (in m^3/s) of the Wheaton River near Carcross in Yukon Territory, Canada. The data consists of 72 exceedances for the years 1958-1984. These data have been analyzed by many authors including Choulakian and Stephens (2001), Nadarajah (2005), Akinsete et al. (2008), Bourguignon et al. (2013), Merovci and Puka (2014), and Chhetri et al. (2017), among others. A summary of the descriptive statistics of the data set is given in Table 2. The analysis of these data with the *Kw*TP is comapared with the results from few commmonly used distributions in the literature. All required computations are carried out using the AdequacyModel script of R-package (Marinho et al. (2016)). In particular we compare the *Kw*TP with the Kumaraswamy Pareto distribution (*Kw*P), the transmuted Pareto distribution (TP), exponentiated Pareto distribution (EP), beta transmuted Pareto distribution (BTP), beta Pareto distribution (BP), and the Pareto distribution (P). Using the method of maximum likelihood estimation, we estimate the distribution parameters. The goodness of fit measures, including the log-likelihood function ($-\ell$), Akaike Information Criterion (AIC), Bayesian Information Criterion (BIC), Consistent Akaike Information Criterion (CAIC), and Hannan-Quinn Information Criteria (HQIC) are obtained for the fitted models.

Table 3 lists the values the MLEs and their standard errors, whereas the values of $-\widehat{\ell}$, *AIC, CAIC, HQIC, BIC* and Kolmogorov (K-S) statistic are given in Table 4. From Table 4, it is noted that the *KwTP* distribution has the lowest values for

Table 3 Estimated parameters and their standard errors for the Wheaton river data

Model	$\hat{a}$	$\hat{b}$	$\hat{\lambda}$	$\hat{\alpha}$	$\hat{\theta}$
KwTP	4.2684	17.0139	-0.3687	0.2003	0.1
	(1.5669)	(12.6727)	(0.5308)	(0.0609)	-
KwP	2.8553	85.8468	-	0.0528	0.1
	(0.3371)	(60.4213)	-	(0.0185)	-
BTP	3.9118	17.3874	-0.8518	0.1159	0.1
	(1.8159)	(11.7365)	(0.2588)	(0.0509)	-
BP	3.1473	85.7508	-	0.0088	0.1
	(0.4993)	(0.0001)	-	(0.0015)	-
TP	1	1	-0.952	0.3490	0.1
	-	-	(0.089)	(0.072)	-
EP	2.8797	1	-	0.4241	0.1
	(0.4911)	-	-	(0.0463)	-
P	1	1	-	0.2438	0.1
	-	-	-	(0.0287)	-

Table 4 The AIC, CAIC, BIC, HQIC and K-S test statistic of the Wheaton river data

Model	Statistics					
	$-\ell(.,x)$	AIC	CAIC	BIC	HQIC	KS
KwTP	254.017	516.034	516.641	525.085	519.634	0.147
BTP	256.577	521.154	521.760	530.204	524.753	0.160
KwP	271.200	548.400	548.753	555.230	551.119	0.170
BP	283.700	573.400	573.753	580.230	576.119	0.175
TP	286.201	576.402	576.575	580.954	578.214	0.287
EP	287.300	578.600	578.774	583.153	580.413	0.199
P	303.100	608.200	608.257	610.477	609.106	0.332

the $-\widehat{\ell}$, *AIC, CAIC, HQIC, BIC,* and K-S statistic. One can employ the likelihood ratio statistic to test the superiority of the *Kw*TP distribution over the other distributions. The plots comparing the *Kw*TP distribution with other distributions are given in Figs. 7, 8 and 9. The values in Table 4, and the plots, all indicate that the *Kw*TP distribution is more superior and fits the data more adequately than the other comparative distributions.

11.2 The Norwegian fire insurance data (in kr)

We apply the *Kw*TP to the Norwegian fire insurance claims data for the years 1988 and 1990. Several authors have analyzed these data. See for example, Mdziniso and Cooray (2017) and related references in the paper. The 1988 Norwegian fire claims data consist of 827 fire insurance losses in thousand Norwegian krones, ranging from 500 to 465,365 thousand Norwegian krones. The 1990 Norwegian fire data consist of 628 fire insurance losses in thousand Norwegian krones, ranging from 500 to 78,537 thousand Norwegian krones. Data for both years are highly positively skewed. Using these data, Mdziniso and

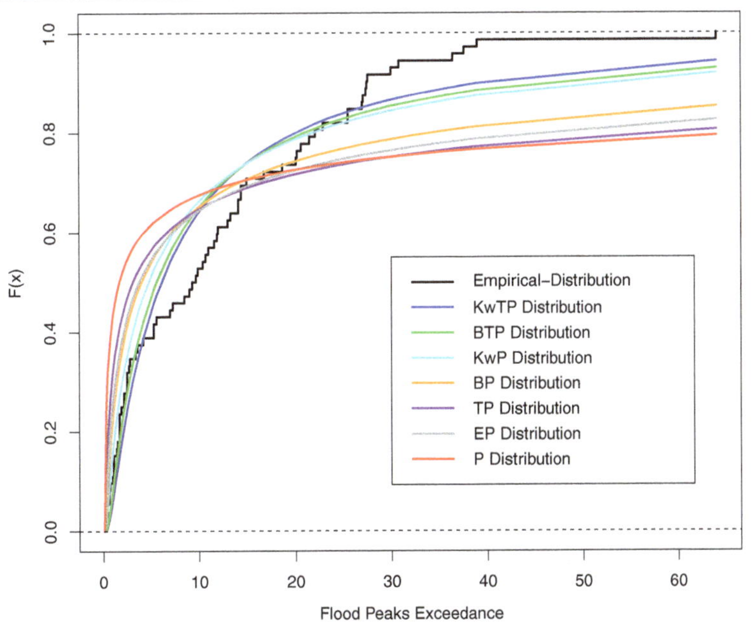

Fig. 7 Fitted CDF of the Wheaton river data

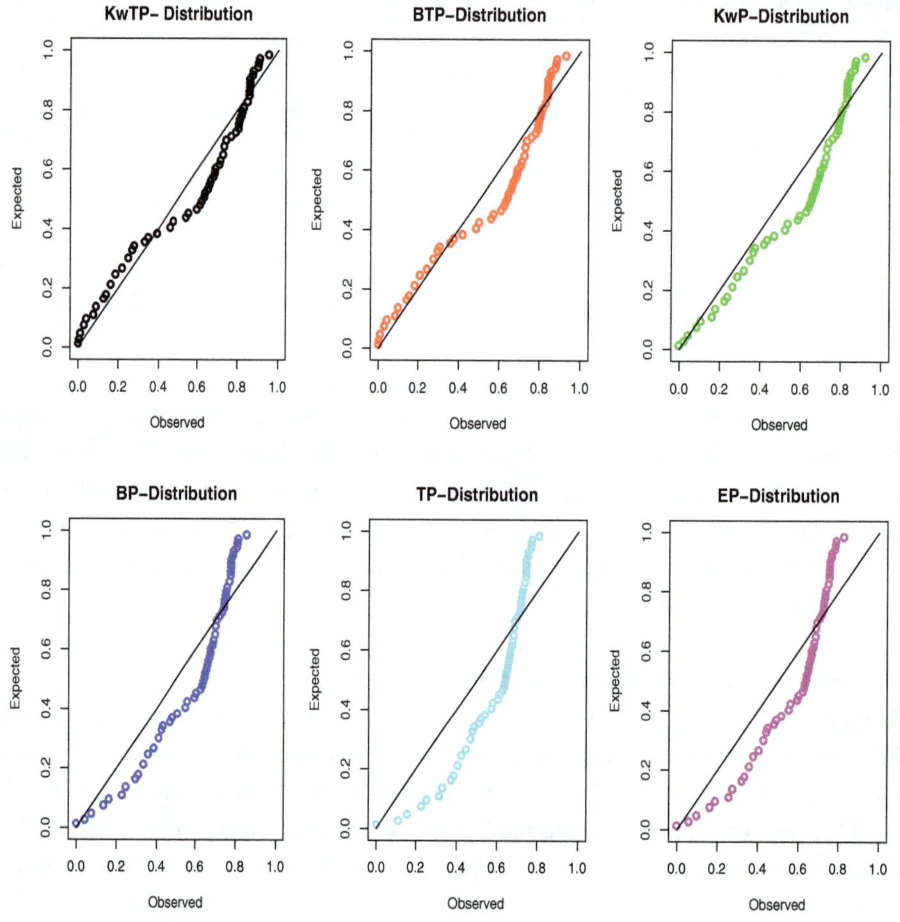

Fig. 8 QQ-plots of the Wheaton river data

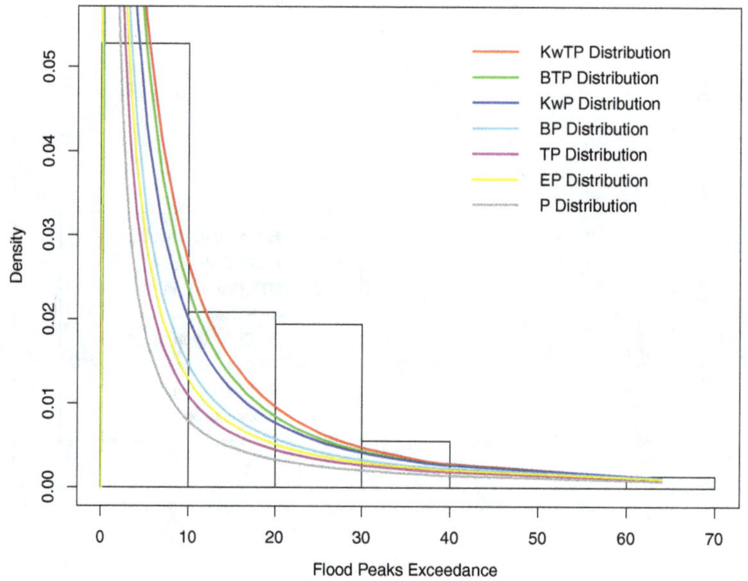

Fig. 9 Fitted PDF of the Wheaton river data

Cooray (2017) compared the performances of the Pareto (P), the 3-paramater generalized Pareto (GP), the odd-Pareto (OP) and its extension (OP*), the 3-parameter Burr, the exponentiated Pareto (EP), the exponentiated odd Pareto (EOP) and the odd generalized Pareto (OGP) distributions.

We provide in Table 5, the parameter estimation of the insurance data for both 1988 and 1990. Comparing the performance of *Kw*TP with the distributions listed in Table 4 of Mdziniso and Cooray (2017) for the 1988 data, we see from the values of the AIC, A-D, and K-S, that *Kw*TP displays superiority over the P, GP, OP, OP*, Burr, and EP distributions. It also performs better than the OGP judging from their K-S values. The same pattern of performances of *Kw*TP and the distributions listed in Table 5 of the paper are observed in the case for the 1990 data. With the exception of the exponentiated odd Pareto (EOP) distribution, the *Kw*TP is a better fit for these sets of data.

12 Simulation study

In this section we perform some simulation study to assess the reliability of the MLEs. An ideal technique for simulating random numbers from the *Kw*TP distribution is the inversion method. For fixed selected combinations of α, λ, a and b in Eq. (17), we generate samples of sizes $n = 5, 10, 20, 50, 100, 200, 500$ and 1000 from the *Kw*TP distribution. We repeated the simulations $N = 1,000$ times and calculated the mean estimates, the root mean square errors (RMSEs) and the mean absolute errors (MAEs). The empirical results obtained using R are given in Table 6. Observe that RMSEs decay as the sample size increases, while the maximum likelihood estimates get closer to their true values.

13 Conclusion

We have proposed in this article, a new distribution that is being referred to as the Kumaraswamy transmuted Pareto (*Kw*TP). The transmuted Pareto distribution is used as a baseline distribution in the Kumaraswamy distribution to construct the *Kw*TP distribution. Many mathematical and statistical properties and special cases of the *Kw*TP are obtained. The estimation of the model parameters is performed by the maximum likelihood method. We compare the distribution with few other distributions in modeling two real datasets. Various statistics indicate that *Kw*TP better fits the Wheaton river data set than other comparative distributions, and majority of its competitive distributions for the Norwegian insurance claims data. We conducted a simulation

Table 5 Norwegian fire insurance claims: Estimated values for *Kw*TP

Parameters				$-\ell$	AIC	BIC	A-D	K-S	p-value
(Year = 1988, $n = 827$)									
$\hat{\alpha}$	$\hat{\lambda}$	$\hat{a}$	$\hat{b}$						
1.0197	-0.7034	0.9698	1.2808	6751.24	13510.49	13529.29	0.46	0.03	0.62
(0.4104)	(0.2148)	(0.1609)	(0.6186)						
(Year = 1990, $n = 628$)									
$\hat{\alpha}$	$\hat{\lambda}$	$\hat{a}$	$\hat{b}$						
0.6310	-0.7285	1.0873	3.1417	5039.91	10087.82	10105.57	0.96	0.04	0.40
(0.2739)	(0.1873)	(0.1669)	(1.8428)						

Table 6 Simulation from *KwTP* of the MLE for $\alpha = 2, \lambda = 0.1, a = 3$, and $b = 2.5$

Sample size (n)	Parameter	Estimate	RMSE	MAE
5	$\hat{\alpha}$	2.0490	0.7727	0.6466
	$\hat{\lambda}$	-0.1342	0.6446	0.5640
	$\hat{a}$	2.7096	1.0869	0.8522
	$\hat{b}$	2.4806	1.1296	0.9987
10	$\hat{\alpha}$	2.0722	0.6983	0.5810
	$\hat{\lambda}$	-0.1046	0.6089	0.5274
	$\hat{a}$	2.7421	0.9994	0.7871
	$\hat{b}$	2.5427	1.0715	0.9389
20	$\hat{\alpha}$	2.0429	0.6157	0.5030
	$\hat{\lambda}$	-0.0228	0.5786	0.5022
	$\hat{a}$	2.8104	0.9036	0.7274
	$\hat{b}$	2.5201	0.9400	0.8048
50	$\hat{\alpha}$	2.0078	0.5674	0.4486
	$\hat{\lambda}$	0.0330	0.5507	0.4692
	$\hat{a}$	2.8363	0.8251	0.6755
	$\hat{b}$	2.4863	0.8682	0.7382
100	$\hat{\alpha}$	1.9644	0.5361	0.4302
	$\hat{\lambda}$	0.0590	0.5288	0.4445
	$\hat{a}$	2.8337	0.7392	0.5914
	$\hat{b}$	2.4907	0.7856	0.6617
200	$\hat{\alpha}$	1.9783	0.5214	0.4246
	$\hat{\lambda}$	0.0628	0.5082	0.4210
	$\hat{a}$	2.8550	0.6567	0.5191
	$\hat{b}$	2.5186	0.7352	0.6141
500	$\hat{\alpha}$	1.9691	0.5515	0.4607
	$\hat{\lambda}$	0.0846	0.4972	0.4057
	$\hat{a}$	2.8452	0.6056	0.4562
	$\hat{b}$	2.5001	0.7121	0.5980
1000	$\hat{\alpha}$	1.9802	0.5694	0.4841
	$\hat{\lambda}$	0.1073	0.4818	0.3830
	$\hat{a}$	2.8629	0.5301	0.3895
	$\hat{b}$	2.4704	0.6748	0.5693

study to assess the performance of the maximum likelihood estimation procedure for estimating the parameters of the *KwTP* distribution. It is expected that the *KwTP* distribution will serve as a better alternative in modeling data sets exhibiting the extreme value properties.

Acknowledgements
The authors would like to thank the Editor and the anonymous reviewers for their valuable comments and suggestions which have greatly improved the quality of the paper.

Authors' contributions
The authors, SBC, AAA, GA and HL with the consultation of each other carried out this work and drafted the manuscript together. All authors read and approved the final manuscript.

Competing interests
The authors declare that they have no competing interests.

Author details
[1]Department of Mathematical Sciences, Florida Atlantic University, Bacon Raton, FL, USA. [2]Department of Mathematics, Marshall University, Huntington, WV, USA. [3]Department of Mathematics, Statistics and Computer Science, Purdue University Northwest, Hammond, IN, USA.

References

Afify, AZ, Cordeiro, GM, Yousof, HM, Alzaatreh, A, Nofal, ZM: The Kumaraswamy transmuted-g family of distributions: Properties and applications. J. Data Sci. **14**, 245–270 (2016)
Afify, AZ, Nofal, ZM, Butt, NS: Transmuted complementary Weibull geometric distribution. Pak. J. Stat. Oper. Res. **4**, 435–454 (2014)
Akinsete, A, Famoye, F, Lee, C: The beta-Pareto distributions. Statistics. **42**(6), 547–563 (2008)
Akinsete, AA, Famoye, F, Lee, C: The Kumaraswamy - geometric distribution. J. Stat. Distrib. Appl. **1**, 1–21 (2014)
Aryal, GR, Tsokos, CP: On the transmuted extreme value distribution with application. Nonlinear Anal. Theory Methods and Appl. **7**, 1401–1407 (2009)
Aryal, GR, Tsokos, CP: Transmuted Weibull distribution: A generalization of the Weibull probability distribution. Eur. J. Pur. Appl. Math. **2**(2), 89–102 (2011)
Ashour, SK, Eltehiwy, MA: Transmuted Lomax distribution. Am. J. Appl. Math. Stat. **1**(6), 121–127 (2013)
Bourguignon, MB, Silva, RB, Zea, LM, Cordeiro, GM: The Kumaraswamy Pareto distribution. J. Stat. Theory Appl. **12**(2), 129–144 (2013)
Bowley, AL: Elements of Statistics. 4th edn. Charles Scribner, New York, NY (1920)
Chhetri, S, Long, H, Aryal, G: The beta transmuted Pareto distribution: Theory and applications. J. Stat. Appl. Prob. **6**(2), 243–258 (2017)
Choulakian, V, Stephens, MA: Goodness-of-fit for the generalized Pareto distribution. Technometrics. **43**, 478–484 (2001)
Cordeiro, GM, de Castro, M: A new family of generalized distributions. J. Stat. Comput. Simul. **81**, 883–898 (2011)
Cordeiro, GM, Ortega, EMM, Nadarajah, S: The Kumaraswamy Weibull distribution with application to failure data. J. Frankl. Inst. **347**, 1399–1429 (2010)
de Pascoa, MAR, Ortega, EMM, Cordeiro, GM: The Kumaraswamy generalized gamma distribution with application in survival analysis. Stat Methodol. **8**, 411–433 (2011)
Elbatal, I: The Kumaraswamy exponentiated Pareto distribution. Econ. Qual. Control. **28**(1), 1–8 (2013)
Eugene, N, Lee, C, Famoye, F: The beta-normal distribution and its applications. Commun. Stat. Theory Methods. **31**, 497–512 (2002)
Gradshteyn, IS, Ryzhik, IM: Table of Integrals, Series, and Products. 3rd edn. Academic Press, New York (2000)
Hady, AE, Ebraheim, N: Exponentiated transmuted Weibull distribution: A generalization of the Weibull distribution. Int. Sch. Sci. Res. Innov. **8**(6), 903–911 (2014)
Hussian, MA: Transmuted exponentiated gamma distribution: A generalization of the exponentiated gamma probability distribution. Appl. Math. Sci. **8**(27), 1297–1310 (2014)
Jones, MC: Kumaraswamy's distribution: A beta-type distribution with some tractability advantages. Stat Methodol. **6**(1), 70–81 (2009)
Kenny, JF, Keeping, ES: Mathematics of Statistics. 3rd edn. D. Van Nostrand Co., New Jersey (1962)
Khan, MS, King, R: Transmuted modified Weibull distribution: A generalization of the modified Weibull probability distribution. Eur. J. Pur. Appl. Math. **6**(1), 66–88 (2013)
Khan, MS, King, R, Hudson, IL: Transmuted Kumaraswamy distribution. Stat. Transit. **17**(2), 183–210 (2016)
Kumaraswamy, P: A generalized probability density function for double-bounded random processes. Hydrology. **46**, 79–88 (1980)
Marinho, PRD, Bourguignon, M, Dias, CRB: R Package - Adequacymodel (2016). https://cran.r-project.org/web/packages/AdequacyModel/AdequacyModel.pdf
Mdziniso, NC, Cooray, K: Odd Pareto families of distributions for modelling loss payment data. Scand. Actuar. J. 1–22 (2017). http://dx.doi.org/10.1080/03461238.2017.1280527
Merovci, F, Puka, L: Transmuted Pareto distribution. ProbStat Forum. **07**, 1–11 (2014)
Moors, JJA: A quantile alternative for kurtosis. J. R. Stat. Series D Stat. **37**(1), 25–32 (1988)
Nadarajah, S: Exponentiated Pareto distribution. Stat. J. Theor. Appl. Stat. **39**(3), 255–260 (2005)
Owokolo, EA, Oguntade, PE, Adejumo, AO: Performance rating of the transmuted exponential distribution: an analytical approach. SpringerPlus. **4**, 818 (2015). doi:10.1186/s40064-015-1590-6
Saulo, H: The Kumaraswamy Birnbaum-Saunders distribution. J. Stat. Theory Pract. **6**(4), 745–759 (2012)
Shams, TM: The Kumaraswamy-generalized Lomax distribution. Middle East J. Sci. Res. **17**(5), 641–646 (2013)
Shaw, WT, Buckley, IRC: The alchemy of probability distributions: beyond Gram-Charlier & Cornish-Fisher expansions, and skew-normal or kurtotic-normal distributions. Conference on Computational Finance, IMA, 6670 (2007). Research Report
Shaw, WT, Buckley, IRC: The alchemy of probability distributions: beyond Gram-Charlier expansions, and a skew-kurtotic-normal distribution from a rank transmutation map. Conference on Computational Finance, IMA, 0901–0434 (2009). Research Report
Tahir, MH, Cordeiro, GM: Compounding of distributions: a survey and new generalized classes. J. Stat. Distrib. Appl. **3**, 183–210 (2016)
Yousof, HM, Afify, AZ, Alizadeh, M, Butt, NS, Hamedani, GG, Ali, MM: The transmuted exponentiated generalized-g family of distributions. Pak. J. Stat. Oper. Res. **4**, 441–464 (2015)

7

On $(p_1, \ldots, p_k)$-spherical distributions

Wolf-Dieter Richter

Correspondence:
wolf-dieter.richter@uni-rostock.de
Institute of Mathematics University of Rostock, Rostock, Germany

Abstract

The class of $(p_1, \ldots, p_k)$-spherical probability laws and a method of simulating random vectors following such distributions are introduced using a new stochastic vector representation. A dynamic geometric disintegration method and a corresponding geometric measure representation are used for generalizing the classical χ^2-, t- and F-distributions. Comparing the principles of specialization and marginalization gives rise to an alternative method of dependence modeling.

Keywords: $(p_1, ..., p_k)$-power exponential distribution, simulation of $(p_1, ..., p_k)$-spherical uniform distribution, sign-invariant distribution, $(p_1, ..., p_k)$-spherical surface content measure, $(p_1, ..., p_k)$-spherical radius variable, matrix times vector stochastic representation, dynamic disintegration, $(p_1, ..., p_k)$-spherical coordinates, generalized χ^2-, t- and F-distributions, dependence modeling, specialization vs. marginalization, specialization Copula density, poly Beta function, angular Beta density

Mathematics Subject Classification (2000): 60E05, 62E15, 60D05, 14J29, 28A50, 28A75

1 Introduction

A basic notion from the theory of spherical probability laws is that of the stochastic basis which is a random vector following the uniform distribution on the Euclidean unit sphere in the k-dimensional Euclidean space $\mathbb{R}^k$, see e.g. Fang et al. (1990). In this monograph, multivariate uniform distributions are introduced in an algebraic way without referring to any type of surface measure. Numerous authors deal with the (singular) uniform distribution by considering the density of its $k-1$-dimensional marginal distribution, an approach which will, however, not further be discussed, here. Instead, the point of view of uniformity of which is the speech here is to define it by having a constant Radon-Nikodym derivative with respect to the Euclidean surface content measure. This geometric view onto the class of spherical distributions is the background of the corresponding geometric measure representation (2) in Richter (1991). This representation extends the one given in (3) in Richter (1985) for the Gaussian law and was exploited later on in a series of papers on probabilities of large deviations and on various statistical distributions. For a related survey see e.g. Richter (2015b).

Many authors studied more general or modified multivariate distribution classes. Just for getting an impression of this research area taking into account different points of view we refer to the more recent papers Field and Genton (2006), Arnold et al. (2008), Kamiya et al. (2008), Balkema et al. (2010), Balkema and Nolde (2010), Richter (2014), Richter (2015a) and Nolan (2016) as well as the references given there.

A detailed geometric description of uniformity of the stochastic basis of $l_{k,q}$-spherical distributions, $q > 0, k \in \{2, 3, \ldots\}$, is given in Richter (2009). The mathematical background there is based upon both a study of the different notions of Euclidean and $l_{k,q}$-surface content measures on $l_{k,q}$-spheres and on using suitable coordinates for evaluating the uniform measure of given subsets of spheres. The coordinates mentioned were introduced in Richter (2007) solving a long standing problem apparently conclusively treated as insolvable in Szablowski (1998).

It is a natural next step of research to consider random vectors having p-spherical uniform distributions with positive components of $p = (p_1, \ldots, p_k)$. The studies of (limit laws in) high risk scenarios in Balkema and Embrechts (2007) and of natural image patches in Sinz et al. (2009), e.g., deliver well motivating examples from this direction. A stochastic matrix times vector and a dynamic geometric measure representation were proved for the corresponding two-dimensional case in Richter (2017). More generally, the necessities of flexible probabilistic modeling in the era of big statistical data make it desirable to further study the class of p-spherical or $l_{k,p}$-symmetric distributions. For technical reasons, we deal exclusively with the case

$$p_i \neq p_j \text{ for } i \neq j, \tag{1}$$

here. As a consequence, distributions studied here are not invariant w.r.t. the class of orthogonal transformations or at least w.r.t. particular rotations but still appear to be sign-invariant meaning invariance w.r.t. multiplication with sign matrixes. While sample schemes for identically distributed variables automatically imply exchangeability of all variables in the present paper this property is excluded due to assumption (1). Distributions considered here are therefore fully outside the scope of the spherical distribution part and in consequence of the remaining parts of the classical monograph by Fang et al. (1990) and numerous work following it.

The present paper is structured as follows. The class of p-spherical uniform distributions is introduced in a geometrically motivated way in Section 2 and extended to the class of p-spherical or $l_{k,p}$-symmetric distributions in Section 3.1. The Sections 3.2 and 3.3 deal with a geometric measure representation and a combination of the principles of specialization and marginalization in dependence modeling, respectively.

2 The class of p-spherical uniform distributions

Let us consider the functional

$$|x|^{(p)} = \frac{|x_1|^{p_1}}{p_1} + \ldots + \frac{|x_k|^{p_k}}{p_k}, \; x = (x_1, \ldots, x_k)^T \in \mathbb{R}^k$$

where the vector $p = (p_1, \ldots, p_k)$ consists of pairwise different components throughout this paper. The set

$$B_p(r) = \left\{ x \in \mathbb{R}^k : |x|^{(p)} \leq r \right\}$$

will be called the p-ball with p-spherical radius parameter $r > 0$. Clearly, the "unit ball" $B_p = B_p(1)$ is not a norm or antinorm ball, r is not a radius in the sense of Euclidean or any $l_{k,q}$-geometry and the Minkowski functional of $B_p(r)$ is not homogeneous. But if we would allow, for a meantime ignoring assumption (1), to put $p_1 = \ldots = p_k = q \geq 1$ or $p_1 = \ldots = p_k = q \in (0, 1)$ then $B_p(r)$ would be a convex norm ball or a radially concave antinorm ball with norm or antinorm radius $(qr)^{1/q}$, respectively. In these cases the notation $B_p(r)$ would coincide with the notation $B_q\left((qr)^{1/q}\right)$ in Richter (2014); Richter (2015a) where q is just a scalar while p is a k-dimensional vector, here. For the notions of antinorm and radial concavity we refer to Moszyńska and Richter (2012). The functional $x \to |x|^{(p)}$ is invariant w.r.t. multiplication with sign matrixes, that is $|Sx|^{(p)} = |x|^{(p)}$ if S is a diagonal matrix with entries $s_1, ..., s_k$ which can be arbitrarily chosen from $\{-1, 1\}$. The topological boundary $S_p(r)$ of $B_p(r)$ is called the p-sphere with p-spherical radius parameter r. We call $S_p(r)$ (positive) matrix-homogeneous meaning that it allows the representation $S_p(r) = D_p(r)S_p$ where

$$D_p(r) = diag\left(r^{\frac{1}{p_1}}, ..., r^{\frac{1}{p_k}}\right)$$

is a diagonal matrix and $S_p = S_p(1)$ denotes the "unit sphere". Note that $B_p(r_1) \subset B_p(r_2)$ if $r_1 < r_2$ and

$$B_p(r) = \bigcup_{\varrho=0}^{r} S_p(\varrho).$$

Let μ and $\mathfrak{B}^k$ denote the Lebesgue measure and the Borel σ-field in $\mathbb{R}^k$, respectively. Assume the random vector X follows the uniform distribution on $\mathfrak{B}(B_p) = \mathfrak{B}^k \cap B_p$, that is

$$P(X \in M) = \frac{\mu(M)}{\mu(B_p)}, M \in \mathfrak{B}(B_p),$$

and let $R^{(p)} = |X|^{(p)}$. What can we say then about the distribution of the random vector

$$U_p = D_p\left(\frac{1}{R^{(p)}}\right)X?$$

The answer to this question is basic for disclosing the main message of this paper and will be given below. The sets

$$CPC_p(A) = \{D_p(r)x : x \in A, r > 0\}$$

and

$$Se_p(A, r) = CPC_p(A) \cap B_p(r)$$

defined for $A \in \mathfrak{B}^k \cap S_p = \mathfrak{B}(S_p)$ are called D_p-transformed central projection cone and D_p-transformed ball sector, respectively. For evaluating the volume of the latter type of sets we shall use the following coordinates.

Definition 1 *Let $\mathfrak{p} > 0$ be a parameter and $M_k = (0, \infty) \times M_k^*$ where $M_k^* = [0, \pi)^{\times(k-2)} \times [0, 2\pi)$. The $(p, \mathfrak{p})$-spherical coordinate transformation $Pol_{p,\mathfrak{p},k} : M_k \to \mathbb{R}^k$ with*

$$(x_1, ..., x_k)^T = Pol_{p,\mathfrak{p},k}(r, \varphi_1, ..., \varphi_{k-1})$$

is defined by

$$
\begin{aligned}
x_1 &= (p_1 r)^{\frac{1}{p_1}} sign(\cos_{\mathfrak{p}} \varphi_1)|\cos_{\mathfrak{p}} \varphi_1|^{\frac{\mathfrak{p}}{p_1}},\\
x_2 &= (p_2 r)^{\frac{1}{p_2}} sign(\cos_{\mathfrak{p}} \varphi_2)(\sin_{\mathfrak{p}} \varphi_1 |\cos_{\mathfrak{p}} \varphi_2|)^{\frac{\mathfrak{p}}{p_2}},\\
x_3 &= (p_3 r)^{\frac{1}{p_3}} sign(\cos_{\mathfrak{p}} \varphi_3)(\sin_{\mathfrak{p}} \varphi_1 \sin_{\mathfrak{p}} \varphi_2 |\cos_{\mathfrak{p}} \varphi_3|)^{\frac{\mathfrak{p}}{p_3}},\\
&\vdots\\
x_{k-1} &= (p_{k-1} r)^{\frac{1}{p_{k-1}}} sign(\cos_{\mathfrak{p}} \varphi_{k-1})(\sin_{\mathfrak{p}} \varphi_1 \cdot \ldots \cdot \sin_{\mathfrak{p}} \varphi_{k-2} |\cos_{\mathfrak{p}} \varphi_{k-1}|)^{\frac{\mathfrak{p}}{p_{k-1}}},\\
x_k &= (p_k r)^{\frac{1}{p_k}} sign(\sin_{\mathfrak{p}} \varphi_{k-1})(\sin_{\mathfrak{p}} \varphi_1 \cdot \ldots \cdot \sin_{\mathfrak{p}} \varphi_{k-2} |\sin_{\mathfrak{p}} \varphi_{k-1}|)^{\frac{\mathfrak{p}}{p_k}}.
\end{aligned}
\tag{2}
$$

Here, the q-generalized trigonometric functions $\sin_q \varphi = \frac{\sin\varphi}{N_q(\varphi)}$ and $\cos_q \varphi = \frac{\cos\varphi}{N_q(\varphi)}$ with $N_q(\varphi) = (|\sin\varphi|^q + |\cos\varphi|^q)^{1/q}$, $q > 0$ are introduced in Richter (2007) and used in studying and geometrically representing generalized spherical power exponential distributions in a series of papers starting from Richter (2009).

Remark 1 *The map $Pol_{p,\mathfrak{p},k}$ is almost one-to-one and its inverse is given by*

$$
r = \sum_{i=1}^{k} \frac{|x_i|^{p_i}}{p_i}, \tag{3}
$$

$$
\varphi_i = \arccos_p[\, sign(x_i) \left(\frac{|x_i|^{p_i}/p_i}{\sum_{j=1}^{k} |x_j|^{p_j}/p_j} \right)^{1/\mathfrak{p}}], i = 1, ..., k-2 \tag{4}
$$

and

$$
\varphi_{k-1} = \arctan \left[sign\left(\frac{x_{k-1}}{x_k}\right) \left(\frac{|x_k|^{p_k}/p_k}{|x_{k-1}|^{p_{k-1}}/p_{k-1}} \right)^{1/\mathfrak{p}} \right]. \tag{5}
$$

Evaluating the Jacobian of the transformation $Pol_{p,\mathfrak{p},k}$, the volume of the D_p-transformed ball sector satisfies

$$
\mu(Se_p(A,r)) = \frac{1}{\frac{1}{p_1} + ... + \frac{1}{p_k}} r^{\frac{1}{p_1}+...+\frac{1}{p_k}} \pi_p^*(A) \tag{6}
$$

where

$$
\pi_p^*(A) = \int\limits_{Pol^{*-1}_{p,\mathfrak{p},k}(A)} J_k^*(\varphi) d\varphi
$$

with $J_k^*(\varphi)d\varphi$ being equal to

$$
\mathfrak{p}^{k-1} p_1^{\frac{1}{p_1}-1} \cdot ... \cdot p_k^{\frac{1}{p_k}-1} \prod_{i=1}^{k-1} |\cos_{\mathfrak{p}} \varphi_i|^{\frac{\mathfrak{p}}{p_i}-1} |\sin_{\mathfrak{p}} \varphi_i|^{\frac{\mathfrak{p}}{p_k}+...+\frac{\mathfrak{p}}{p_{i+1}}-1} \frac{d\varphi_i}{N_{\mathfrak{p}}^2(\varphi_i)}. \tag{7}
$$

Here and below, the exponent of $|\sin_{\mathfrak{p}} \varphi_{k-1}|$ is defined to be $\frac{\mathfrak{p}}{p_k} - 1$ and

$$
Pol^*_{p,\mathfrak{p},k}(\varphi) = Pol_{p,\mathfrak{p},k}(1,\varphi), \varphi = (\varphi_1, ..., \varphi_{k-1})^T
$$

is the restriction of the function $Pol_{p,\mathfrak{p},k}$ to radius parameter 1.

Remark 2 *If the expression in (7) is equivalently rewritten by substituting* $x_i = \cos\varphi_i, i = 1, ..., k$ *then* $\pi_p^*(A)$ *can be represented as*

$$\pi_p^*(A) = \mathfrak{p}^{k-1} p_1^{\frac{1}{p_1}-1} \cdot ... \cdot p_k^{\frac{1}{p_k}-1} \left(\int\limits_{M^+} + \int\limits_{M^-} \right) J_k^{\circledcirc}(x_1, ..., x_{k-1}) d(x_1, ..., x_{k-1}) \tag{8}$$

where

$$J_k^{\circledcirc}(x_1, ..., x_{k-1}) = \prod_{i=1}^{k-1} \frac{|x_i|^{\frac{\mathfrak{p}}{p_i}-1} \left(1 - x_i^2\right)^{\frac{1}{2}\left(\frac{\mathfrak{p}}{p_k}+...+\frac{\mathfrak{p}}{p_{i+1}}\right)-1}}{\left(|x_i|^{\mathfrak{p}} + \left(1 - x_i^2\right)^{\mathfrak{p}/2}\right)^{\frac{1}{p_i}+\frac{1}{p_{i+1}}+...+\frac{1}{p_k}}}$$

and

$$M^{+(-)} = \left\{ (x_1, ..., x_{k-1})^T : \left(x_1, ..., x_{k-1}, +(-)\left(1 - \sum_{j=1}^{k-1} |x_j|^{p_j}\right)^{1/p_k} \right)^T \in A \right\}.$$

The following definition is well motivated by the fact that for the case $p = (q, \dots, q)$ (not considered here) the $l_{k,q}$-surface content measure is similarly introduced in Richter (2009) and proved to be equivalent to the differential-geometric definition. Much more general results of this type of equivalence are proved for norm and antinorm spheres in Richter (2015a).

Definition 2 *Let* $f_A(r) = \mu(Se_p(A, r))$ *for* $r > 0$ *and* $A \in \mathfrak{B}(S_p)$. *We call*

$$O_p(D_p(r)A) = f_A'(r)$$

the p-spherical surface content of $D_p(r)A$ *or its* S_p*-surface content, for short.*

It follows from this definition and equation (6) that

$$O_p(A) = f_A'(1) = \pi_p^*(A).$$

Let us emphasize again that differently from what is assumed in a broad literature p is a vector, here. In particular, as because

$$Pol_{p,\mathfrak{p},k}^{*-1}(S_p) = M_k^*$$

and

$$\int\limits_0^{\pi/2} (\cos_{\mathfrak{p}} \varphi)^{a-1} (\sin_{\mathfrak{p}} \varphi)^{b-1} \frac{d\varphi}{N_{\mathfrak{p}}^2(\varphi)} = \frac{B\left(\frac{a}{\mathfrak{p}}, \frac{b}{\mathfrak{p}}\right)}{\mathfrak{p}} \tag{9}$$

there holds

$$O_p(S_p) = 2^k B\left(\frac{1}{p_1}, ..., \frac{1}{p_k}\right) \prod_{i=1}^{k} p_i^{\frac{1}{p_i}-1}. \tag{10}$$

Here,

$$B(x_1, ..., x_k) = \frac{\Gamma(x_1) \cdot ... \cdot \Gamma(x_k)}{\Gamma(x_1 + ... + x_k)}, x_i > 0, i = 1, ..., k$$

denotes the poly Beta function where $\Gamma(x) = \int_0^\infty t^{x-1}e^{-t}dt, x > 0$ is the Gamma function, and $B(.,.)$ is the classical Beta function. Moreover,

$$O_p(D_p(r)A) = \pi_p^*(A)r^{\frac{1}{p_1}+...+\frac{1}{p_k}-1}. \tag{11}$$

Definition 3 *The density*

$$f^*_{\mathfrak{p},a,b}(\varphi) = \frac{\mathfrak{p}}{B\left(\frac{a}{\mathfrak{p}}, \frac{b}{\mathfrak{p}}\right)} \frac{(\cos_{\mathfrak{p}}\varphi)^{a-1}(\sin_{\mathfrak{p}}\varphi)^{b-1}}{N^2_{\mathfrak{p}}(\varphi)}, \varphi \in (0, \pi/2)$$

is called angular Beta density with parameters $\mathfrak{p} > 0$, $a > 0$ *and* $b > 0$.

Remark 3 *The notion of* S_p*-surface content of* $D_p(r)A$ *is different from the notion of Euclidean surface content of* $D_p(r)A$ *(unless for the case* $p = (2, ..., 2)$ *which, however, because of assumption (1) is not allowed to appear, here).*

Definition 4 *The p-spherical uniform probability law on the Borel* σ*-field* $\mathfrak{B}(S_p)$ *is defined by*

$$\omega_p(A) = \frac{O_p(A)}{O_p(S_p)}.$$

This definition corresponds to Definition 8 in Richter (2007) where, however, p is a scalar while it is a k-dimensional vector, here. Similarly, the notation $U_p = D_p\left(\frac{1}{R^{(p)}}\right)X$ introduced above is closely connected to that given in the same paper where p is a scalar. However, while $R^{(p)} = \frac{|X_1|^{p_1}}{p_1} + ... + \frac{|X_k|^{p_k}}{p_k}$ denotes a certain "mixed-power-of-radius", here, not being the power of a norm or antinorm, in Richter (2007) the symbol R actually means a norm or antinorm. The proof of the following theorem is analogous to that of Theorem 2 in Richter (2017) and will therefore be omitted, here.

Theorem 1 *(a) The random vector* U_p *follows the p-spherical uniform distribution,* $U_p \sim \omega_p$*, is independent of* $R^{(p)}$*, and* $R^{(p)}$ *has the following density with respect to the Lebesgue measure on the real line*

$$\left(\frac{1}{p_1} + ... + \frac{1}{p_k}\right) r^{\frac{1}{p_1}+...+\frac{1}{p_k}-1} I_{[0,1)}(r)dr. \tag{12}$$

(b) If, vice versa, ξ *and* W *are independent where* ξ *has density (12) and* $W \sim \omega_p$ *then* $\eta = D_p(\xi) \cdot W$ *is uniformly distributed on the unit ball* B_p.

We are now in a position to disclose the basic message of this paper as follows: matrix-multiplication of a p-spherical uniformly distributed random vector U_p by $D_p(R)$ where U_p and the random variable $R \geq 0$ are independent generates the world of distributions being of main interest, here.

Remark 4 *If the random variable* Y *is uniformly distributed on* $(0, 1)$ *then* $Z_p = Y^{1/\left(\frac{1}{p_1}+...+\frac{1}{p_k}\right)}$ *follows the density (12), that is* $Z_p \overset{d}{=} R^{(p)}$*. Thus* $R^{(p)}$ *can be simulated by* Z_p.

Theorem 2 *Let the random vector* X *be uniformly distributed on the unit p-ball* B_p*. Then the p-spherical radius variable* $R^{(p)}$ *and the p-spherical angles* $\Phi_1, ..., \Phi_{k-1}$ *of* X *are* *independent and the angle* Φ_i *follows the angular Beta density with parameters* $\mathfrak{p}$, a_i *and* b_i, $f_{\mathfrak{p},a_i,b_i}$, *where* $a_i = \frac{\mathfrak{p}}{p_i}, b_i = \frac{\mathfrak{p}}{p_k} + ... + \frac{\mathfrak{p}}{p_{i+1}}, i = 1, ..., k-2$ *and* $a_{k-1} = \frac{\mathfrak{p}}{p_{k-1}}, b_{k-1} = \frac{\mathfrak{p}}{p_k}$.

Proof According to Definition 1, the vector U_p allows the representation

$$U_p = \begin{pmatrix} (p_1)^{\frac{1}{p_1}} |\cos_{\mathfrak{p}} \Phi_1|^{\frac{\mathfrak{p}}{p_1}} S_1 \\ (p_2)^{\frac{1}{p_2}} |\sin_{\mathfrak{p}} \Phi_1 \cos_{\mathfrak{p}} \Phi_2|^{\frac{\mathfrak{p}}{p_2}} S_2 \\ \vdots \\ (p_k)^{\frac{1}{p_k}} |\sin_{\mathfrak{p}} \Phi_1 \sin_{\mathfrak{p}} \Phi_2 \cdot \ldots \cdot \sin_{\mathfrak{p}} \Phi_{k-2} \sin_{\mathfrak{p}} \Phi_{k-1}|^{\frac{\mathfrak{p}}{p_k}} S_k \end{pmatrix} \tag{13}$$

where the signature vector $S = (S_1, \ldots, S_k)^T$ is independent of $\Phi_1, \ldots, \Phi_{k-1}$ and uniformly distributed in $\{-1, 1\}^{\times k}$. It follows from the evaluation of the Jacobian of the transformation $Pol_{p,\mathfrak{p},k}$ that the vector $\left(R^{(p)}, \Phi_1, ..., \Phi_{k-1}\right)^T$ has the density

$$f_p(r, \varphi) = r^{\frac{1}{p_1} + ... + \frac{1}{p_k} - 1} J_k^*(\varphi_1, ..., \varphi_{k-1})$$

where J_k^* satisfies representation (7). For $i = 1, ..., k-1$, the independent angles Φ_i thus have the densities

$$f_{p,\mathfrak{p},i}(\varphi) = \frac{\mathfrak{p}}{B\left(\frac{1}{p_i}, \frac{1}{p_k} + \ldots + \frac{1}{p_{i+1}}\right)} \frac{|\cos_{\mathfrak{p}} \varphi|^{\frac{\mathfrak{p}}{p_i} - 1} |\sin_{\mathfrak{p}} \varphi|^{\frac{\mathfrak{p}}{p_k} + ... + \frac{\mathfrak{p}}{p_{i+1}} - 1}}{N_{\mathfrak{p}}^2(\varphi)}. \tag{14}$$

Now, Definition 3 applies. □

Remark 5 *If* Φ_i *follows the density in (14) then* $Y = |\cos_{\mathfrak{p}} \Phi_i|^{\mathfrak{p}} \sim B(l, m)$ *where* $l = \frac{1}{p_i}, m = \frac{1}{p_{i+1}} + ... + \frac{1}{p_k}$. *Let* $(V_{0,1}, V_{0,2})$ *be uniformly distributed in* $(0, 1) \times (0, 1)$, *then there holds*

$$P(Y \in B) = P\left(\frac{V_{0,1}^{1/l}}{V_{0,1}^{1/l} + V_{0,2}^{1/m}} \in B | V_{0,1}^{1/l} + V_{0,2}^{1/m} \leq 1\right), B \in \mathfrak{B} \cap [0, 1].$$

This allows to simulate Y by an acceptance-rejection method, see equation (A5) and algorithm A.1, step 2 in Kalke and Richter (2013). Thus, methods for simulating vectors U_p *and X being p-spherical uniformly distributed on* S_p *and uniformly distributed on* B_p, $U_p \sim U(S_p)$ *and* $X \sim U(B_p)$, *respectively, can now be established as follows.*

Simulation Algorithm 1[p-spherical uniform distribution in $S_p, p = (p_1, ..., p_k)^T$]

Step 1 For $i \in \{1, \ldots, k\text{-}1\}$, simulate $(V_{i,1}, V_{i,2}) \sim U([0, 1] \times [0, 1])$ until $V_{i,1}^{p_i} + V_{i,2}^{1/\left(\frac{1}{p_{i+1}} + ... + \frac{1}{p_k}\right)} \leq 1$.

Step 2 Calculate $W_i = \frac{V_{i,1}^{p_i}}{V_{i,1}^{p_i} + V_{i,2}^{1/\left(\frac{1}{p_{i+1}} + ... + \frac{1}{p_k}\right)}}, i = 1, ..., k-1$.

Step 3 Simulate independently $(S_1, \ldots, S_k) \sim U\left(\{-1, +1\}^{\times k}\right)$.

Step 4 Calculate $U_{p,i} = S_i \left(\prod_{j=1}^{i-1} (1 - W_j) W_i\right)^{1/p_i}$ for $i = 1, \ldots, k-1$

and $U_{p,k} = S_k \prod_{j=1}^{k} (1 - W_j)^{1/p_k}$.

Step 5 Return $U_p = (U_{p,1}, ..., U_{p,k})^T$.

Simulation Algorithm 2 [Uniform distribution in $B_p, p = (p_1, ..., p_k)^T$]

Step 1 Simulate $U_p \sim U(S_p)$ according to Algorithm 1.
Step 2 Simulate independently: $Y \sim U(0,1)$.
Calculate $R^{(p)} = Y^{1/(\frac{1}{p_1}+...+\frac{1}{p_k})}$.
Step 3 Return $X = R^{(p)} \cdot U_p$

Remark 6 *By symmetry, the distribution center of U_p is $\mathbb{E}(U_p) = 0_k$ where $0_k = (0,...,0)^T \in \mathbb{R}^k$ is the zero vector of the sample space. The vector U_p has uncorrelated components. Using formulas (9), (13) and (14) one can show that the variances of U_p's components are*

$$\mathbb{V}(U_{p,i}) = \gamma_1 \cdot \ldots \cdot \gamma_i \cdot p_i^{\frac{2}{p_i}} \frac{\Gamma(\frac{3}{p_i})\Gamma\left(\frac{1}{p_1}+...+\frac{1}{p_k}\right)}{\Gamma(\frac{1}{p_i})\Gamma\left(\frac{3}{p_i}+\sum\limits_{j\neq i}\frac{1}{p_j}\right)}, i = 1,...,k$$

where $\gamma_1 = ... = \gamma_{k-2} = 2$ and $\gamma_{k-1} = \gamma_k = 4$.

Remark 7 *(a) Let us call*

$$sm_p(A) = \frac{\mu(Se_p(A,1))}{\mu(B_p)}$$

the D_p-transformed-sector measure on $\mathfrak{B}(S_p)$ (or, more precisely, the uniform probability measure of the D_p-transformed sector $Se(A,1)$ of B_p). It follows from the obvious equations

$$\frac{O_P(A)}{O_P(S_p)} = \frac{f_A'(1)}{f_{S_p}'(1)} = \frac{\pi_p^*(A)}{\pi_p^*(S_p)} = \frac{\mu(Se_p(A,1))}{\mu(Se_p(S_p,1))} = \frac{\mu(Se_p(A,1))}{\mu(B_p)}$$

that

$$\omega_p(A) = sm_p(A).$$

Thus, the p-spherical uniform probability law ω_p can also be called a D_p-transformed-sector measure. For an interpretation of ω_p as cone measure see (e) and Remark 9 (b).

(b) According to Remark 3, the notion of p-spherical uniform distribution on $\mathfrak{B}(S_p)$ is different from the notion of uniform distribution with respect to Euclidean surface content (unless for $p_1 = ... = p_k \in \{1,2,\infty\}$).

(c) Fine properties of the Euclidean surface content measure defined on the Borel σ-field of the Euclidean unit sphere, a precursor of the S_p-surface content measure O_p considered here, are exploited by the author in the eighties and nineties of the last century in a series of papers on large deviations. A main idea in the background of those work is the development and application of a generalized method of indivisibles extending a classical approach by Cavalieri and Torricelli, see Richter (1985), Richter (2015b) and Günzel et al. (2012).

(d) It is a challenging problem to find a differential-geometric interpretation of O_p as it was found in Richter (2009) for the $l_{k,q}$-surface content measure and in Richter (2015a) for norm and antinorm spheres. This problem was first stated for the two-dimensional case in Richter (2017).

(e) *Several authors who study uniform distributions on (non-D_p-transformed) generalized spheres make use of the notion cone measure instead of sector measure and rely on the last representation in (a), that is on relating volumes to each other, see e.g. Naor and Romik (2003) and Barte et al. (2005) and a series of follow up papers.*

(f) *Only a few days before finishing the present paper, Amir Ahmadi-Javid kindly let me know his joint article Ahmadi-Javid and Moeini (2019) where the authors follow another way of considering a uniform distribution on $\mathfrak{B}(S_p)$. They start, in Definition 2.3, with a random vector being uniformly distributed in a parallelepiped and, referring to the work of Schechtman and Zinn (1990), Rachev and Rüschendorf (1991), Song and Gupta (1997), Liang and Ng (2008), Harman and Lacko (2010) and Lacko and Harman (2012), later make use of a common (non-dynamical) notion of cone measure (just like the one mentioned in (e)) for studying a certain type of uniform distributions on $\mathfrak{B}(S_p)$. A closer comparison with the method presented here, where we start with a uniform distribution on a ball B_p and continue with a dynamically transformed-cone measure, would be of interest for future work.*

3 The class of *p*-spherical distributions

A random vector distributed according to the p-spherical uniform distribution builds the stochastic basis of any p-spherical distributed random vector considered in Section 3.1. Examples of light and heavy distribution centers and tails are possible. A geometric measure representation and its applications allow studying exact distributions of generalized χ^2-, t- and F-statistics in Section 3.2. The final Section 3.3 gives a sketch of an alternative approach to describing dependence of random variables following one-dimensional specializations of k-dimensional distributions instead of marginal distributions.

3.1 Definitions and Examples

Definition 5 *Let the random vector U_p follow the p-spherical uniform distribution on the Borel σ-field $\mathfrak{B}(S_p)$, $U_p \sim \omega_p$, and R be a nonnegative random variable having cumulative distribution function (cdf) F and characteristic function (cf) ϕ and being independent of U , then*

$$X = D_p(R)U_p \tag{15}$$

is said to follow the p-spherical distribution $\Phi_p^{cdf(F)} = \Phi_p^{cf(\phi)}$. The vector U_p is called the p-spherical uniform basis and R the generating variate of X, and (15) a stochastic representation of X. The distribution of X will alternatively be denoted $\Phi_p^{pdf(f)}$ if R has probability density function (pdf) f.

Remark 8 *If* $\mathbb{E}\left(R^{(p)\frac{1}{\min\{p_1,\dots,p_k\}}}\right)$ *is finite then, due to Remark 6,* $\mathbb{E}(X) = 0_k$ *and if* $\mathbb{E}\left(R^{(p)\frac{2}{\min\{p_1,\dots,p_k\}}}\right) < \infty$ *then* $\mathbb{V}(X_i) = \mathbb{E}\left(R^{(p)\frac{2}{p_i}}\right)\mathbb{V}(U_{p,i}), i = 1, \dots, k$. *For the derivation of moments in the case* $p_1 = \dots = p_k$ *we refer to Arellano-Valle and Richter (2012).*

Theorem 3 *The characteristic function of a p-spherically distributed random vector X satisfying representation (15) can be written as*

$$\phi_X(t) = \int_0^\infty \phi_{U_p}(D_p(r)t)P^R(dr),\ t \in R^k$$

where P^R and ϕ_{U_p} denote the distribution law of R and the characteristic function of U_p, respectively.

Proof Because of the diagonal structure of $D_p(r)$ we have

$$\phi_X(t) = \mathbb{E}e^{i(t,D_p(R)U_p)} = \mathbb{E}e^{i(D_p(R)t,U_p)}.$$

If $\mathbb{E}(Y|R)$ denotes the conditional expectation of Y given $R = r$ then

$$\phi_X(t) = \mathbb{E}\mathbb{E}\left(e^{i(D_p(r)t,U_p)}|R = r\right) = \int_0^\infty \phi_{U_p}(D_p(r)t)P^{R|U_p}(dr)$$

from where the result follows by independence of U_p and R. □

Corollary 1 *(a) The distribution of a p-spherically distributed random vector X is uniquely determined by the distribution of its generating variate R.*

(b) If a p-spherically distributed random vector X has a density, then it is of the form $f_X = \varphi_{g;p}$,

$$\varphi_{g;p}(x) = C(g;p)g\left(|x|^{(p)}\right), x \in \mathbb{R}^n$$

where $g : [0,\infty) \to [0,\infty)$ is a density generating function (dgf) satisfying

$$0 < I(g;p) = \int_0^\infty r^{\frac{1}{p_1}+\ldots+\frac{1}{p_k}-1} g(r)dr < \infty,$$

and the normalizing constant allows the factorization

$$\frac{1}{C(g;p)} = I(g;p)O_p(S_p).$$

This density is invariant w.r.t. multiplication with sign matrices, or sign-invariant or sign-symmetric. For a general class of symmetric distributions we refer to Arellano-Valle et al. (2002) and Arellano-Valle and del Pino (2004).

Both this result and the following definition transfer earlier statements from Richter (2014) to the present case. The following definition adopts notation in Müller and Richter (2016) and is aimed to make the notion of dgf unique.

Definition 6 *A dgf g satisfying the equation*

$$I(g;p)O_p(S_p) = 1 \tag{16}$$

is called density generator (dg) of a continuous p-spherical distribution.

Example 1 *(a) The dg of the p-spherical Kotz type distribution having parameters $M > 1 - \frac{1}{p_1} - \ldots - \frac{1}{p_k}$ and β and γ from $(0,\infty)$ is*

$$g^{(p)}_{Kt;M,\beta,\gamma}(r) = C^{(p)}_{Kt;M,\beta,\gamma} r^{M-1} e^{-\beta r^\gamma} I_{(0,\infty)}(r)$$

where

$$C^{(p)}_{Kt;M,\beta,\gamma} = \frac{\gamma\beta^{\left(M-1+\frac{1}{p_1}+...+\frac{1}{p_k}\right)/\gamma}}{\Gamma\left(\left(M-1+\frac{1}{p_1}+...+\frac{1}{p_k}\right)/\gamma\right)2^k B\left(\frac{1}{p_1},...,\frac{1}{p_k}\right)\prod_{i=1}^{k} p_i^{1/p_i-1}}.$$

The p-spherical Kotz type pdf with parameters M, β, γ is therefore

$$\varphi^{(p)}_{Kt;M,\beta,\gamma}(x) = C^{(p)}_{Kt;M,\beta,\gamma}\left(\frac{|x_1|^{p_1}}{p_1}+...+\frac{|x_k|^{p_k}}{p_k}\right)^{M-1} e^{-\beta\left(\frac{|x_1|^{p_1}}{p_1}+...+\frac{|x_k|^{p_k}}{p_k}\right)^{\gamma}}. \qquad (17)$$

(b) Particularly, the dg of the p-spherical power exponential distribution is

$$g^{(p)}_{PE}(r) = \left(\prod_{i=1}^{k} C_i\right) e^{-r} I_{(0,\infty)}(r)$$

and one may write then

$$\Phi^{(p)}(dx) = \left(\prod_{i=1}^{k} C_i\right) \exp\left\{-|x|^{(p)}\right\} dx_1...dx_k$$

to denote the k-dimensional p-power exponential density where $C_i = \frac{p_i^{1-1/p_i}}{2\Gamma(1/p_i))}$, $i = 1,...,k$ are individual normalizing constants. In this case the random vector X consists of independent components, and according to (a) there holds

$$\prod_{i=1}^{k} C_i = C^{(p)}_{Kt;1,1,1}.$$

Example 2 *The dg of the p-spherical Pearson Type VII distribution having parameters $M > \max\left\{1, \frac{1}{p_1}+...+\frac{1}{p_k}\right\}$ and $\nu > 0$ is*

$$g^{(p)}_{PT7;M,\nu}(r) = C^{(p)}_{PT7;M,\nu}\left(1+\frac{r}{\nu}\right)^{-M} I_{(0,\infty)}(r)$$

where

$$C^{(p)}_{PT7;M,\nu} = \frac{\Gamma(M)\prod_{i=1}^{k} p_i^{1-1/p_i}}{2^k \nu^{\frac{1}{p_1}+...+\frac{1}{p_k}} \cdot \Gamma\left(M-\frac{1}{p_1}-...-\frac{1}{p_k}\right)\Gamma\left(\frac{1}{p_1}\right)\cdot...\cdot\Gamma\left(\frac{1}{p_k}\right)}.$$

The p-spherical Pearson Type VII density with parameters M and ν is therefore

$$\varphi^{(p)}_{PT7;M,\nu}(x) = C^{(p)}_{PT7;M,\nu}\left(1+\frac{1}{\nu}\left(\frac{|x_1|^{p_1}}{p_1}+...+\frac{|x_k|^{p_k}}{p_k}\right)\right)^{-M}. \qquad (18)$$

3.2 Geometric measure representation

The following theorem is a geometric-measure theoretic counterpart to Theorem 3. Its proof follows the line of author's earlier work in Richter(1985, 1991, 2014, 2017). The main aim of this section is to present first applications of the geometric measure representation extending the classical Helmert-Pearson χ^2-, Gosset alias Student t- and Fisher F-distributions.

Let $\Phi^{(p)}_g$ denote the continuous p-spherical distribution law having dg g.

Theorem 4 *For every* $B \in \mathfrak{B}^k$,

$$\Phi_g^{(p)}(B) = \frac{1}{I(g;p)} \int_0^\infty r^{\frac{1}{p_1}+...+\frac{1}{p_k}-1} g(r)\mathfrak{F}_p(B,r)dr.$$

Let a random vector X follow the p-spherical distribution law with dg g, $X \sim \Phi_{g;p}$, and $T = T(X)$ be any statistic. The statistic T satisfies $T < \lambda$ if and only if the outcome of X belongs to the sub-level set

$$B_T(\lambda) = \left\{x \in \mathbb{R}^k : T(x) < \lambda\right\},$$

thus

$$P(T < t) = \frac{1}{I(g;p)} \int_0^\infty r^{\frac{1}{p_1}+...+\frac{1}{p_k}-1} g(r)\mathfrak{F}_p(B_T(\lambda),r)dr. \tag{19}$$

Example 3 ***Chi-p statistic*** *Let*

$$T(X) = |X|^{(p)}$$

denote the p-spherical radius variable of the random vector X, that is $T(X) = R^{(p)}$, *then the ipf of the set* B_T *satisfies*

$$\mathfrak{F}_p(B_T(\lambda),r) = I_{(0,\lambda)}(r), r > 0.$$

The Chi-$(g;p)$ *pdf is therefore*

$$f(\lambda) = \lambda^{\frac{1}{p_1}+...+\frac{1}{p_k}-1} g(\lambda)I_{(0,\infty)}(\lambda). \tag{20}$$

Differently from the $\chi^2(k)$*-distribution where the parameter k corresponds to the dimension of a subspace of the sample space, here, the parameter* $\frac{1}{p_1}+...+\frac{1}{p_k}$ *itself does not allow interpretation as dimension of a linear space or a linear subspace of the sample space, but its number of summands k does.*

Note that the density of T was dealt with for the particular dg of the generalized Gaussian law in Taguchi (1978). For $\frac{1}{p_1}+...+\frac{1}{p_k}=\frac{k}{2}$ *the pdf in (20) is called the g-generalization of the Chi-square density with k degrees of freedom (d.f.) in Richter (1991) (note that there holds* $p_1 = \ldots = p_k$ *but (16) is not assumed to be satisfied there). For more partial cases and statistical applications of this distribution, see Richter (2007, 2009, 2016).*

If we specify $g = g_{PE}^{(p)}$ *in (19), see Example 1(b), then*

$$f_{R^{(p)}}(\lambda) = \lambda^{\frac{1}{p_1}+...+\frac{1}{p_k}-1} e^{-\lambda} \prod_{i=1}^k \frac{p_i^{1-1/p_i}}{2\Gamma(1/p_i)}.$$

In this case we have

$$\mathbb{E}R^{(p)} = \Gamma\left(\frac{1}{p_1}+...+\frac{1}{p_k}+1\right)\prod_{i=1}^k \frac{p_i^{1-1/p_i}}{2\Gamma(1/p_i)}$$

and

$$\mathbb{E}(R^{(p)2}) = \Gamma\left(\frac{1}{p_1}+...+\frac{1}{p_k}+2\right)\prod_{i=1}^k \frac{p_i^{1-1/p_i}}{2\Gamma(1/p_i)}$$

as well as

$$\mathbb{E}(R^{(p)\frac{2}{p_i}}) = \Gamma\left(\frac{3}{p_i} + \sum_{j\neq i}\frac{1}{p_j}\right)\prod_{j=1}^{k}\frac{p_j^{1-1/p_j}}{2\Gamma(1/p_j)}.$$

It follows from Remarks 6 and 8 that

$$\mathbb{V}(X_i) = \frac{\gamma_1\cdot\ldots\cdot\gamma_i}{2^k B\left(\frac{1}{p_1},\ldots,\frac{1}{p_k}\right)}\frac{\Gamma\left(\frac{3}{p_i}\right)}{\Gamma\left(\frac{1}{p_i}\right)}p_i^{1+\frac{1}{p_i}}\prod_{j\neq i}p_j^{1-\frac{1}{p_j}}. \tag{21}$$

Similarly, if we put $g = g^{(p)}_{Kt;M,\beta,\gamma}$ *or* $g = g^{(p)}_{PT7;M,\nu}$ *in (19) then* $f_{R^{(p)}}$ *and* $\mathbb{V}(X_i)$ *will be correspondingly specified.*

Example 4 ***Fisher-p statistic*** *Let the vectors* $X^{(1)T} = (X_1,\ldots,X_m)$ *and* $X^{(2)T} = (X_{m+1},\ldots,X_k)$ *be sub-vectors of* $X = (X_1,\ldots,X_m,X_{m+1},\ldots,X_k)^T$ *and* $p^{(1)} = (p_1,\ldots,p_m)^T$, $p^{(2)} = (p_{m+1},\ldots,p_k)^T$ *be sub-vectors of the shape-tail parameter vector* $p = (p_1,\ldots,p_m,p_{m+1},\ldots,p_k)^T$, *and assume that* $X \sim \Phi_{g;p}$. *We consider the F-p statistic*

$$T(X) = \frac{|X^{(1)}|_{(p_1)}/m}{|X^{(2)}|_{(p_2)}/(k-m)}$$

and recognize that $T(x) = T(D_p(\gamma)x)$ *for all* $\gamma > 0$. *Roughly spoken,* B_T *has the curved cone-type property*

$$D_p(\gamma)B_T(\lambda) = B_T(\lambda), \lambda > 0. \tag{22}$$

Thus the ipf of the set B_T *does not depend on r and equation (19) shows that*

$$P(T < \lambda) = \mathfrak{F}_p(B_T(\lambda), 1).$$

Making use of the coordinate transformations

$$Pol_{p^{(1)},\mathrm{p},m} : (r_1,\varphi_1,\ldots,\varphi_{m-1}) \to x^{(1)}$$

and

$$Pol_{p^{(2)},\mathrm{p},m} : (r_2 = \cos_1\varphi_{m+1},\ldots,\varphi_{k-1}) \to x^{(2)}$$

instead of the coordinate transformations $SPH_{p,1}$ *and* $SPH_{p,2}$ *used in Richter (2009), and*

$$r_1 = \cos_1\varphi, r_2 = \sin_1\varphi, 0 \leq \varphi < \pi/2,$$

and further following the line of the proof of Theorem 6 there, we get

$$\mathfrak{F}_p(B_T(\lambda),1) = \frac{1}{B\left(\frac{1}{p_1}+\ldots+\frac{1}{p_m},\frac{1}{p_{m+1}}+\ldots+\frac{1}{p_k}\right)} \cdot \int_{\operatorname{arc\,cot}\left(\frac{m\lambda}{k-m}\right)}^{\pi/2}\frac{(\cos\varphi)^{\frac{1}{p_1}+\ldots+\frac{1}{p_m}-1}(\sin\varphi)^{\frac{1}{p_{k+1}}+\ldots+\frac{1}{p_k}-1}}{N_1(\varphi)^{\frac{1}{p_1}+\ldots+\frac{1}{p_k}}}d\varphi.$$

Taking the derivative shows that the pdf of statistic T is

$$f_{m,k-m}(\lambda) = \frac{\left(\frac{m}{k-m}\right)^{\frac{1}{p_1}+\ldots+\frac{1}{p_m}}\lambda^{\frac{1}{p_1}+\ldots+\frac{1}{p_m}-1}}{B\left(\frac{1}{p_1}+\ldots+\frac{1}{p_m},\frac{1}{p_{m+1}}+\ldots+\frac{1}{p_k}\right)\left(1+\left(\frac{m\lambda}{k-m}\right)\right)^{\left(\frac{1}{p_1}+\ldots+\frac{1}{p_k}\right)}}. \tag{23}$$

We call the pdf in (23) the Fisher-p density with $(m, k-m)$ d.f. or $F_{m,k-m}(p)$-density, for short.

Example 5 ***Student-p statistic*** *Let the t-p statistic be defined by*

$$T(X) = \frac{X_1}{\left(|X^{(2)}|^{(p_2)}/(k-1)\right)^{1/p_1}}.$$

The pdf of T is called Student p-density with $k-1$ d.f. or $t_{k-1}\left(\frac{1}{p_1}, \ldots, \frac{1}{p_k}\right)$-density, for short:

$$\frac{d}{dt}P(T < t) = \frac{p_1}{2} f_{1,k-1}(t^{p_1})t$$

$$= \frac{p_1}{2(k-1)^{\frac{1}{p_1}} B\left(\frac{1}{p_1}, \frac{1}{p_2} + \ldots + \frac{1}{p_k}\right)\left(1 + \frac{|t|^{p_1}}{k-1}\right)}.$$

Note that, as in Example 4, B_T is a D_p-transformed-cone or curved-transformed-cone type set satisfying (22).

Remark 9 *(a) Because Fisher-p and Student-p distributions do not depend on the dg, the F-p and t-p statistics are called g-robust. For a study of g-sensitivity and g-robustness of certain statistics see Ittrich et al. (2000) and for a study of statistics generating curved-transformed-cone type sets, see Ittrich and Richter (2005).*

(b) Let $B_T(\lambda)$ be a curved-transformed-cone type set satisfying (22) and put $A = B_T(\lambda) \cap S_p$. By Theorem 4, $\Phi_g^{(p)}(B_T(\lambda)) = \omega_p(A)$. It is reasonable therefore to call $C^{(p)}(A) = \omega_p(A)$ the D_p-transformed-cone measure of $A \in \mathfrak{B}(S_p)$.

(c) We recall that several representations of Student distributed statistics were given in Richter (1995). In particular, the two facts are exploited there that the ipf of the cone $\{T < \lambda\}$ does not depend on its radius variable and that the multivariate standard Gaussian law is invariant w.r.t. orthogonal transformations, together leading to g-robustness of Fisher's and Student's statistics. Due to assumption (22), we observe in the present situation that the ipf of $B_T(t)$ does not depend on the generalized radius variable and we observe invariance of Fisher's and Student's statistics w.r.t. any transformation $D_p(r)$ where $r > 0$ is used to prove g-robustness of T. Thus, if an arbitrary statistic T generates sub-level sets satisfying assumption (22) then such statistic is g-robust.

3.3 Dependence modeling: specialization vs. marginalization

Let $1 \le m < k, 1 \le i_1 < i_2 < \ldots < i_m \le k$. We assume that

$$\varphi_{m,Kt}(x_{i_1}, \ldots, x_{i_m}) = C_{m,Kt}\left(\frac{|x_{i_1}|^{p_{i_1}}}{p_{i_1}} + \ldots + \frac{|x_{i_m}|^{p_{i_m}}}{p_{i_m}}\right)^{M-1} \cdot e^{-\beta\left(\frac{|x_{i_1}|^{p_{i_1}}}{p_{i_1}} + \ldots + \frac{|x_{i_m}|^{p_{i_m}}}{p_{i_m}}\right)}$$

and

$$\varphi_{m,PT7}(x_{i_1}, \ldots, x_{i_m}) = C_{m,PT7}\left(1 + \frac{1}{\nu}\left(\frac{|x_{i_1}|^{p_{i_1}}}{p_{i_1}} + \ldots + \frac{|x_{i_m}|^{p_{i_m}}}{p_{i_m}}\right)\right)^{-M}$$

are suitably normalized densities and call them m-dimensional specializations of the Kotz and Pearson type densities $\varphi^{(p)}_{Kt;M,\beta,\gamma}$ and $\varphi^{(p)}_{PT7;M,\nu}$, respectively. It is well known that marginal densities are not of the same type as specializations, in general. For the well known possibilities and problems of finding marginal densities of elliptically contoured distributions we refer to Fang et al. (1990).

Imagine now each of k experimenters observe another random variable, let them combine their (possibly dependent) observations by a vector and describe this vector by a joint cdf $F^{(k)}$ (possibly including dependence). In hindsight, did the experimenters construct the multivariate cdf $F^{(k)}$ starting from the specializations $F_1, ..., F_k$ of $F^{(k)}$ or from the marginal cdfs $F_1^*, ..., F_k^*$ of $F^{(k)}$? In other words, are the experimenters searching for a multivariate cdf $F^{(k)}$ such that their original observations follow marginal distributions or specializations of $F^{(k)}$? For an illustration, in the case $p_1 = ... = p_k$ and at hand of certain stock exchange indices, see Müller and Richter (2016).

A common way of studying dependence among components of random vectors makes use of marginal distributions and copulas. Here, we approach dependence by comparing a vector density with the product of all its one-dimensional specializations. To this end, let

$$c_{sp}(x) = J(x)/P(x), x \in \mathbb{R}^k$$

where $P(x) = \prod_{i=1}^{k} f_i(x_i)$ and J is the joint density which combines $f_1, \ldots, f_k$ by a certain dependence construction.

For comparison, let $x \to c^{(p)}(x)$, F_i and $g_i, i = 1, ..., k$ denote the Copula density, the marginal cdfs and pdfs of the distribution law $\Phi_g^{(p)}$, respectively. Then

$$c^{(p)}(F_1(x_1), ..., F_k(x_k)) = \varphi_g^{(p)}(x) / \prod_{i=1}^{k} g_i(x_i). \tag{24}$$

The following definition is therefore well motivated.

Definition 7 *We call c_{sp} the specialization copula density.*

Example 6 *Each of the functions*

$$f_i(x_i) = \frac{\gamma \beta^{\frac{1}{\gamma}(M-1+\frac{1}{p_i})}}{2 p_i^{\frac{1}{p_i}-1} \Gamma\left(\frac{1}{\gamma}\left(M-1+\frac{1}{p_i}\right)\right)} \left(\frac{|x_i|^{p_i}}{p_i}\right)^{M-1} e^{-\beta\left(\frac{|x_i|^{p_i}}{p_i}\right)^{\gamma}}, i = 1, ..., k \tag{25}$$

is a one-dimensional specialization of the Kotz type density (17) meaning that, vice versa, $\varphi_{Kt;M,\beta,\gamma}$ in (17) generalizes f_i, $i = 1, \ldots, k$. In other words, the function in (17) is thought being build by a certain dependence construction applied to $f_1, ..., f_k$. Thus,

$$c_{sp,Kt}(x) = \varphi_{Kt;M,\beta,\gamma}^{(p)}(x) / \prod_{i=1}^{k} f_i(x_i)\, x \in \mathbb{R}^k.$$

and $\varphi_{Kt;M,\beta,\gamma}^{(p)}$ allows the representation

$$\varphi_{Kt;M,\beta,\gamma}^{(p)}(x) = c_{sp,Kt}(x) \prod_{i=1}^{k} f_i(x_i) \tag{26}$$

where the specialization copula density is explicitly given by

$$c_{sp,Kt}(x) = C \cdot \left(\frac{\sum_{i=1}^{k} \frac{|x_i|^{p_i}}{p_i}}{\prod_{i=1}^{k} \frac{|x_i|^{p_i}}{p_i}}\right)^{M-1} \frac{\exp\left\{\beta \sum_{i=1}^{k} \left(\frac{|x_i|^{p_i}}{p_i}\right)^{\gamma}\right\}}{\exp\left\{\beta \left(\sum_{i=1}^{k} \frac{|x_i|^{p_i}}{p_i}\right)^{\gamma}\right\}}$$

with

$$C = \frac{\beta^{\frac{(k-1)(1-M)}{\gamma}} \Gamma\left(\frac{M-1+\frac{1}{p_1}}{\gamma}\right) \cdot \ldots \cdot \Gamma\left(\frac{M-1+\frac{1}{p_k}}{\gamma}\right)}{\gamma^{k-1} B\left(\frac{1}{p_1}, \ldots, \frac{1}{p_k}\right) \Gamma\left(\frac{M-1+\frac{1}{p_1}+\ldots+\frac{1}{p_k}}{\gamma}\right)}.$$

Clearly, searching for marginal densities is not as easy as determining specializations, here, and the corresponding copula density has not such nice structure.

Example 7 *Each of the functions*

$$f_i(x_i) = \frac{\Gamma(M)}{2\Gamma\left(1+\frac{1}{p_i}\right)\Gamma\left(M-\frac{1}{p_i}\right)(\nu p_i)^{\frac{1}{p_i}}}\left(1+\frac{|x_i|^{p_i}}{\nu p_i}\right)^{-M}$$

is a one-dimensional specialization of the Pearson Type VII pdf in (18)*. The pdf* $\varphi^{(p)}_{PT7;M,\nu}$ *allows the representation*

$$\varphi^{(p)}_{PT7;M,\nu}(x) = c_{sp,PT7}(x) \prod_{i=1}^{k} f_i(x_i)$$

where the dependence function is

$$c_{sp,PT7}(x) = C \cdot \frac{\prod\limits_{i=1}^{k}\left(1+\frac{|x_i|^{p_i}}{\nu p_i}\right)^{-M}}{\left(1+\sum\limits_{i=1}^{k}\frac{|x_i|^{p_i}}{\nu p_i}\right)^{-M}}$$

with

$$C = p_1 \cdot \ldots \cdot p_k \frac{B\left(M-\frac{1}{p_1}, \ldots, M-\frac{1}{p_{k-1}}, M-\frac{1}{p_k}\right)}{B(\underbrace{M, \ldots, M}_{k-1}, M-\frac{1}{p_1}-\ldots-\frac{1}{p_1})}.$$

As in the preceding example, the specialization copula density has a nice structure and is explicitly given.

Remark 10 *The univariate q-generalized normal distribution or q-power exponential distribution has been parameterized in different ways in the literature, for a recent survey see Dytso et al. (2018). For different purposes, any of these parameterizations can be used to derive modified representations of the distributions considered in this paper.*

4 Model extensions: a concluding remark

Although Definition 5 deals with the whole class of p-spherical distributions later consideration is concentrated on continuous p-spherical distributions. To finally widen again the view we refer to Remark 1 in Richter (2015a) where a way is described to derive new distributions from the elements of a given class of distributions by restricting the region of definition of such distributions. The following definition sums up that for the present situation.

Definition 8 *Let* $\Sigma \in \mathfrak{B}(S_p)$ *with* $O_p(\Sigma) > 0$ *and*

$$\omega_\Sigma(A) = \frac{O_p(A)}{O_p(\Sigma)}, \; A \in \mathfrak{B}(S_p) \cap \Sigma = \mathfrak{B}(\Sigma).$$

Then ω_Σ is called p-uniform distribution on $\mathfrak{B}(\Sigma)$.

The extension of the whole class of p-spherical distributions follows accordingly.

It might be a further task of future work to extend the class of q-spherical processes, $q > 0$, introduced in Müller and Richter (2019) to a class of p-spherical processes, $p = (p_1, ..., p_k) \in (0, \infty)^{\times k}$.

Acknowledgement
The author is grateful to the Referees for their valuable hints.

Authors' contributions
The author read and approved the final manuscript.

Funding
Not applicable.

Competing interests
The author declares no competing interests.

References

Ahmadi-Javid, A., Moeini, A: Uniform distributions and random variate generation over generalized l_p-balls and spheres. J. Stat. Plann. Infer. **201**, 1–19 (2019)

Arellano-Valle, R., del Pino, G: From symmetric to asymmetric distributions: a unified approach. In: Genton M (ed.) Skew-Elliptical Distributions and Their Applications: A Journey Beyond Normality. Chapman and Hall/CRC, Boca Raton, (2004)

Arellano-Valle, R., Richter, W.-D.: On skewed continuous $l_{n,p}$-symmetric distributions. Chil. J. Stat. **3**(2), 193–212 (2012)

Arellano-Valle, R., del Pino, G., San Martin, E: Definition and probabilistic properties of skew distributions. Stat. Probab. Lett. **58**, 111–121 (2002)

Arnold, B., Castillo, E., Sarabia, J.: Multivariate distributions defined in terms of contours. J. Stat. Plann. Infer. **138**, 4158–4171 (2008)

Balkema, A., Nolde, N.: Asymptotic independence for unimodal densities. Adv Appl Prob. **42**, 411–432 (2010)

Balkema, A., Embrechts, P., Nolde, N.: Meta densities and the shape of their sample clouds. J. Multivar. Anal. **101**, 1738–1754 (2010)

Balkema, G., Embrechts, P.: High Risk Scenarios and Extremes. European Mathematical Society, Zürich (2007)

Barte, F., Gudon, O., Mendelson, S., Naor, A.: A probabilistic approach to the geometry of the l^{n_p}-ball. Ann Probab. **33**, 480–513 (2005)

Dytso, A., Bustin, R., Poor, H., Shamai, S.: Analytical properties of generalized gaussian distributions. J. Stat. Distrib. Appl. **5**(6), 1–40 (2018)

Fang, K. T., Kotz, S., Ng, K. W.: Symmetric multivariate and related distributions, London (1990)

Field, C., Genton, M.: The multivariate g-and-h Distribution. Technometrics. **48**(1), 104–111 (2006)

Günzel, T., Richter, W.-D., Scheutzow, S., Schicker, K., Venz, J.: Geometric approach to the skewed normal distribution. J. Stat. Plann. Infer. **142**(12), 3209–3224 (2012). https://doi.org/10.1016/j.jspi.2012.06.009

Harman, R., Lacko, V.: On decomposition algorithms for uniform sampling from n-Spheres and n-balls. J. Multivar. Anal. **101**, 2297–2304 (2010)

Ittrich, C., Richter, W.-D.: Exact tests and confidence regions in nonlinear regression. Statistics. **39**, 13–42 (2005)

Ittrich, C., Krause, D., Richter, W.-D.: Probabilities and large quantiles of noncentral generalized chi-square distributions. Statistics. **34**, 53–101 (2000)

Kalke, S., Richter, W.-D.: Simulation of the p-generalized Gaussian distribution. J. Stat. Comput. Simul. **83**(4), 639–665 (2013). https://doi.org/10.1080/00949655.2011.631187

Kamiya, H., Takemura, A., Kuriki, S.: Star-shaped distributions and their generalizations. J. Stat. Plann. Infer. **138**, 3429–3447 (2008)

Lacko, V., Harman, R.: A conditional distribution approach to uniform sampling on spheres and balls in l_p spaces. Metrika. **75**, 939–951 (2012)

Liang, J., Ng, K.: A method for generating uniformly scattert points on the l_p-norm unit sphere and its application. Metrika. **68**, 83–98 (2008)

Moszyńska, M., Richter, W.-D.: Reverse triangle inequality. Antinorms and semi-antnorms. Stud. Sci. Math. Hung. **49:1**, 120–138 (2012). https://doi.org/10.1556/SScMath.49.2012.1.1192

Müller, K., Richter, W.-D.: Exact distributions of order statistics of dependent random variables from $l_{n,p}$-symmetric sample distributions $n \in \{3, 4\}$. Depend Model. **4**, 1–29 (2016). https://doi.org/10.1515/demo-2016-0001

Müller, K., Richter, W.-D.: On p-generalized elliptical random processes. J Stat Distrib Appl 1:20. **6:1**, 1–37 (2019). https://doi.org/10.1186/s40488-019-0090-6

Naor, A., Romik, D.: Projecting the surface measure of the sphere of l_p^n. Annals de l'Institut Henri Poincare (B) Probab Stat. **39**, 241–261 (2003)

Nolan, J. P.: An r package for modeling and simulating generalized spherical and related distributions. J Stat Distrib Appl. **3**(1), 14 (2016)
Rachev, S., Rüschendorf, L.: Approximate independence of distributions on spheres and their stability properties. Ann Probab. **19**, 1311–1337 (1991)
Richter, W.-D.: Laplace-Gauss integrals, Gaussian measure asymptotic behavior and probabilities of moderate deviations. Zeitschrift für Analysis und ihre Anwendungen. **4**(3), 257–267 (1985)
Richter, W.-D.: Eine geometrische Methode in der Stochastik. Rostock Math Kolloq. **44**, 63–72 (1991)
Richter, W.-D.: A geometric approach to the Gaussian law. In: Mammitzsch, Schneeweiß (ed.) Symposia Gaussiana. Conf. B, pp. 25–45. Walter de Gruyter & Co., Berlin, (1995)
Richter, W.-D.: Generalized spherical and simplicial coordinates. J. Math. Anal. Appl. **335**, 1187–1202 (2007). https://doi.org/10.1016/j.jmaa.2007.03.047
Richter, W.-D.: Continuous $l_{n,p}$-symmetric distributions. Lith Math. J. **49**(1), 93–108 (2009). https://doi.org/10.1007/s10986-009-9030-3
Richter, W.-D.: Geometric disintegration and star-shaped distributions. J. Stat. Distrib. Appl. **1**(20), 1–24 (2014). https://doi.org/10.1186/s40488-014-0020-6
Richter, W.-D.: Convex and radially concave contoured distributions. J. Probab. Stat., 1–12 (2015a). https://doi.org/10.1155/2015/165468
Richter, W.-D.: Norm contoured distributions in R^2. In: Lecture notes of Seminario Interdisciplinare di Matematica. Vol. XII, pp. 1–12. Potenza: Seminario Interdisciplinare di Matematica (S.I.M.), Italy, (2015b)
Richter, W.-D.: Exact inference on scaling parameters in norm and antinorm contoured sample distributions. J. Stat. Distrib. Appl. **3**(8), 1–16 (2016). https://doi.org/10.1186/s40488-016-0046-z
Richter, W.-D.: The class of (p,q)-spherical distributions with an extension of the sector and circle number functions. Risks 2017. **3**(40) (2017). https://doi.org/103390/risks5030040
Schechtman, G., Zinn, J.: Concentration on the l_{p^n} ball. Elsevier Science, Amsterdam (1990). https://doi.org/10.1016/j.sigpro.2009.07.015
Sinz, F., Simoncelli, E., Bethge, M.: Hierarchical modeling of local image features through L_p-nested symmetric distributions. In: Twenty-Third Annual Conference on Neural Information Processing Systems, pp. 1–9, (2009)
Song, D., Gupta, A. K.: l_p-norm uniform distributions. Proc. Am. Math. Soc. **125**(2), 595–601 (1997)
Szablowski, P. J.: Uniform distributions on spheres in finite dimensional l_α and their generalizations. J. Multivar. Anal. **64**, 103–107 (1998)
Taguchi, T.: On a generalization of Gaussian distribution. Ann. Inst. Stat. Math. **30**, 211–242 (1978). https://doi.org/10.1007/BF02480215

8

A nonparametric approach for quantile regression

Mei Ling Huang[1*] and Christine Nguyen[2]

*Correspondence:
mhuang@brocku.ca
[1]Department of Mathematics & Statistics, Brock University, St. Catharines, Ontario L2S 3A1, Canada
Full list of author information is available at the end of the article

Abstract

Quantile regression estimates conditional quantiles and has wide applications in the real world. Estimating high conditional quantiles is an important problem. The regular quantile regression (QR) method often designs a linear or non-linear model, then estimates the coefficients to obtain the estimated conditional quantiles. This approach may be restricted by the linear model setting. To overcome this problem, this paper proposes a direct nonparametric quantile regression method with five-step algorithm. Monte Carlo simulations show good efficiency for the proposed direct QR estimator relative to the regular QR estimator. The paper also investigates two real-world examples of applications by using the proposed method. Studies of the simulations and the examples illustrate that the proposed direct nonparametric quantile regression model fits the data set better than the regular quantile regression method.

Keywords: Conditional quantile, Goodness-of-fit, Gumbel's second kind of bivariate exponential distribution, Nonparametric kernel density estimator, Nonparametric regression, Weighted loss function

AMS 2010 Subject Classifications: primary: 62G32; secondary: 62J05

1 Introduction

It is important to study quantile regression to estimate high conditional quantiles in real-world events Koenker (2005). Some extreme events can cause damages to society: stock market crashes, pipeline failures, large flooding, wildfires, pollution, earth quakes and hurricanes. We wish to estimate high conditional quantiles of a random variable y with cumulative distribution function (c.d.f.) $F(y)$ given a variable vector, $\mathbf{x} = (x_1, x_2, \ldots, x_d)$, and $\mathbf{x}_p = (1, x_1, x_2, \ldots, x_d)^T \in R^p$ where $p = d + 1$. The τth conditional linear quantile is defined by

$$Q_y(\tau|\mathbf{x}) = Q_y(\tau|x_1, x_2, \ldots, x_d) = F^{-1}(\tau|\mathbf{x}),\ 0 < \tau < 1. \tag{1}$$

The traditional quantile regression is concerned with the estimation of the τth conditional quantile regression (QR) of y for given $\mathbf{x}$ which often sets a linear model as

$$Q_y(\tau|\mathbf{x}) = \mathbf{x}_p^T \beta(\tau) = \beta_0(\tau) + \beta_1(\tau)x_1 + \cdots + \beta_d(\tau)x_d, 0 < \tau < 1, \tag{2}$$

where $\beta(\tau) = (\beta_0(\tau), \beta_1(\tau), \beta_2(\tau), \ldots, \beta_d(\tau))^T$.

For linear model *(2)*, we estimate the coefficient $\beta(\tau) = (\beta_0(\tau), \beta_1(\tau), \beta_2(\tau), \ldots, \beta_d(\tau))^T \in R^p$ from a random sample $\{(y_i, \mathbf{x}_i),\ i = 1, \ldots, n\}$, where $\mathbf{x}_{pi} = (1, x_{i1}, x_{i2}, \ldots, x_{id})^T$ is the p-dimensional design vector and y_i is the univariate response variable from a continuous distribution with a c.d.f. $F(y)$. Koenker and Bassett (1978) proposed an L_1-weighted loss function to obtain estimator $\widehat{\beta}(\tau)$ by solving

$$\widehat{\beta}(\tau) = \arg \min_{\beta(\tau)\in R^p} \sum_{i=1}^{n} \rho_\tau(y_i - \mathbf{x}_{pi}^T\beta(\tau)),\ 0 < \tau < 1, \tag{3}$$

where ρ_τ is a loss function, namely

$$\rho_\tau(u) = u(\tau - I(u < 0)) = \begin{cases} u(\tau - 1), u < 0; \\ u\tau,\ u \geq 0. \end{cases}$$

The linear quantile regression problem can be formulated as a linear program

$$\min_{(\beta(\tau),\mathbf{u},\mathbf{v})\in R^p\times R_+^{2n}} \{\tau\mathbf{1}_n^T\mathbf{u} + (1-\tau)\mathbf{1}_n^T\mathbf{v}|\mathbf{X}\beta(\tau) + \mathbf{u} - \mathbf{v} = \mathbf{y}\},$$

where $\mathbf{1}_n^T$ is an n-vector of 1s, $\mathbf{X}$ denotes the $n\times p$ design matrix, and $\mathbf{u}, \mathbf{v}$ are $n \times 1$ vectors with elements of $u_i, v_i,\ i = 1, \ldots, n$, respectively (Koenker, 2005).

In recent years, studies are looking for efficiency improvements of estimator *(3)* (Yu et al. 2003; Wang and Li 2013; Huang et al. 2015; Huang and Nguyen 2017). The regular linear quantile regression *(2)* needs the estimator $\widehat{\beta}(\tau)$ in *(3)* for the estimated conditional quantile curves. But this estimated conditional quantile curve may be restricted under the model setting.

Many studies have used nonparametric method of quantile regression in recent years, for example, Chaudhuri (2003), Yu and Jones (1991), Hall et al. (1999) and Yu et al. (2003). Chapter 7 in Keoker (2005) proposed a local polynomial quantile regression (LPQR), and other methods. Also we can see detailed discussions on theory, methodologies and applications in Li and Racine (2007) and Cai (2013).

In order to overcome the limitation of the model setting in *(2)* in this paer we propose a direct nonparametric quantile regression method which uses the ideas of nonparametric kernel density estimation and nonparametric kernel regression. The proposed method is not only different from most other existing nonparametric quantile regression methods, it also overcome thecrossing problem of estimating quantile curves. We like to see if the new method has an improvement relative to the regular linear quantile regression and other nonparametric quantile regression methods, we will do two studies in this paper:

1. Monte Carlo simulations will be performed to confirm the better efficiency of the new direct QR estimator relative to the regular QR estimator and a nonparametric LPQR.
2. The new proposed method will be applied to two real-world examples of extreme events and compared with the linear model in Huang and Nguyen (2017).

In Section 2, we propose a direct nonparametric quantile regression estimator. A relative measure of comparing goodness-of-fit for quantile models is given in Section 3. In Section 4, the results of Monte Carlo simulations generated from Gumbel's second kind of bivariate exponential distribution Gumbel (1960) show that the proposed direct method produces high efficiencies relative to existing linear QR and LPQR methods. In Section 5, the regular linear quantile regression and the proposed direct quantile regression are applied to two real-life examples: the Buffalo snowfall and CO_2 emission examples in Huang and Nguyen (2017). The study of these examples illustrate that the proposed direct nonparametric quantile regression model fits the data better than the existing linear quantile regression method.

2 Proposed direct nonparametric quantile regression

In this paper, for generality, we ignore the idea of the linear model *(2)*. We obtain a direct estimator for true conditional quantile in *(1)*:

$$\widehat{Q}_y(\tau|\mathbf{x}) = \widehat{Q}_y(\tau|x_1, x_2, \ldots, x_d) = \widehat{F}^{-1}(\tau|\mathbf{x}),$$

by using local conditional quantile estimator $\xi_i(\tau|\mathbf{x}_i) = Q_y(\tau|\mathbf{x}_i)$ based the ith point of given random sample, $\{(y_i, \mathbf{x}_i), i = 1, \ldots, n\}$, for $\mathbf{x}_i = (x_{1i}, x_{2i}, \ldots, x_{di})^T$.

We construct the following a five-step algorithm of a direct nonparametric quantile regression:

Step 1: Estimate the conditional density of y for given $\mathbf{x} = (x_1, x_2, \ldots, x_d)$ using a kernel density estimation method (Silverman 1986; Scott 2015):

$$\widehat{f}(y|\mathbf{x}) = \frac{\widehat{f}(y, \mathbf{x})}{\widehat{g}(\mathbf{x})}, \tag{4}$$

where $\widehat{f}(y, \mathbf{x})$ is an estimator of the joint density of y and $\mathbf{x}$, and $\widehat{g}(\mathbf{x})$ is an estimator of the marginal density of $\mathbf{x}$.

A d-dimensional kernel density estimator from a random sample $\mathbf{X}_i = (X_{1i}, X_{2i}, \ldots, X_{di})$, $i = 1, 2, \ldots, n$, from a population $\mathbf{x} = (x_1, x_2, \ldots, x_d)$ for joint density $g(\mathbf{x})$,is given by

$$\widehat{g}(\mathbf{x}) = \frac{1}{nh^d} \sum_{i=1}^{n} K\left\{\frac{\mathbf{x} - \mathbf{X}_i}{h}\right\},$$

where $h > 0$ is the bandwidth and the kernel function $K(\mathbf{x})$ is a function defined for d-dimensional $\mathbf{x} = (\mathbf{x}_1, \mathbf{x}_2, \ldots, \mathbf{x}_d)$ which satisfies $\int_{R^d} K(\mathbf{x})d\mathbf{x} = 1$.

Fukunaga (1972) suggested using

$$\widehat{g}(\mathbf{x}) = \frac{(\det \mathbf{S})^{-1/2}}{nh^d} \sum_{i=1}^{n} k\left\{\frac{(\mathbf{x} - \mathbf{X}_i)^T \mathbf{S}^{-1} (\mathbf{x} - \mathbf{X}_i)}{h^2}\right\},$$

where $\mathbf{S}$ is the sample covariance matrix of the data, K is the normal kernel, the function k is

$$k(u) = \left(\frac{1}{2\pi}\right)^{d/2} \exp\left(-\frac{u}{2}\right), \quad k(\mathbf{x}^T\mathbf{x}) = K(\mathbf{x}) = (2\pi)^{-d/2} \exp\left(-\frac{1}{2}\mathbf{x}^T\mathbf{x}\right).$$

A plug-in selector of the bandwidth $h > 0$ will be given by (Silverman 1986, p. 85) as

$$h_{opt} = \left\{\int t^2 K(t) dt\right\}^{-2/(d+2)} \left\{\int K(t)^2 dt\right\}^{1/(d+4)} \left\{\int \left(\nabla^2 g(\mathbf{x})\right)^2 d\mathbf{x}\right\}^{-1/(d+4)} n^{-1/(d+4)}, \tag{5}$$

If a multivariate normal kernel is used for smoothing the normal distribution data with unit variance,

$$h_{opt} = \left\{\frac{4}{d+2}\right\}^{1/(d+4)} n^{-1/(d+4)}.$$

Step 2: Estimate the conditional c.d.f. of y given $\mathbf{x}$:

$$\widehat{F}(y|\mathbf{x}) = \int_{-\infty}^{y} \widehat{f}(y|\mathbf{x}) dy.$$

Step 3: Estimate the local conditional quantile function $\xi(\tau|\mathbf{x})$ of y given $\mathbf{x}$ by inverting an estimated conditional c.d.f. $\widehat{F}(y|\mathbf{x})$.

$$\widehat{\xi}(\tau|\mathbf{x}) = \widehat{Q_y}(\tau|\mathbf{x}) = \inf\{y : \widehat{F}(y|\mathbf{x}) \geq \tau\} = \widehat{F}^{-1}(\tau|\mathbf{x}).$$

It is difficult to compute a global inverse function $\widehat{\xi}(\tau|\mathbf{x})$ of the kernel estimated conditional c.d.f. $\widehat{F}(y|\mathbf{x})$ which has many terms. To avoid the the computational global difficulties, we estimate the local conditional quantile point $\xi_i(\tau|\mathbf{x}_i)$ of y given $\mathbf{x}_i$ by inverting $\widehat{F}(y|\mathbf{x}_i)$ at the ith data point $(y_i, \mathbf{x}_i)$:

$$\widehat{\xi_i}(\tau|\mathbf{x}_i) = \widehat{Q_y}(\tau|\mathbf{x}_i) = \inf\{y : \widehat{F}(y|\mathbf{x}_i) \geq \tau\} = \widehat{F}^{-1}(\tau|\mathbf{x}_i), \quad i = 1, 2, \ldots, n. \tag{6}$$

Thus, we have n points $\left(\mathbf{x}_i, \widehat{\xi_i}(\tau|\mathbf{x}_i)\right),\ i = 1, 2, \ldots, n$.

Step 4: We propose a direct nonparametric quantile regression estimator for the τth conditional quantile curve of $\mathbf{x}$ by using Nadaraya-Watson (NW) nonparametric regression estimator (Scott, 2015, p. 242) on $\left(\mathbf{x}_i, \widehat{\xi_i}(\tau|\mathbf{x}_i)\right),\ i = 1, 2, \ldots, n$:

$$Q_D(\tau|\mathbf{x}) = \widehat{\xi}(\tau|\mathbf{x}) = \frac{\sum_{i=1}^{n} K_{\mathbf{h}}\{\mathbf{x} - \mathbf{X}_i\}\widehat{\xi_i}(\tau|\mathbf{x}_i)}{\sum_{j=1}^{n} K_{\mathbf{h}}\{\mathbf{x} - \mathbf{X}_j\}} = \sum_{i=1}^{n} W_{h_{\mathbf{x}}}(\mathbf{x}, \mathbf{X}_i)\widehat{\xi_i}(\tau|\mathbf{x}_i), \quad 0 < \tau < 1, \tag{7}$$

where $W_{h_{\mathbf{x}}}(\mathbf{x}, \mathbf{X}_i)$ is called an equivalent kernel, and $\mathbf{h} = (h_1, \ldots, h_d)$,

$$W_{h_{\mathbf{x}}}(\mathbf{x}, \mathbf{X}_i) = \frac{K_{\mathbf{h}}\{\mathbf{x} - \mathbf{X}_i\}}{\sum_{j=1}^{n} K_{\mathbf{h}}\{\mathbf{x} - \mathbf{X}_j\}}, \quad i = 1, 2, \ldots, n,$$

where

$$K_{\mathbf{h}}\{\mathbf{x} - \mathbf{X}_i\} = \frac{1}{nh_1 \ldots h_d} \prod_{j=1}^{d} K\left(\frac{x - x_{ij}}{h_j}\right), \quad i = 1, \ldots, n,$$

where K is the kernel function, and $h_j > 0$ is the bandwidth for the j th dimension.

The new point of *(7)* is that it uses Step 3's *(6)* numerical results: n points $\left(\mathbf{x}_i, \widehat{\xi_i}(\tau|\mathbf{x}_i)\right),\ i = 1, 2, \ldots, n$, to estimate a conditional mean curve of the τth quantile function based on these n points, then smoothes these n points out.

In this paper, for the kernel regression, we use K which is the standard normal kernel. Similar as formula *(5)*, we use the optimal bandwidth for the jth dimension (Silverman 1986, p.40),

$$h_{j,opt} = \left\{\int t^2 K(t)dt\right\}^{-2/5} \left\{\int K(t)^2 dt\right\}^{1/5} \left\{\int \left(\nabla^2 \widehat{g_j}(x_j)\right)^2 d\mathbf{x}_j\right\}^{-1/5} n^{-1/5}, \quad j = 1, \ldots, d, \tag{8}$$

where $\widehat{g_j}(x_j)$ is the estimated the jth dimensional marginal density of x_j in $\mathbf{x} = (x_1, x_2, \ldots, x_d)$, n is the sample size of the random sample in *(4)*.

Step 5: Check all procedures, and make any necessary adjustments.

3 Comparison of goodness-of-fit on quantile regression models

In order to compare the regular QR estimator in *(3)* and the direct nonparametric QR estimator in *(7)*, we extend the idea of measuring goodness-of-fit by Koenker and Machado (1999). We suggest using a Relative $R(\tau),\ 0 < \tau < 1$, which is defined as

$$\textit{Relative } R(\tau) = 1 - \frac{V_D(\tau)}{V_R(\tau)}, \quad -1 \le R(\tau) \le 1, \quad \text{where} \tag{9}$$

$$V_D(\tau) = \sum_{y_i \ge Q_D(\tau|\mathbf{x}_i)} \frac{\tau}{n} \left|y_i - Q_D(\tau|\mathbf{x}_i)\right| + \sum_{y_i < Q_D(\tau|\mathbf{x}_i)} \frac{(1-\tau)}{n} \left|y_i - Q_D(\tau|\mathbf{x}_i)\right|,$$

where $Q_D(\tau|\mathbf{x}_i)$ is obtained by *(7)*, and

$$V_R(\tau) = \sum_{y_i \ge \mathbf{x}_i^T \widehat{\beta}(\tau)} \frac{\tau}{n} \left|y_i - \mathbf{x}_i^T \widehat{\beta}(\tau)\right| + \sum_{y_i < \mathbf{x}_i^T \widehat{\beta}(\tau)} \frac{(1-\tau)}{n} \left|y_i - \mathbf{x}_i^T \widehat{\beta}(\tau)\right|,$$

where $\widehat{\beta}(\tau)$ is given by *(3)*.

4 Simulations

For investigating the proposed direct nonparametric quantile regression estimator in *(7)*, in this Section, Monte Carlo simulations are performed. We generate m random samples with size n each from the second kind of Gumbel's bivariate exponential distribution Gumbel (1960) which has a non-linear conditional quantile function of y given x in *(11)*. It has c.d.f. $F(x, y)$ and density function $f(x, y)$ in *(10)*:

$$F(x,y) = (1 - e^{-x})(1 - e^{-y})(1 + \alpha e^{-(x+y)}), \ x \ge 0, \ y \ge 0, \ \alpha > 0, \tag{10}$$

$$f(x,y) = e^{-(x+y)}(1 + \alpha(2e^{-x} - 1)(2e^{-y} - 1)), \ x \ge 0, \ y \ge 0, \ \alpha > 0.$$

The conditional density of y for given x is

$$f(y|x) = e^{-y}(1 + \alpha(2e^{-x} - 1)(2e^{-y} - 1)), \ x \ge 0, \ y \ge 0, \ \alpha > 0.$$

The conditional c.d.f. of y for given x is

$$F(y|x) = e^{-y}(\alpha(2e^{-x} - 1)(1 - e^{-y}) - 1) + 1, \ x \ge 0, \ y \ge 0, \ \alpha > 0.$$

The true τth conditional quantile function of y given x of *(10)* is

$$\xi(\tau|x) = Q_y(\tau|x) = \ln\left(\frac{2\alpha(2e^{-x} - 1)}{\alpha(2e^{-x}-1) - 1 + \sqrt{(\alpha(2e^{-x}-1)+1)^2 - 4\alpha\tau(2e^{-x} - 1)}}\right), \tag{11}$$
$$x \ge 0, \ \alpha > 0, \ 0 < \tau < 1.$$

Letting $\alpha = 1$, the c.d.f. in *(10)* is in Fig. 1.

We use three quantile regression methods:

1. The regular quantile regression $Q_R(\tau|x)$ estimation based on *(3)*:

$$Q_R(\tau|x) = \widehat{\beta}_0(\tau) + \widehat{\beta}_1(\tau)x. \quad 0 < \tau < 1 \tag{12}$$

2. The first-order linear polynomials Quantile Regression (LPQR) $Q_{LP}(\tau|x)$ (Chaudhuri 1991, Keoker 2005, Yu and Jones 1998), for z in a neighborhood of x,

$$Q_{LP}(\tau|x) = \widehat{a}_0(\tau, x) + \widehat{a}_1(\tau, x)(z - x). \quad 0 < \tau < 1, \tag{13}$$

where

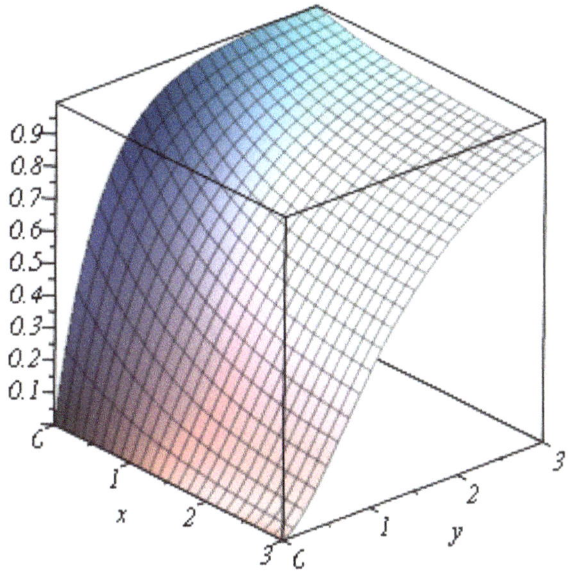

Fig. 1 The c.d.f. of Gumbel's Second kind of bivariate exponential distribution with $\alpha = 1$

$$\widehat{\mathbf{a}}(\tau, x) = \arg \min_{\beta(\tau) \in R^p} \sum_{i=1}^{n} \rho_\tau (y_i - a_0(\tau, x) - a_1(\tau, x)(x_i - x)) K\left(\frac{x - x_i}{h}\right), \quad 0 < \tau < 1,$$

here $\mathbf{a}(\tau, x) = (a_0(\tau, x), a_1(\tau, x))^T$, h and K are the bandwidth and kernel function. the LPQR can be computed by the R package 'quantreg' Koenker (2018).

3. The direct nonparametric quantile regression $Q_D(\tau|x)$ estimation based on *(7)*

$$Q_D(\tau|x) = \sum_{i=1}^{n} W_{h_\mathbf{x}}(\mathbf{x}, \mathbf{X}_i)\widehat{\xi}_i(\tau|x_i), \quad 0 < \tau < 1, \tag{14}$$

where $\widehat{\xi}_i(\tau|x_i)$ is obtained by *(6)*, $W_{h_\mathbf{x}}(\mathbf{x}, \mathbf{X}_i)$ is given by *(7)*.

For each method, we generate size $n = 100$, $m = 100$ samples. $Q_{R,i}(\tau|x)$, $Q_{LP,i}(\tau|x)$ and $Q_{D,i}(\tau|x)$, $i = 1, 2, \ldots, m$, are estimated in the ith sample. Let $\alpha = 1$ in *(11)*. Then the true τth conditional quantile is

$$\xi(\tau|x) = Q_y(\tau|x) = \ln\left(\frac{2e^{-x} - 1}{e^{-x} - 1 + \sqrt{e^{-2x} - \tau(2e^{-x} - 1)}}\right), \; x \geq 0, \; \alpha > 0, \; 0 < \tau < 1. \tag{15}$$

The simulation mean squared errors (SMSEs) of the estimators *(12)*, *(13)* and *(14)* are:

$$SMSE(Q_R(\tau|x)) = \frac{1}{m}\sum_{i=1}^{m}\int_0^N (Q_{R,i}(\tau|x) - Q_y(\tau|x))^2 dx; \tag{16}$$

$$SMSE(Q_{LP}(\tau|x)) = \frac{1}{m}\sum_{i=1}^{m}\int_0^N (Q_{LP,i}(\tau|x) - Q_y(\tau|x))^2 dx, \tag{17}$$

$$SMSE(Q_D(\tau|x)) = \frac{1}{m}\sum_{i=1}^{m}\int_0^N (Q_{D,i}(\tau|x) - Q_y(\tau|x))^2 dx, \tag{18}$$

where the true τth conditional quantile $Q_y(\tau|x)$ is defined in *(15)*. N is a finite x value such that the c.d.f. in *(10)* $F(N, N) \approx 1$. We take $N = 6$ and the simulation efficiencies (SEFFs) are given by

Table 1 Simulation Mean Square Errors (SMSEs) and Efficiencies (SEFFs) of Estimating $Q_y(\tau|x), m = 100, n = 100, N = 6$.

τ	0.95	0.96	0.97	0.98	0.99
$SMSE(Q_R(\tau\|x))$	22.091	26.632	28.982	42.725	73.340
$SMSE(Q_{LP}(\tau\|x))$	8.160	9.667	11.074	15.080	23.734
$SMSE(Q_D(\tau\|x))$	5.161	6.630	6.552	8.850	11.596
Efficiency					
$SEFF(Q_{LP}(\tau\|x))$	**2.7072**	**2.7449**	**2.6171**	**2.8332**	**3.0901**
$SEFF(Q_D(\tau\|x))$	**4.2804**	**4.0169**	**4.4234**	**4.8278**	**6.3246**

$$SEFF(Q_{LP}(\tau|x)) = \frac{SMSE(Q_R(\tau|x))}{SMSE(Q_{LP}(\tau|x))}, \quad SEFF(Q_D(\tau|x)) = \frac{SMSE(Q_R(\tau|x))}{SMSE(Q_D(\tau|x))},$$

where $SMSE(Q_R(\tau|x))$, $SMSE(Q_{LP}(\tau|x))$ and $SMSE(Q_D(\tau|x))$ are defined in *(16), (17)* and *(18),* respectively.

Table 1 shows that all of the $SEFF(Q_D(\tau|x))$ are larger than 1 when $\tau = 0.95, \ldots, 0.99$.

Figure 2 compares the $SMSE(Q_R(\tau|x)), SMSE(Q_{LP}(\tau|x))$ with the $SMSE(Q_D(\tau|x))$ for $\tau = 0.95, \ldots, 0.99$. It demonstrates that all $SMSE(Q_D(\tau|x))$ have smaller values than both $SMSE(Q_{LP}(\tau|x))$ and $SMSE(Q_R(\tau|x))$, thus, the simulation results show that the proposed estimator $Q_D(\tau|x)$ is more efficient relative to the regular linear estimator $Q_R(\tau|x)$ and nonparametric local polynomial estimator $Q_D(\tau|x)$.

Next, we compare $Q_D(\tau|x)$ and $Q_R(\tau|x)$ in Figs. 3 and 4.

Figure 3 shows the boxplots of $Q_R(\tau|x)$ and $Q_D(\tau|x)$ for $\tau = 0.95, 0.97$, and 0.99.(The true conditional quantiles are in blue line). The $Q_D(\tau|x)$ has much smaller variance than $Q_R(\tau|x)s$.

Figure 4 shows the average curves of the 100 estimated $\tau = 0.95$th quantile curves of $Q_R(\tau|x)$ (in blue dash line) and that of $Q_D(\tau|x)$ (in red solid). The average $Q_D(\tau|x)$ curve is much closer than $Q_R(\tau|x)$ to the true quantile curve (in green dash).

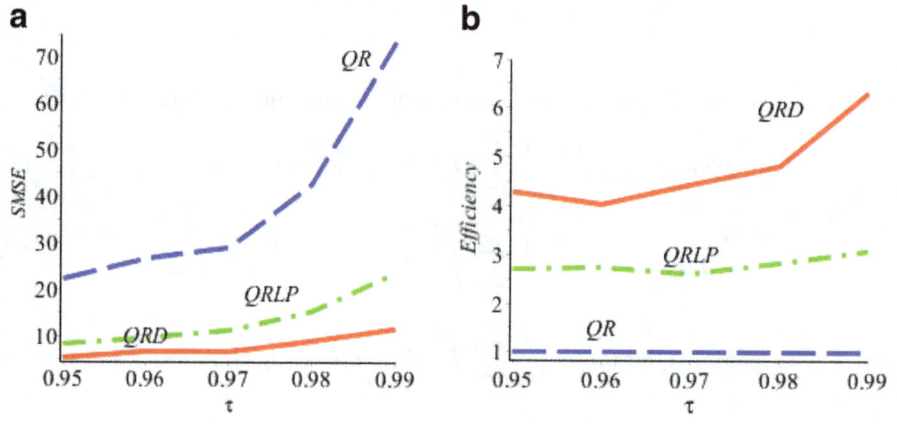

Fig. 2 **a** $SMSE(Q_D(\tau))$ is the red solid line, $SMSE(Q_{LP}(\tau))$ is the green dash-dot line, $SMSE(Q_R(\tau|x))$ is the blue dash line. **b** $SEFF(Q_D(\tau|x))$ is the red solid line, $SEFF(Q_{LP}(\tau|x))$ is the green dash-dot line,$SEFF(Q_R(\tau|x)) \equiv 1$ is blue dash line

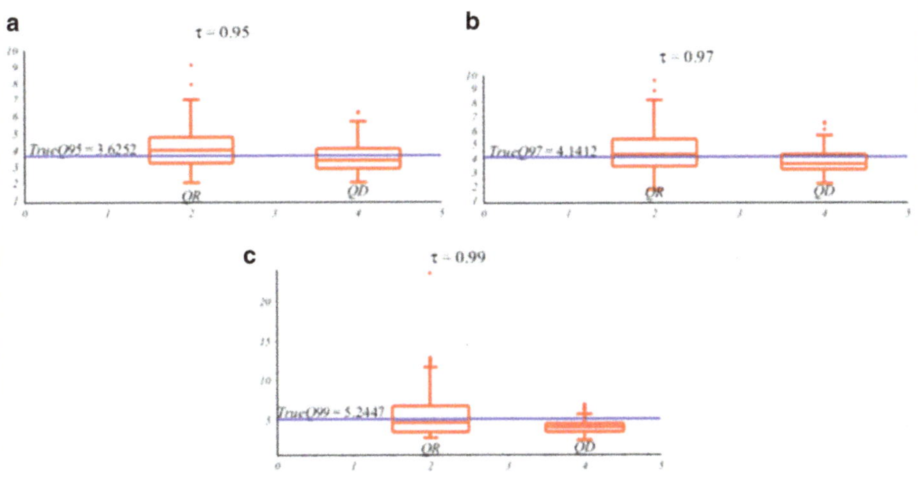

Fig. 3 Box plots for (**a**) $\tau = 0.95$th quantile curves; (**b**) $\tau = 0.97$th quantile curves; (**c**) $\tau = 0.99$th quantile curves. The true conditional quantile lines are in blue

From the overall results of the simulation, we can conclude that Table 1 and Figs. 2, 3, and 4 show that for $\tau = 0.95, \ldots, 0.99$, the proposed direct estimator $Q_D(\tau|x)$ in *(7)* is more efficient relative to the regular regression $Q_R(\tau|x)$ in *(2)* and a nonparametric LPQR in *(13)*.

5 Real examples of applications

In this section, we apply the following two regression models to the Buffalo snowfall and CO_2 emission examples in Huang and Nguyen (2017):

1. The regular quantile regression $Q_R(\tau|\mathbf{x})$ in model *(2)* using estimator $\widehat{\beta}(\tau)$ in *(3)*;
2. The direct nonparametric quantile regression $Q_D(\tau|\mathbf{x})$ in *(7)*.

5.1 Buffalo snowfall example

Huang and Nguyen (2017) used the following linear second order polynomial quantile regression model for this example (National Weather Service Forecast Office 2017):

$$Q_y(\tau|x) = \beta_0(\tau) + \beta_1(\tau)x + \beta_2(\tau)x^2,$$

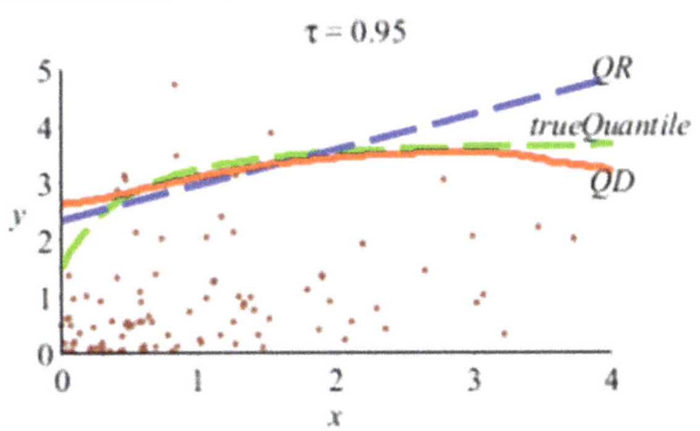

Fig. 4 In $n = 100$, $m = 100$, $\tau = 0.95$ simulations, the true Quantile-green dash; average regular QR-blue dash; average direct QD-red solid

Table 2 Buffalo Daily Snowfalls (cm) at High Quantiles Using Q_R and Q_D

	$\tau = 0.97$		$\tau = 0.99$	
Temperature (°C)	Q_R	Q_D	Q_R	Q_D
-15	37.38	25.49	105.46	60.64
-10	33.19	30.23	87.95	62.98
-5	30.98	33.33	72.08	56.54
0	30.73	29.89	57.86	54.56
5	32.47	33.27	45.29	52.39
10	36.17	37.34	34.36	43.04

where y represents the total snowfall (cm) and x represents the maximum temperature ($°C$).

In this paper we use the proposed five-step algorithm in Section 2 to obtain the new direct nonparametric quantile estimator $Q_D(\tau|\mathbf{x})$ in *(7)*. We compare the new estimator $Q_D(\tau|\mathbf{x})$ with the regular quantile estimator $Q_R(\tau|\mathbf{x})$ in Huang and Nguyen (2017). Table 2 and Fig. 5 show the difference of values of two estimators. Figure 5a, b and c show the scatter plot of the daily snowfall vs. maximum temperature with the fitted Q_R, and Q_D

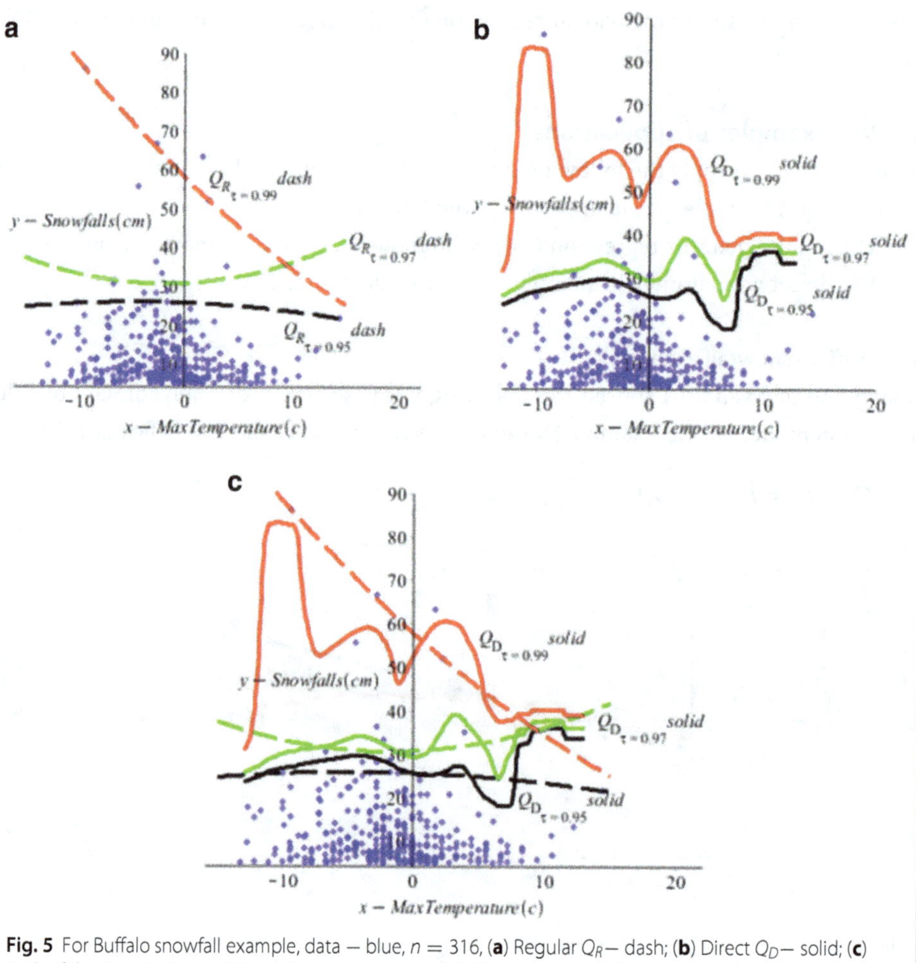

Fig. 5 For Buffalo snowfall example, data – blue, $n = 316$, (**a**) Regular Q_R– dash; (**b**) Direct Q_D– solid; (**c**) Both of the Regular Q_R and Direct Q_D in a plot at $\tau = 0.95$ in black, $\tau = 0.97$ in green and $\tau = 0.99$ in red

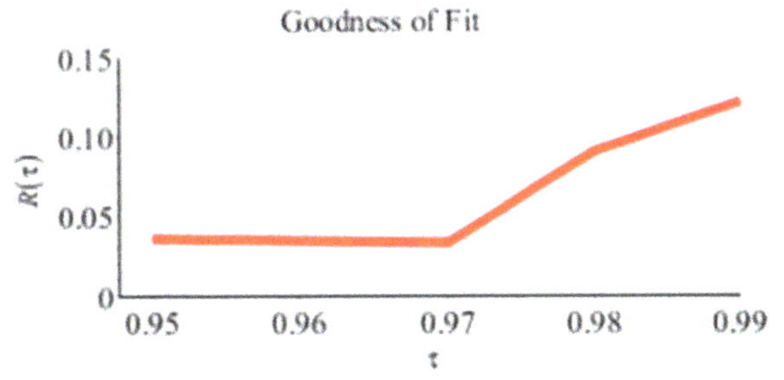

Fig. 6 Relative R_τ of Q_D relative to Q_R for the Buffalo snowfall example

quantile curves at $\tau = 0{,}95$, 0.97 and 0.99. It is interesting to see that the Q_D curves appear to follow the data patterns closer than the Q_R curves.

Table 2 lists the estimated Buffalo snowfall quantile values at a given maximum temperature for $\tau = 0.97$ and 0.99. It demonstrates that when quantiles are at high τ, the Q_D gives greater variety of snowfall predictions than the Q_R. The relationship of snowfall and max-temperature is not necessarily linear.

Figure 6 and Table 3 show the values of the Relative $R(\tau)$ in *(9)* for given $\tau = 0.95, \ldots, 0.99$. We note that $R(\tau) > 0$ which means that $V_D(\tau) < V_R(\tau)$ and Q_D is a better fit to the data than Q_R.

Figure 5c shows that the proposed direct nonparametric quantile regression Q_D predicts that for moderate temperatures, such as $5°C$ to $10°C$, it is likely to have smaller but varied snowfalls in Buffalo than the regular Q_D predicts. For temperature over $10°C$, the Q_D predicts a much higher value snow amount than the regular Q_R predicts. On another side, for very low temperatures, such as $-15°C$ to $0°C$, the Q_D and Q_R both predict more likely to have extreme heavy snowfalls that may cause damage. Thus prediction of heavy snowfalls is related to cold weather forecasts. But the prediction snowfalls related to temperature from the Q_D is not as a simple linear relationship as Q_R predicts. We also note that lots of snow occurred between -5$°C$ to $0°C$; the predictions form the Q_D are reflecting this fact and give varied predictions.

5.2 CO_2 emission example

Huang and Nguyen (2017) used the linear quantile regression model for this example:

$$Q_y(\tau|x_1, x_2) = \beta_0(\tau) + \beta_1(\tau)x_1 + \beta_2(\tau)x_2,$$

where y represents CO_2 emission (tonnes) per capita, x_1 represents ln of gross domestic product (GPD) (US \$), per capita and x_2 represents ln of electricity consumption (E.C.) (kilowatts) per capita (Carbon Dioxide Information Analysis Centre (2017)).

Similar as in the Buffalo Snowfall example in Subsection 5.1, we use the proposed five-step algorithm in Section 2 to obtain the new direct nonparametric quantile estimator

Table 3 Relative $R(\tau)$ Values for the Buffalo Snowfall Example

	$\tau = 0.95$	$\tau = 0.96$	$\tau = 0.97$	$\tau = 0.98$	$\tau = 0.99$
Relative $R(\tau)$	0.0359	0.0346	0.0324	0.0903	0.1206

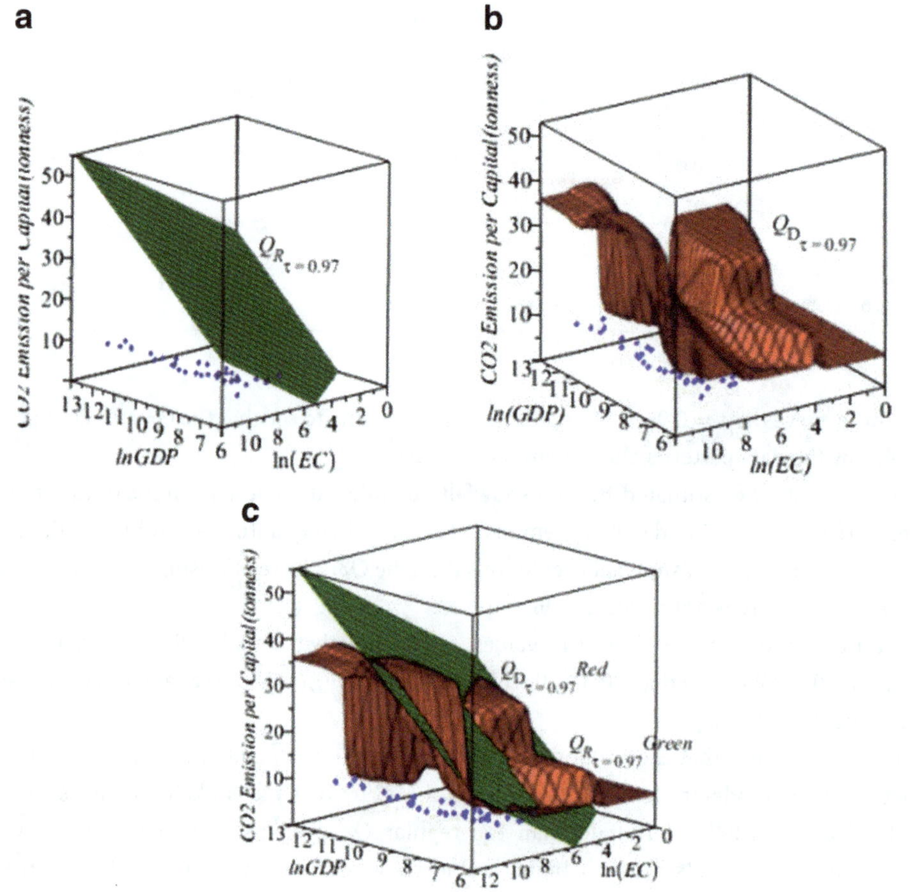

Fig. 7 3D Plots for CO_2 Emission, data – blue, $n = 123$, (**a**) Regular Q_R– green at $\tau = 0.97$; (**b**) Direct Q_D– red at $\tau = 0.97$; (**c**) Regular Q_R–green and Direct Q_D–red in a plot at $\tau = 0.97$

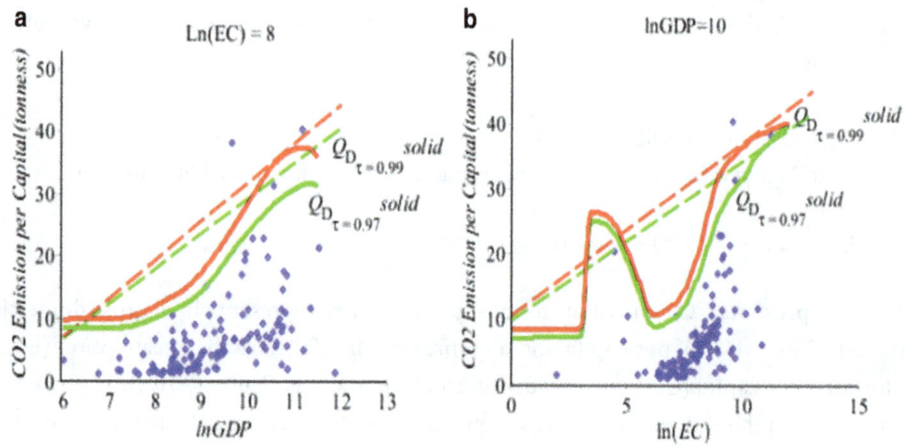

Fig. 8 2D plots for CO_2 Emission, data – blue, $n = 123$, (**a**) Regular Q_R (in dash) and direct Q_D (in solid) of the CO_2 emission vs ln(GDP) when the country's E.C. is 2980.96 kilowatts at $\tau = 0.97$ (green) and 0.99 (red). (**b**) Regular Q_R (in dash) and direct Q_D (in solid) of the CO_2 emission vs ln(E.C.) when the country's GDP is \$13,359.73 at $\tau = 0.97$ (green) and 0.99 (red)

Table 4 CO_2 Emission per capita at high quantiles given ln(GDP) estimators Q_R and Q_D

	$\tau = 0.97$	
ln of GDP per capita ($)	Q_R	Q_D
7.5	15.2181	8.8737
8	18.0437	10.1949
8.5	20.8693	11.7828
9	23.6950	14.4143
9.5	26.5206	19.0458
10	29.3462	24.0338
10.5	32.1718	27.9596
11	34.9975	31.1097
11.5	37.8231	30.7696
12	40.6487	31.2366

2980.96 Kilowatts of Electricity Consumed per capita

$Q_D(\tau|\mathbf{x})$ in (7). We compare the new estimator $Q_D(\tau|\mathbf{x})$ with the regular quantile estimator $Q_R(\tau|\mathbf{x})$ in Huang and Nguyen (2017). Figures 7, 8 and Tables 4, 5 show the differences of the values of two estimators. Figure 7a shows the 3D scatter plot of CO_2 emission vs ln(GDP) and ln(EC) with the fitted regular Q_R surface at $\tau = 0.97$. Figure 7b shows the 3D scatter plot of CO_2 emission vs ln(GDP) and ln(EC) with the fitted direct Q_D surface at $\tau = 0.97$. Figure 7c shows the 3D scatter plot with both the regular Q_R (green) and direct Q_D (red) quantile surfaces of CO_2 emission vs the ln(GDP) and ln(E.C.) at $\tau = 0.97$. It is interesting to see the difference between the Q_R and Q_D quantile surfaces.

We may see the Q_R and Q_D quantile curves more cleanly in 2D plots. Figure 8a shows the 2D scatter plot of CO_2 emission vs ln(GDP) when the country's E.C. is 2980.96 kilowatts with the fitted regular Q_R and direct Q_D curves at at $\tau = 0.97$. Figure 8b shows the 2D scatter plot of CO_2 emission vs ln(E.C.) when the country's GDP is $13,359.73 with the fitted regular Q_R and direct Q_D curves at at $\tau = 0.97$. We note that the Q_R and Q_D quantile regression curves appear to fit the data. In general, the Q_D curves follow the data patterns closer than Q_R quantile lines, and the Q_D produces different estimated CO_2 emissions than the Q_R estimated at high quantiles. In Fig. 7, it is interesting to see that the Q_D conditional quantile surfaces are not linear as the linear planes of the Q_R.

Tables 4 and 5 provide details of the estimated high quantiles about countries' CO_2 emission at $\tau = 0.97$ when the countries consume 2980.96 kilowatts of electricity and have a GDP of $13,359.73, respectively.

Table 5 CO_2 emission per capita at high quantiles given ln(E.C.) estimators Q_R and Q_D

ln of Electricity Consumption	$\tau = 0.97$	
per capita (kilowatts)	Q_R	Q_D
0	6.9775	7.1919
2	11.8632	7.2759
4	16.7490	24.6924
6	21.6348	9.5560
8	26.5206	15.9569
10	31.4064	31.5634
12	36.2921	39.6481

GDP per capita of 13, 359.73

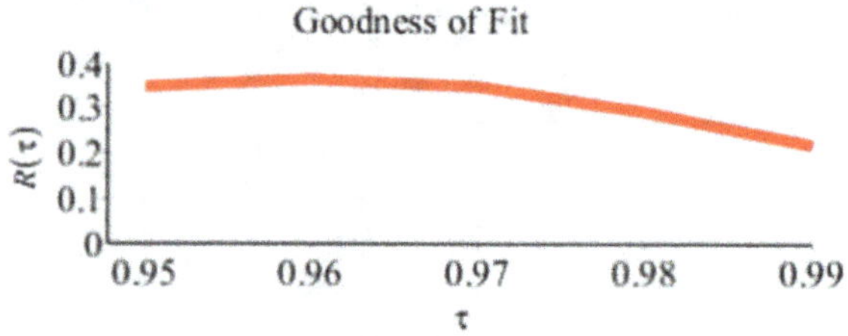

Fig. 9 Relative $R(\tau)$ of Q_D relative to Q_R for the CO_2 emission example

Figure 9 and Table 6 show the Relative $R(\tau)$ in (9), for $\tau = 0.95, \ldots, 0.99$. All values of Relative $R(\tau)$ are larger than 0, which signifies that $V_D(\tau) < V_R(\tau)$ and it also suggests that the direct quantile regression estimator Q_D is a better fit to the CO_2 emission data than the regular quantile regression estimator Q_R.

Over all, it is interesting to see that the proposed direct estimator Q_D gave more variety of predictions than the Q_R on CO_2 emissions relative to gross domestic product and amounts of electricity produced. The relationships are not necessarily linear and model free. We expect that the predictions from Q_D may be more reasonable. The predictions may benefit prevention of further damages of CO_2 emissions to the environment.

6 Conclusions

After the above studies, we can conclude:

1. This paper proposes a new direct nonparametric quantile regression method which is model free. It uses nonparametric density estimation and nonparametric regression techniques to estimate high conditional quantiles. The paper provides a computational five-step algorithm which overcomes the limitations of the estimation in the linear quantile regression model and some other nonparametric quantile regression methods.

2. The Monte Carlo simulation works on the second kind of Gumbel's bivariate exponential distribution which has a nonlinear conditional quantile function. The simulation is different from the bivariate Pareto distribution which has a linear conditional quantile function, in Huang and Nguyen (2017). The simulation results confirm that the proposed new method is more efficient relative to the regular quantile regression estimators and a local polynomial nonparametric estimator.

3. The proposed new direct nonparametric quantile regression can be used to predict extreme values of snowfall and CO_2 emission examples in Huang and Nguyen (2017). The proposed direct quantile regression Q_D estimator gives a variety of predictions which fits data very well. The prediction of relationships are not simply just linear. We expect that the predictions from Q_D may be more reasonable than the regular quantile regression predictions. The new estimator may benefit prevention of further damages of the extreme events to human and the environment.

4. The proposed direct nonparametric quantile regression provides an alternative way for quantile regression. Further studies on the details of this method are suggested.

Table 6 Relative $R(\tau)$ values for CO_2 emission example

	$\tau = 0.95$	$\tau = 0.96$	$\tau = 0.97$	$\tau = 0.98$	$\tau = 0.99$
Relative $R(\tau)$	0.3480	0.3612	0.3494	0.2895	0.2151

Acknowledgements
We are grateful for the comments of the reviewers and editor. They have helped us to improve the paper. This research is supported by *the Natural Science and Engineering Research Council of Canada (NSERC) grant MLH, RGPIN-2014-04621*. We deeply appreciate the work and suggestions of Ramona Rat and Jenny Tieu which helped to improve the paper.

Authors' contributions
The authors MLH and CN carried out this work and drafted the manuscript together. Both authors read and approved the final manuscript.

Competing interests
The authors declare that they have no competing interests.

Author details
[1]Department of Mathematics & Statistics, Brock University, St. Catharines, Ontario L2S 3A1, Canada. [2]Apotex Inc., Toronto, Ontario M9L 1T9, Canada.

References
Carbon Dioxide Information Analysis Center (2017). http://www.cdiac.ornl.gov. Accessed 20 Oct 2014
Cai, Z: Applied Nonparametric Econometrics. Wang Yanan Institute for Studies in Economics, Xiamen University, China (2013)
Chaudhuri, P: Nonparametric estimates of regression quantile and their local Bahadur representation. Ann. Stat. **2**, 760–777 (1991)
Fukunaga, K: Introduction to Statistical Pattern Recognition. Academic press, New York (1972)
Gumbel, EJ: Bivariate exponential distributions. J. Am. Stat. Assoc. **55**, 698–707 (1960)
Hall, P, Wolff, RCL, Yao, Q: Methods for estimating a conditional distribution. J. Am. Stat. Assoc. **94**, 154–163 (1999)
Huang, ML, Nguyen, C: High quantile regression for extreme events. J. Stat. Distrib. Appl. **4**(4), 1–20 (2017)
Huang, ML, Xu, X, Tashnev, D: A weighted linear quantile regression. J. Stat. Comput. Simul. **85**(13), 2596–2618 (2015)
Koenker, R: Quantile regression. Cambridge University Press, New York (2005)
Koenker, R. Package 'guantreg': Quantile Regression (2018). R Package, Version 5.35 (Available from https://www.r-project.org). Accessed 23 Apr 2018
Koenker, R, Bassett, GW: Regression Quantiles. Econometrica. **46**, 33–50 (1978)
Koenker, R, Machado, JAF: Goodness of fit and related inference processes for quantile regression. J. Am. Stat. Assoc. **96**(454), 1296–1311 (1999)
Li, Q, Racine, JS: Nonparametric Econometrics-Theory and Practice. Prinston University Press, Oxford (2007)
National Weather Service Forecast Office (2017). www.weather.gov/buf. Accessed 22 Sept 2014
Scott, DW: Multivariate Density Estimation, Theory, Practice and Visualization, second edition. John Wiley & Sons, New York (2015)
Silverman, BW: Density estimation for statistics and data analysis. Chapman & Hall, London (1986)
Wang, HJ, Li, D: Estimation of extreme conditional quantile through power transformation. J. Am. Stat. Assoc. **108**(503), 1062–1074 (2013)
Yu, K, Lu, Z, Stander, J: Quantile regression: applications and current research areas. Statistician. **52**(3), 331–350 (2003)
Yu, K, Jones, MC: Local linear regression quantile regression. J. Am. Stat. Assoc. **93**, 228–238 (1998)

9

Meta analysis of binary data with excessive zeros in two-arm trials

Saman Muthukumarana[1*], David Martell[2] and Ram Tiwari[3]

*Correspondence:
Saman.Muthukumarana@umanitoba.ca
[1] Department of Statistics, University of Manitoba, Machray Hall, Winnipeg, Canada
Full list of author information is available at the end of the article

Abstract

We present a novel Bayesian approach to random effects meta analysis of binary data with excessive zeros in two-arm trials. We discuss the development of likelihood accounting for excessive zeros, the prior, and the posterior distributions of parameters of interest. Dirichlet process prior is used to account for the heterogeneity among studies. A zero inflated binomial model with excessive zero parameters were used to account for excessive zeros in treatment and control arms. We then define a modified unconditional odds ratio accounting for excessive zeros in two arms. The Bayesian inference is carried out using Markov chain Monte Carlo (MCMC) sampling techniques. We illustrate the approach using data available in published literature on myocardial infarction and death from cardiovascular causes. Bayesian approaches presented here use all the data, including the studies with zero events and capture heterogeneity among study effects, and produce interpretable estimates of overall and study-level odds-ratios, over the commonly used frequentist's approaches. Results from the data analysis and the model selection also indicate that the proposed Bayesian method, while accounting for zero events, adjusts for excessive zeros and provides better fit to the data resulting in the estimates of overall odds-ratio and study-level odds-ratios that are based on the totality of the information.

Keywords: Dirichlet process, Model selection, Markov chain Monte Carlo, Simulation

1 Introduction

An arm is a standard term for describing clinical trial and it represents a treatment group or a set of subjects. A two-arm study compares a drug with a placebo or drug A with drug B. Sometimes in these studies, the outcome may be binary. A binary outcome is an outcome whose unit can take on only two possible states "0" and "1". For example, outcomes of clinical trials data such as the morbidity and mortality studies are often binary in nature.

The natural distribution for modeling these types of binary data is the binomial distribution given by

$$f(y;p) = \binom{n}{y} p^y (1-p)^{n-y} \text{ for } y = 0, 1, \ldots, n,\ p \in (0,1).$$

The mean and variance for the binomial random variable are $E(Y) = np$ and $Var(Y) = np(1-p)$ respectively. In a two-arm trial with binary outcomes, it is typically assumed that $Y_{T_1}, ..., Y_{T_k}$ and $Y_{C_1}, ..., Y_{C_k}$ are random samples from $Y_{T_i} \sim Bin\left(n_{T_i}, P_{T_i}\right)$ and $Y_{C_i} \sim Bin\left(n_{C_i}, P_{C_i}\right)$ respectively, where k is the number of studies. In a random effects meta analysis of these types of data, the effect size is assumed to vary from study to study. Random effects meta analysis assumes that study effects are a random sample from an underlying relevant distribution of effects, and the combined effect estimates the mean effect of this distribution.

There are a variety of different approaches to analyze these types of data as indicated by some recent literature. See Albert (1995) for various parametrization of binomial models for discrete data within Bayesian settings. Chang et al. (2001) use a mixed effects model to investigate between and within-study variation using rate difference and logit models. Gamalo et al. (2011) propose a Bayesian procedure for testing noninferiority in two-arm studies with a binary primary endpoint that allows the incorporation of historical data on an active control via the use of informative priors but did not consider excessive zeros. Carlin (1992) consider a Bayesian meta-analysis approach for two way contingency table data while Smith et al. (1995) discuss how a full Bayesian analysis can be used to deal with issues in meta-analysis in a natural way using the BUGS language. In this paper, we consider a Bayesian approach for binary data with excessive zeros in two-arm trials. More specifically, we model the excessive zeros using zero inflated binomial distribution and use the Dirichlet process Ferguson (1974) to handle the heterogeneity among studies. There are various zero inflated methods available in the literature. Hall (2000) introduced the framework for count data with many zeros using Poisson and binomial models and likelihood ratio tests based inference for zero inflated Poisson models are discussed in Huang et al. (2014). A Bayesian inference framework for zero inflated Poisson regression models is discussed in Ghosh et al. (2006). A rich class of nonparametric Bayesian priors for study effects and Bayesian nonparametric Polya tree mixture model are developed in Branscum and Hanson (2008) and Burr and Doss (2005).

In Section 2, we describe Bayesian model specification used in the paper. The likelihood function and the priors are described. Study effects have a Dirichlet process prior distribution for capturing heterogeneity among studies. We then obtain posterior summary statistics which describe key features in the model. In particular, posterior expectations are approximated through Markov chain Monte Carlo (MCMC) methods. In Section 3, the model is applied to a large dataset available in the literature Nissen and Wolski (2007). We perform the model selection using the log-pseudo marginal likelihood (LPML) comparing the Binomial and zero-inflated Binomial (ZIB). The results suggest that when the data has a high percentage of observed zeros, ZIB model is a more appropriate model to use. Furthermore, the use of Dirichlet process has advantage over the more commonly used random effects model with normally distributed random effects based on DerSimonian-Laird approach DerSimonian and Laird (1986) or a Bayesian approach using normal priors, in terms of its inherent clustering property resulting in the studies with similar effects to cluster, and thus providing more robust estimates. We also test the approach using simulation studies in Section 4 and study the effect of excessive zeros in the ZIB models. We conclude with a short discussion in Section 5.

2 Model development

Consider two-arm trials with binary outcomes and let $Y_{T_i} \overset{\text{ind}}{\sim} Bin\left(n_{T_i}, P_{T_i}\right)$ and $Y_{C_i} \overset{\text{ind}}{\sim} Bin\left(n_{C_i}, P_{C_i}\right), i = 1, \ldots, k$, where k is the number of studies. Then the joint likelihood $L = L\left(y_{T_1}, \ldots, y_{T_k}, y_{C_1}, \ldots, y_{C_k} | \mu, P_T, P_C\right)$ is

$$L = \prod_{i=1}^{k} \left\{ {}^{n_{T_i}}C_{y_{T_i}} P_{T_i}^{y_{T_i}} (1 - P_{T_i})^{n_{T_i} - y_{T_i}} \right\} \prod_{i=1}^{k} \left\{ {}^{n_{C_i}}C_{y_{C_i}} P_{C_i}^{y_{C_i}} (1 - P_{C_i})^{n_{C_i} - y_{C_i}} \right\}. \tag{1}$$

In random effects meta-analysis formulation, we assume that P_T and P_C follow logistic models, and define
$P_{T_i} = \frac{exp\{\mu+Tr+\alpha_i+e_i\}}{1+exp\{\mu+Tr+\alpha_i+e_i\}}$ and $P_{C_i} = \frac{exp\{\mu+e_i\}}{1+exp\{\mu+e_i\}}$.
That is, $logit(P_{T_i}) = \mu + Tr + \alpha_i + e_i; logit(P_{C_i}) = \mu + e_i, i = 1, \ldots, k$. This gives,

$$logit(P_{T_i}) - logit(P_{C_i}) = Tr + \alpha_i \,; i = 1, \ldots, k,$$

where $logit(p) = log\left(\frac{p}{1-p}\right)$ is the log-odds ratio of p, μ is the intercept, Tr is the treatment effect, α_i and e_i are the study effects and error terms. As proposed by Muthukumarana and Tiwari in Muthukumarana and Tiwari (2016), consider a Bayesian approach and assume that $\{\alpha_i\,; i = 1, \ldots, k\}$ is a sample from a Dirichlet process with concentration parameter ρ and the baseline distribution H. We assume that the baseline distribution H is $N\left(0, \sigma_H^2\right)$. More specifically, we assume that

$$\begin{aligned} \alpha_i &\sim DP(\rho, H) \\ H &\sim N\left(0, \sigma_H^2\right) \\ e_i &\sim N\left(0, \sigma_e^2\right) \\ f(\mu) &\propto \text{constant} \\ Tr &\sim N\left(0, \sigma_{Tr}^2\right) \\ \rho &\sim U[0.1, 1000] \end{aligned} \tag{2}$$

where hyper parameters $\sigma_H^2, \sigma_{Tr}^2$ and σ_e^2 are assumed to be known. We now obtain the posterior characterizations of parameters using Neal's algorithm Neal (2000) using Gibbs sampling as follows.

$$\begin{aligned} f\left(\alpha_c | y_{T_j} : c_j = c\right) \propto & \prod_{j:c_j=c}^{k} \left(\frac{1}{1 + exp\left\{\mu + Tr + \alpha_c + e_j\right\}} \right)^{n_{T_j}} \\ & exp\left\{ \frac{-1}{2\sigma_H^2} \left(\alpha_c - \sigma_H^2 \sum_{j:c_j=c} y_{T_j} \right)^2 \right\} \end{aligned} \tag{3}$$

$$\begin{aligned} f\left(e_i | \underline{y}\right) \propto & \left(\frac{1}{1 + exp\{\mu + Tr + \alpha_i + e_i\}} \right)^{n_{T_i}} \left(\frac{1}{1 + exp\{\mu + e_i\}} \right)^{n_{C_i}} \\ & exp\left\{ \frac{-1}{2\sigma_e^2} \left(e_i - \sigma_e^2 \left(y_{T_i} + y_{C_i}\right)\right)^2 \right\} \end{aligned} \tag{4}$$

$$\begin{aligned} f\left(\mu | \underline{y}\right) \propto & \prod_{j=1}^{k} \left(\frac{1}{1 + exp\left\{\mu + Tr + \alpha_j + e_j\right\}} \right)^{n_{T_j}} \left(\frac{1}{1 + exp\left\{\mu + e_j\right\}} \right)^{n_{C_j}} \\ & exp\left\{ \left(\sum_{j=1}^{k} y_{T_j} + y_{C_j} \right) \mu \right\} \end{aligned} \tag{5}$$

$$f\left(Tr|\underline{y}\right) \propto \prod_{j=1}^{k}\left(\frac{1}{1+exp\left\{\mu+Tr+\alpha_j+e_j\right\}}\right)^{n_{T_j}} \\ exp\left\{\frac{-1}{2\sigma_{Tr}^2}\left(Tr-\sigma_{Tr}^2\sum_{j=1}^{k}y_{T_j}\right)^2\right\} \tag{6}$$

$$f\left(\rho|\underline{y}\right) \propto \rho^{r-1}\left(\rho+k\right)B\left(\rho+1,k\right)I_{[0.1,1000]}\left(\rho\right) \tag{7}$$

Note that the likelihood in (1) does not account for excessive zeros in the data. For this reason, we now consider a zero inflated binomial model for the data as follows.

$$Y_{T_i} \overset{\text{ind}}{\sim} ZIB\left(p_0, n_{T_i}, P_{T_i}\right), \;\; Y_{C_i} \overset{\text{ind}}{\sim} ZIB\left(q_0, n_{C_i}, P_{C_i}\right), i=1,\ldots,k.$$

That is,

$$Y_{T_i} = \begin{cases} 0 & \text{with probability } p_0 \\ Bin\left(n_{T_i}, P_{T_i}\right) & \text{with probability } 1-p_0. \end{cases}$$

Similarly,

$$Y_{C_i} = \begin{cases} 0 & \text{with probability } q_0 \\ Bin\left(n_{C_i}, P_{c_i}\right) & \text{with probability } 1-q_0. \end{cases}$$

This modification brings two more extra parameters to the model and we assume that

$$\begin{aligned} p_0 &\sim Beta\left(a,b\right) \\ q_0 &\sim Beta\left(c,d\right). \end{aligned} \tag{8}$$

where hyper parameters a, b, c and d are assumed to be known. We obtain the the posterior characterizations of parameters under zero inflated binomial likelihood as follows.

$$f\left(\alpha_c|y_{T_j}:c_j=c\right) \propto \prod_{j:c_j=c}^{k}\left[p_0+\left(1-p_0\right)\left(\frac{1}{1+exp\left\{\mu+Tr+\alpha_c+e_j\right\}}\right)^{n_{T_j}}\right]^{u_j} \\ \left[\left(1-p_0\right)\left(\frac{1}{1+exp\left\{\mu+Tr+\alpha_c+e_j\right\}}\right)^{n_{T_j}} exp\left\{\alpha_c y_{T_j}\right\}\right]^{1-u_j} \\ exp\left\{\frac{-1}{2\sigma_H^2}\alpha_c^2\right\} \tag{9}$$

$$f\left(e_i|\underline{y}\right) \propto \left[p_0+\left(1-p_0\right)\left(\frac{1}{1+exp\left\{\mu+Tr+\alpha_i+e_i\right\}}\right)^{n_{T_i}}\right]^{u_i} \\ \left[\left(1-p_0\right)\left(\frac{1}{1+exp\left\{\mu+Tr+\alpha_i+e_i\right\}}\right)^{n_{T_i}} exp\left\{y_{T_i}e_i\right\}\right]^{1-u_i} \\ \left[q_0+\left(1-q_0\right)\left(\frac{1}{1+exp\left\{\mu+e_i\right\}}\right)^{n_{C_i}}\right]^{w_i} \\ \left[\left(1-q_0\right)\left(\frac{1}{1+exp\left\{\mu+e_i\right\}}\right)^{n_{C_i}} exp\left\{y_{C_i}e_i\right\}\right]^{1-w_i} \\ exp\left\{\frac{-1}{2\sigma_e^2}e_i^2\right\} \tag{10}$$

$$f\left(\mu|\underline{y}\right) \propto \prod_{j=1}^{k}\left[p_0+(1-p_0)\left(\frac{1}{1+exp\left\{\mu+Tr+\alpha_j+e_j\right\}}\right)^{n_{T_j}}\right]^{u_j}$$
$$\left[(1-p_0)\left(\frac{1}{1+exp\left\{\mu+Tr+\alpha_j+e_j\right\}}\right)^{n_{T_j}} exp\left\{y_{C_j}\mu\right\}\right]^{1-u_j}$$
$$\left[q_0+(1-q_0)\left(\frac{1}{1+exp\left\{\mu+e_j\right\}}\right)^{n_{C_j}}\right]^{w_j} \tag{11}$$
$$\left[(1-q_0)\left(\frac{1}{1+exp\left\{\mu+e_j\right\}}\right)^{n_{C_j}} exp\left\{y_{C_j}\mu\right\}\right]^{1-w_j}$$

$$f\left(Tr|\underline{y}\right) \propto \prod_{j=1}^{k}\left[p_0+(1-p_0)\left(\frac{1}{1+exp\left\{\mu+Tr+\alpha_j+e_j\right\}}\right)^{n_{T_j}}\right]^{u_j}$$
$$\left[(1-p_0)\left(\frac{1}{1+exp\left\{\mu+Tr+\alpha_j+e_j\right\}}\right)^{n_{T_j}} exp\left\{y_{T_j}Tr\right\}\right]^{1-u_j} \tag{12}$$
$$exp\left\{\frac{-1}{2\sigma^2_{Tr}}Tr^2\right\}$$

$$f\left(\rho|\underline{y}\right) \propto \rho^{r-1}\left(\rho+k\right)B\left(\rho+1,k\right)I_{[0.1,1000]}(\rho) \tag{13}$$

$$f\left(p_0|\underline{y}\right) \propto \left[p_0+(1-p_0)\left(\frac{1}{1+exp\left\{\mu+Tr+\alpha_j+e_j\right\}}\right)^{n_{T_j}}\right]^{u_j}$$
$$\left[(1-p_0)\left(\frac{1}{1+exp\left\{\mu+Tr+\alpha_j+e_j\right\}}\right)^{n_{T_j}} exp\left\{y_{T_j}\left(\mu+Tr+\alpha_j+e_j\right)\right\}\right]^{1-u_j}$$
$$p_0^{a-1}\left(1-p_0\right)^{b-1} \tag{14}$$

$$f\left(q_0|\underline{y}\right) \propto \left[q_0+(1-q_0)\left(\frac{1}{1+exp\left\{\mu+e_j\right\}}\right)^{n_{C_j}}\right]^{w_j}$$
$$\left[(1-q_0)\left(\frac{1}{1+exp\left\{\mu+e_j\right\}}\right)^{n_{C_j}} exp\left\{y_{C_j}\left(\mu+e_j\right)\right\}\right]^{1-w_j} \tag{15}$$
$$q_0^{c-1}\left(1-q_0\right)^{d-1}$$

where $u_j = \begin{cases} 1, & y_{T_j}=0 \\ 0, & y_{T_j}=1 \end{cases}$ and $w_j = \begin{cases} 1, & y_{C_j}=0 \\ 0, & y_{C_j}=1. \end{cases}$

We investigate the suitability of the zero inflated binomial distribution using the log pseudo marginal likelihood (LPML) Gelfand et al. (1992) in Section 4.

3 Data analysis

We illustrate the approach discussed in Section 2 using a published data set on counts of the number of people experiencing myocardial infarction from the use of drugs with an active ingredient "rosiglitazone" Nissen and Wolski (2007). The data used in this section provides information on diabetes patients, 42 diabetes trials having zero events in both arms, and possible heart condition or death resulting from the use of rosiglitazone. Rosiglitazone is a treatment used to treat patients with type 2 diabetes. The data provide information on diabetes patients, 42 diabetes trials, and possible heart condition or death resulting from the use of rosiglitazone. Rosiglitazone is a treatment for diabetes widely used in treating patients with type 2 diabetes. We separately apply the model on myocardial infarction and death from cardiovascular based on these 42 studies. There were 86 myocardial infarctions in the rosiglitazone group and 72 in the control group. There were 39 deaths from cardiovascular causes in the rosiglitazone group and 22 in the control group. Note that the percentages of observed zeros from the 42 studies in the treatment and control arms for myocardial infarction are 23% and 57% respectively. Similar percentages for cardiovascular causes are 50% and 80% respectively. We set the hyper parameters as $a = b = c = d = 1, \sigma_H^2 = 2, \sigma_{Tr}^2 = 2$ and $\sigma_e^2 = 2$. Note that the choice of these values result sufficiently diffuse priors in the range of logit scale of primary parameters. We implement the models developed in Section 2 using R. The results are based on a MCMC simulation with a burn-in period of 1000 iterations followed by 30,000 iterations using thinning of 5. We use the data from the 42 studies, without stratifying them into small and large studies, as the purpose of the proposed work is an illustration of the method and not in in-depth analysis of the data by using different methods or by slicing and dicing the data. The posterior box plots of study effects under two models on myocardial infarction are given in Figs. 1 and 2. The advantage of using DP prior is the flexibility and also the ability to cluster studies appropriately. The clustering is based on the values assigned to each study effects based on their posterior distributions, which are approximated using MCMC. Those studies that share the same study effects will be considered to belong to the same group. Note that there were 5 clusters in myocardial infarction and 4 clusters in cardiovascular causes based on study effects. To evaluate the performance between Binomial and ZIB models, we use the LPML which is based on Conditional Predictive Ordinates(CPO). A detailed discussion of the CPO statistic and its applications to model selection can be found in Geisser (1993) and Gelfand and Dey (1994). The LPML is computed as $\sum_{i=1}^{k} logp(y_i|y_{-i})$ where y_{-i} denotes the observation vector y with the ith

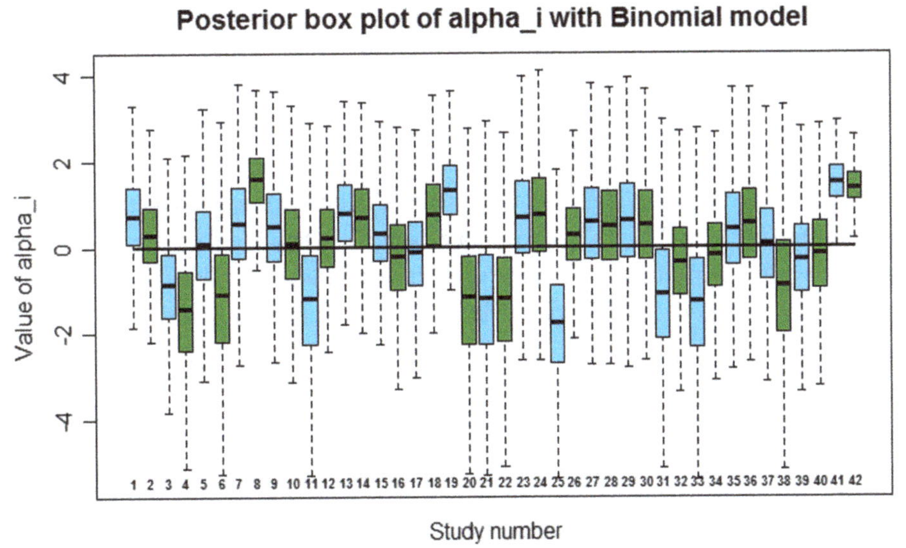

Fig. 1 Posterior summaries of study effects on Myocardial Infarction

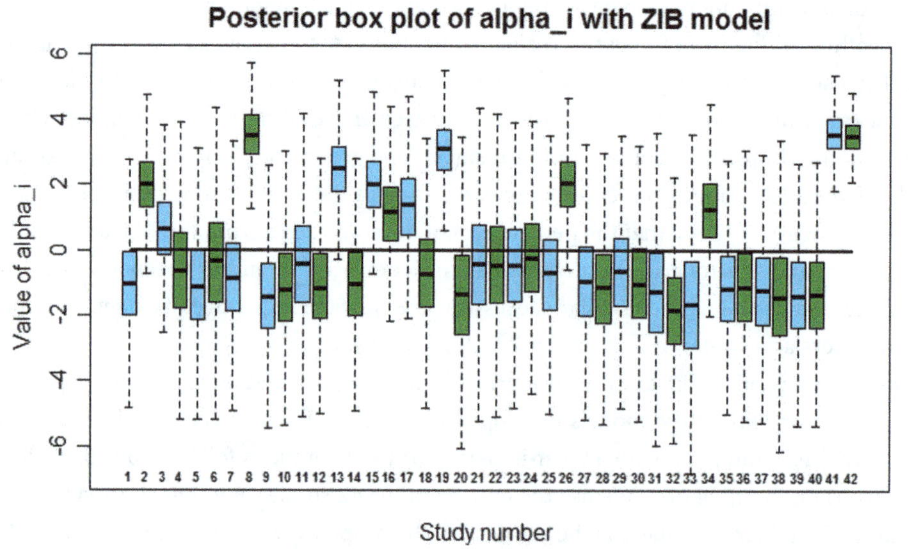

Fig. 2 Posterior summaries of study effects on Myocardial Infarction

observation deleted. The model with larger value of LPML is preferred. The estimates of μ, Tr and the LPML values are given in Table 1. The LPML prefers binomial model over the ZIB model and the two models estimate the parameter Tr differently.

We now investigate the study effects on death from cardiovascular causes. The posterior box plots of study effects under two models on death from cardiovascular causes are given in Figs. 3 and 4. The plots indicate that ZIB model is capable in capturing the heterogeneity of study effects. The estimates of μ, Tr and the LPML values are given in Table 1. In this case, the LPML strongly prefers ZIB model over the binomial model. This is in agreement with the fact that there are large amount of excessive zeros on death from cardiovascular causes relative to myocardial infarction.

A summary of estimates of odds ratios under Binomial, zero inflated Binomial and DerSimonian- Laird random effects models are given in Figs. 5 and 6. For myocardial infarction, DerSimonian- Laird random effects model gives an overall odds ratio of 1.29 with a 95% confidence interval of (0.9, 1.85). On the other hand, Binomial and zero inflated Binomial models provide an overall summary of odds ratio of 1.04 (0.98, 1.1) and 1.07 (0.97, 1.17) respectively. These estimates and 95% credible intervals for cardiovascular causes are 1.2 (0.64, 2.24), 1.03 (0.97, 1.09) and 1.13 (0.93, 1.33) respectively. It is clear that our approach provides overall odds ratios estimates that are slightly lower than that from DerSimonian- Laird overall estimate. Note that DerSimonian-Laird approach is based on the non-zero studies. Also note that Binomial and zero inflated Binomial models identify more heterogeneous study effects than DerSimonian- Laird random effects

Table 1 Parameter estimates with each model along with LPML

	Myocardial infarction		Cardiovascular causes	
Parameter	Binomial model	ZIB model	Binomial model	ZIB model
μ	0.0394 (0.0277)	0.0709 (0.0503)	0.0709 (0.0503)	0.1235 (0.0876)
Tr	-1.1989 (0.3945)	-3.3612 (0.4870)	-3.3612 (0.4870)	-4.5339 (0.5707)
LPML	-173.5474	-179.7584	-156.3964	-125.0447

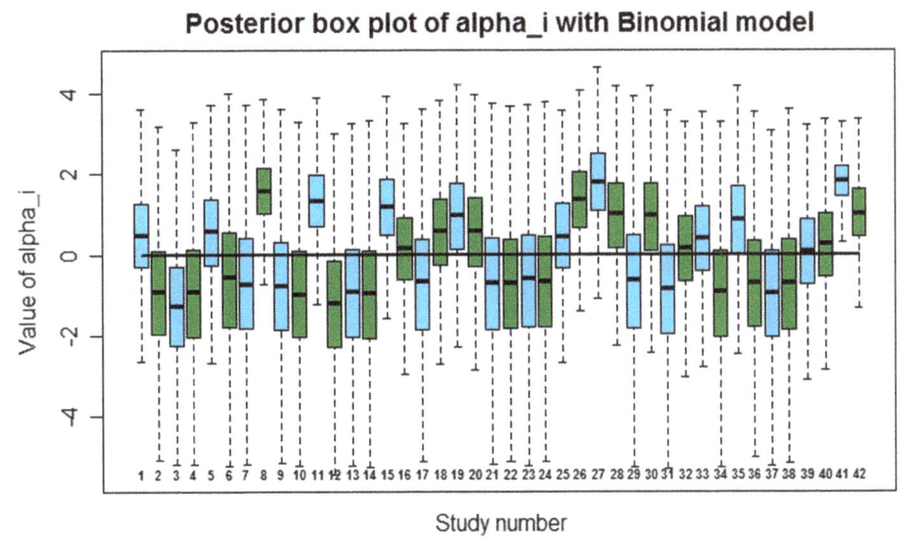

Fig. 3 Posterior summaries of study effects on death from cardiovascular causes

model. According to Figs. 5 and 6, we notice that zero inflated Binomial model identifies more heterogeneous effects than Binomial model while Binomial model identifies more heterogeneous effects than DerSimonian- Laird approach. DerSimonian- Laird estimated random effects variances are zero for both scenarios and this suggests that our approach is superier than DerSimonian- Laird random effects model when there is heteregeniety among studies and LPML model selection criteria will choose the best model in terms of prediction ability.

We now examine the effects of zero inflated parameters p_0 and q_0 on the analysis. The graphical posterior summaries of p_0 and q_0 on myocardial infarction and cardiovascular causes are given in Figs. 7, 8, 9 and 10. In addition, the numerical posterior summaries of p_0 and q_0 are given in Table 2. It is clear that the posterior distributions of p_0 and q_0

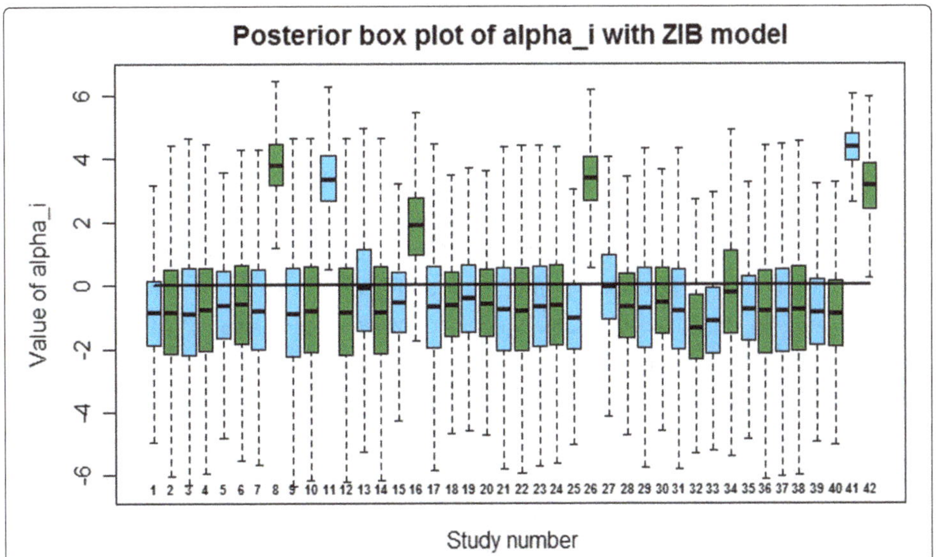

Fig. 4 Posterior summaries of study effects on death from cardiovascular causes

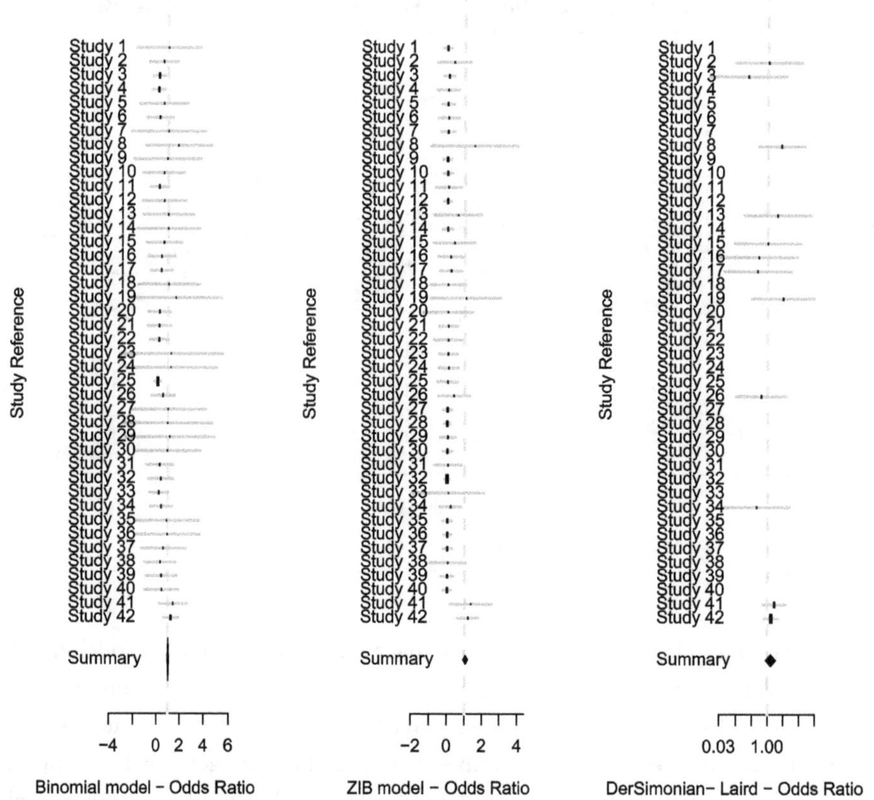

Fig. 5 95% C.I. for odds ratios on Myocardial Infarction

and their numerical summaries for myocardial infarction and cardiovascular causes make sense with respect to the percentages of zeros in the data. We also consider a $Beta(0.5, 0.5)$ prior on p_0 and q_0 in order to investigate the prior sensitivity. The numerical posterior summaries of p_0 and q_0 under $Beta(0.5, 0.5)$ prior are given in Table 3. We notice a magnitude change in estimates of p_0 and q_0 in this case but the estimates of primary parameters μ and Tr are very close indicating that odds ratios are not sensitive to the choice of prior settings. This indicates that inference on p_0 and q_0 will be sensitive to the choice of priors so one should select these priors carefully based on application specific apriori knowledge on zero inflated parameters.

It is important to look at some convergence assessment plots related to the MCMC simulation as this is a high dimensional problem. The trace plot, histogram and autocorrelation plot of μ under binomial model on Myocardial Infarction are given in Fig. 11. The trace plot appears to stabilize immediately and hence provides no indication of lack of convergence in the Markov chain. The autocorrelation plot also appears to dampen quickly. Trace plots of study effects on Myocardial Infarction are given in Fig. 12. The trace plots of study effects on death from cardiovascular causes indicate similar behavior. Similar plots were obtained for all of the parameters under each model and provide the evidence of the convergence of the Markov chains.

Note that one can also assign a simpler parametric normal prior on study effects α_i in place of the DP prior. We now re-analyze the data assuming that study effects are arising from a $N\left(0, \sigma_H^2\right)$ prior distribution. We remark that this is the baseline distribution of the

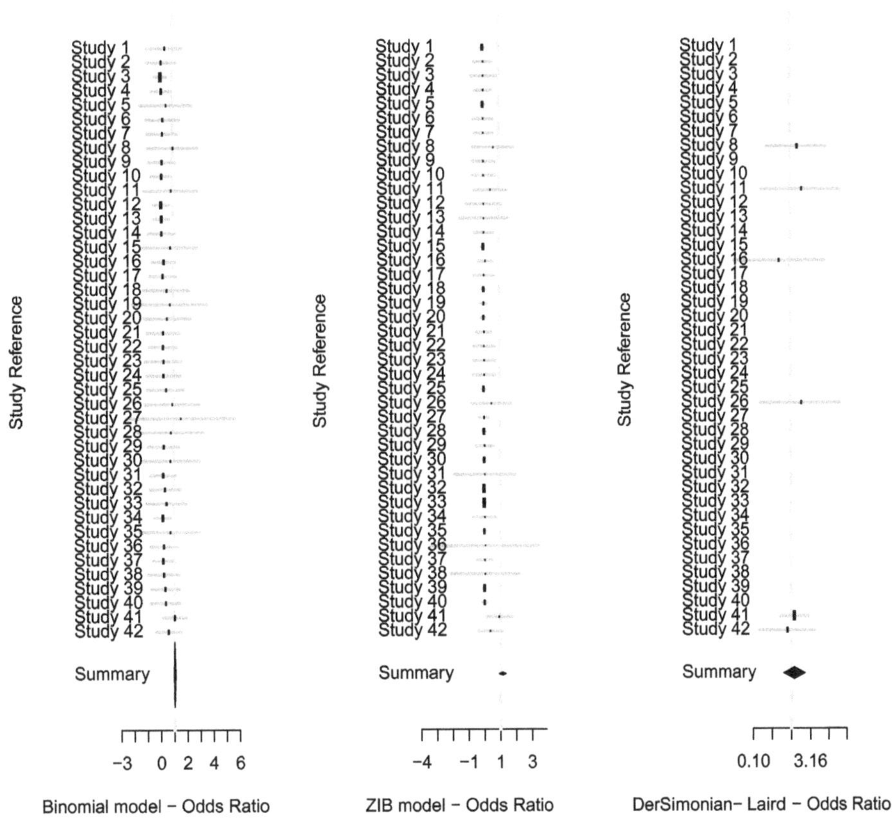

Fig. 6 95% C.I. for odds ratios on death from cardiovascular causes

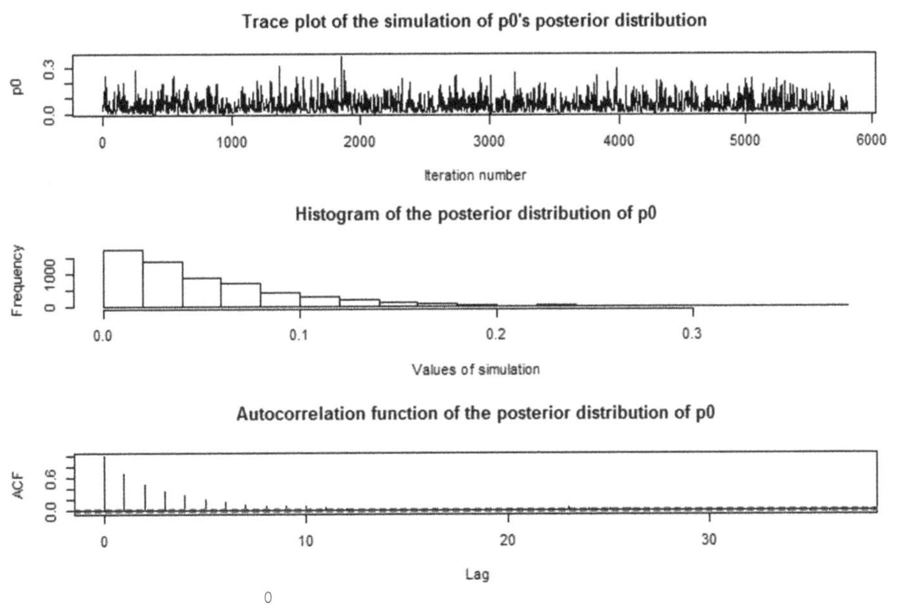

Fig. 7 Posterior summaries of p^0 on Myocardial Infarction

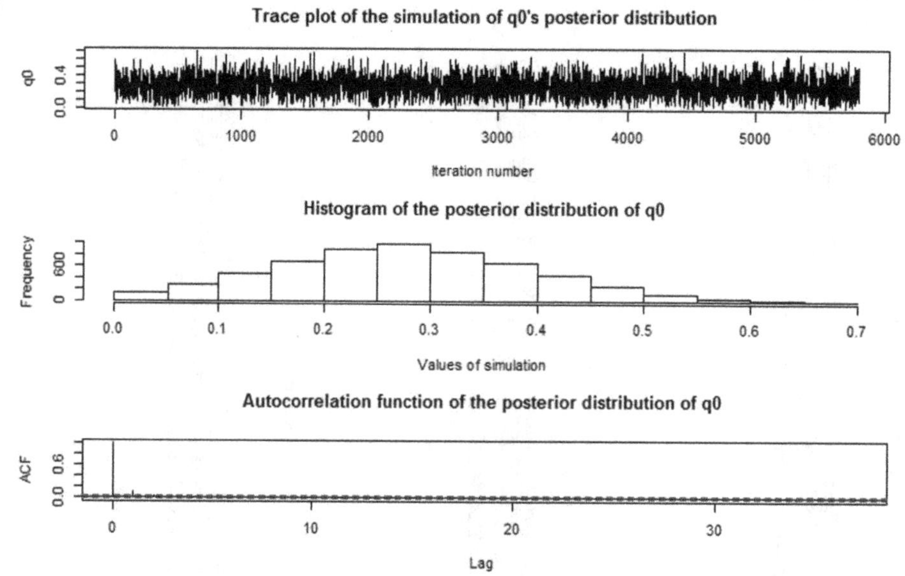

Fig. 8 Posterior summaries of q_0 on Myocardial Infarction

DP prior in (2). In this case, forest plots of odds ratios for each model are given in Fig. 13. The estimates of primary parameters of interest and LPML values are given in Table 4. The LPML model selection criteria clearly indicates that the DP prior in (2) is superior than the conventional parametric prior.

Note that the overall decision to assess the safety should be based on p_0, q_0 and the overall odds ratio (OR). For example, the treatment can be declared is to be safer than the control, if $OR \leq 1$, and $p_0 > q_0$. Also notice that estimates of (p_0, q_0) are independent of the odds ratio because the counts cannot be in "true" zero arms and "Binomial" arms. We combine the two metrics, conditional OR and (p_0, q_0), to come up with an overall

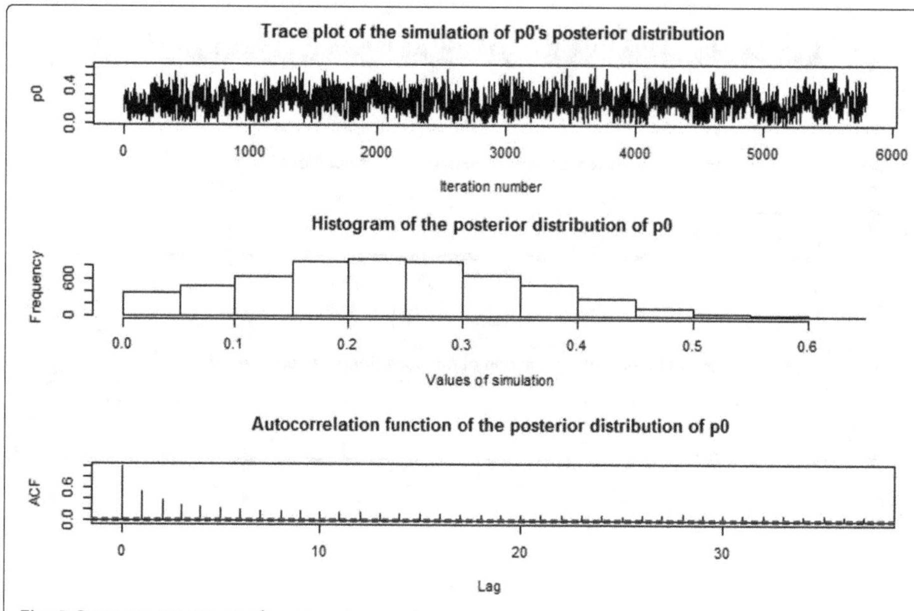

Fig. 9 Posterior summaries of p_0 on cardiovascular causes

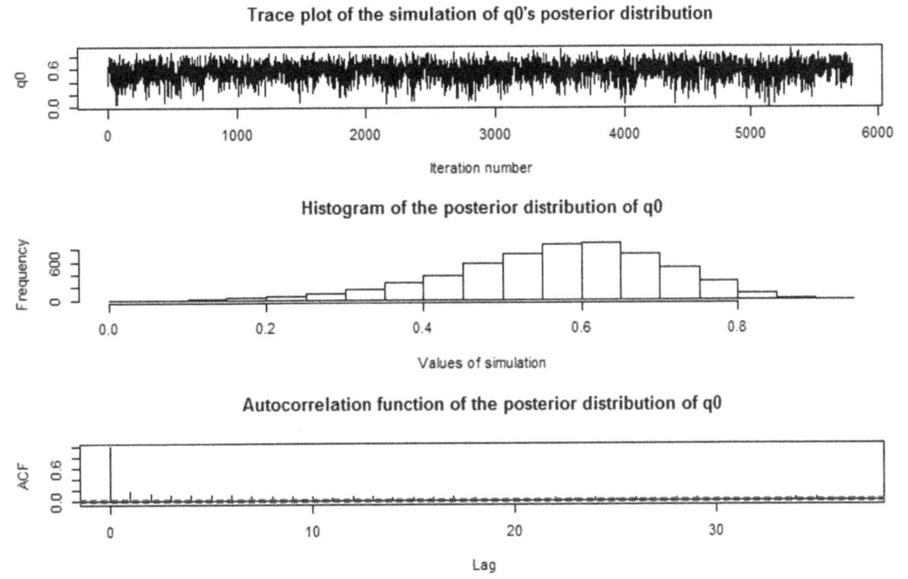

Fig. 10 Posterior summaries of q_0 on cardiovascular causes

unconditional odds ratio. We define it to be modified odds ratio= OR$\times(1-p_0)/(1-q_0)$. Note that when $p_0 = q_0$, modified odds ratio is same as OR. If $p_0 > q_0$, this adjusts OR, by multiplying by a factor less than 1, and if $p_0 < q_0$, it adjust OR by multiplying by a factor > 1. This factor, $h(p_0, q_0) = (1-p_0)/(1-q_0)$ is the ratio of probabilities of observing Bernoulli counts in the two arms, and can be considered as odds for observing Bernoulli counts in the two arms. In frequentist setup, $h(\hat{p}_0, \hat{q}_0)$ is independent of $\hat{\mu}$, and hence independent of conditional odds ratio. In fact, $\hat{p}_0$ and $\hat{q}_0$ converge to p_0 and q_0 with probability 1, and hence $h(\hat{p}_0, \hat{q}_0)$ also converge to $h(p_0, q_0)$ with probability 1, as h is a continuous function (from Slutsky's theorem). So, the estimated modified odds ratio is a consistent estimator for unconditional odds ratio defined as $OR \times (1-p_0)/(1-q_0)$. We provide the estimates of the modified odds ratio for various models in Table 5. As estimate of p_0 is less than q_0 for both examples (Myocardial Infarction and cardiovascular causes), the modified OR values are higher than the corresponding OR values.

4 Results from simulation studies

To understand the role of p_0 and q_0 in the model, different simulation studies were carried out. For this purpose, we generate random ZIB values with empirical binomial parameters. We first generate 42 pairs of independent binary, 0 and 1, variables from Bernoulli (p_0) and Bernoulli (q_0) where p_0 and q_0 are from the set of values $\{(0.1, 0.1), ..., (0.9, 0.9)\}$. We then assign the true-zeros at the places with 1s, and generate binomial outcomes from $B(\bar{n}_T, \hat{P}_{T_i})$ and from $B(\bar{n}_C, \hat{P}_{C_i})$, where $\bar{n}_T, \bar{n}_C, \hat{P}_{T_i}$ and $\hat{P}_{C_i}$ are empirical estimates. Then, MCMC sampling scheme described in Section 2 was carried out using R to obtain the

Table 2 Posterior mean and standard deviation (in parentheses) of p_0 and q_0

Parameter	Myocardial infarction	Cardiovascular causes
p_0	0.0495 (0.044)	0.231 (0.117)
q_0	0.27 (0.118)	0.566 (0.137)

Table 3 Posterior mean and standard deviation (in parentheses) of p_0 and q_0 under $Beta(0.5, 0.5)$ prior distribution

Parameter	Myocardial infarction	Cardiovascular causes
p_0	0.0254 (0.036)	0.179 (0.125)
q_0	0.247 (0.125)	0.538 (0.161)
μ	0.073 (0.051)	0.118 (0.084)
Tr	-3.357 (0.444)	-4.707 (0.632)

posterior estimate of p_0 and q_0. This was done 1000 times for each pair to obtain the mean and standard error of each estimate. For various scenarios of excessive zeros, the results are given in Table 6. The results indicate that when true values of p_0 is small and the observed values of zeros in the simulated data in treatment arm (control arm) is also small (large), the estimated values of p_0 and q_0 are also small (large), whereas when the values of p_0 and q_0 are large the simulated data has large proportion of zeros in both the arms, this results in large estimated values of p_0 and q_0. In both the situations, the estimated values of p_0 and q_0 are in conformity with the observed percentages of zeros in the simulated data. The estimates of p_0 and q_0 remain high in spite of their true choices from the parameter values. Note that our primary interest is on alphas and on treatment arm not on the control arm, so we may not need to investigate q0 very well as it can be trated as nuisance parameter. In practice, one should have a very good apriori knowledge of q0 which will allow to assign an informative prior as it is reflecting the zeros in the control arm. This indicates that the use of ZIB is more appropriate when there are excessive zeros in the data.

5 Discussion

Binary data naturally arise in clinical trials in health sciences. In some cases, they arise with excessive zeros. In this paper, we have provided a random effects meta analysis

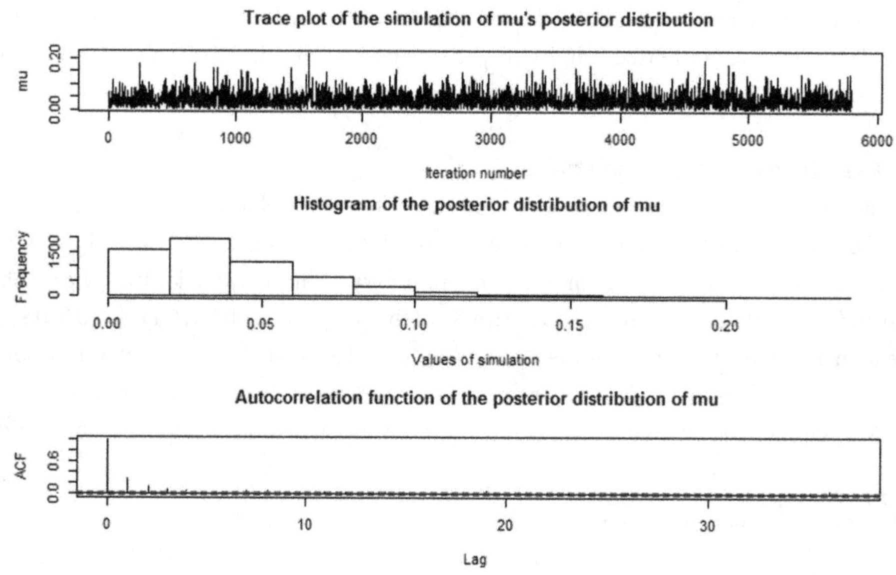

Fig. 11 The trace plot, histogram and autocorrelation plot of μ under binomial model on Myocardial Infarction

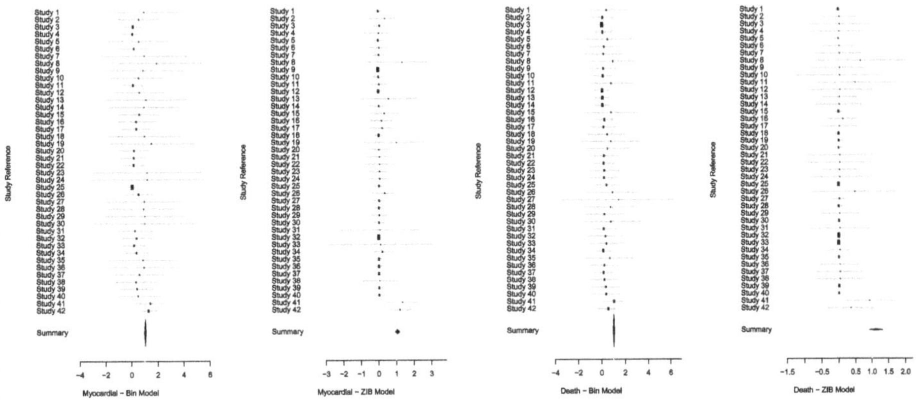

Fig. 12 Trace plots of some study effects on Myocardial Infarction under binomial model

approach for binary data with excessive zeros in two-arm trials. The suitability of the binomial and zero inflated binomial model was assessed in the presence of Dirichlet process as the prior for the study effects. The approach can be used as a template for meta analysis of binary data and a user may choose the proper model using log pseudo marginal likelihood. We have shown that our approach is superior than DerSimonian- Laird random

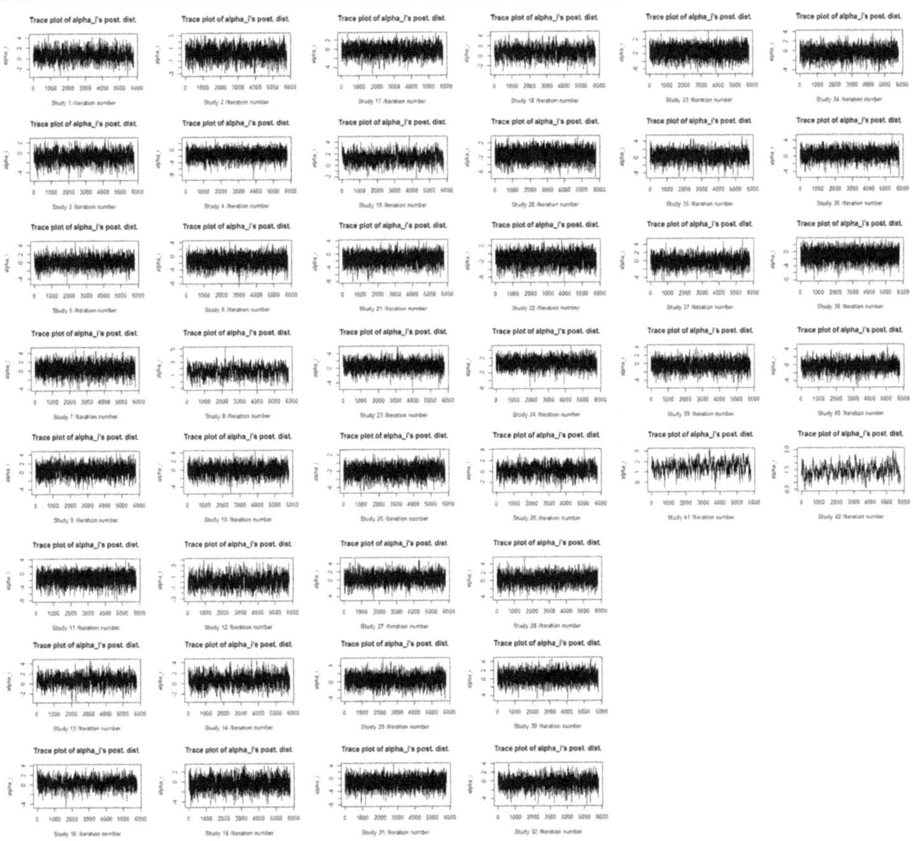

Fig. 13 Forest plots of odds ratios for models under normal prior on study effects

Table 4 Parameter estimates for various models under $N\left(0, \sigma_H^2\right)$ prior on study effects

Model	μ	Tr	LPML	Overall odds ratio with 95% C.I.
Myocardial - Bin	0.039 (0.028)	-1.148 (0.369)	-179.5883	1.04 (0.979, 1.101)
Myocardial - ZIB	0.073 (0.052)	-3.311 (0.435)	-182.5765	1.07 (0.961, 1.194)
Cardiovascular - Bin	0.036 (0.025)	-1.892 (0.427)	-161.3744	1.04 (0.983, 1.094)
Cardiovascular - ZIB	0.123 (0.089)	-4.585 (0.568)	-125.5402	1.13 (0.917, 1.356)

effects model when there is heterogeneity among studies and LPML model selection criteria can be used to selection the best model among the Bayesian models (not including DerSimonian-Laird model) for a given data set.

The Bayesian approaches discussed in this paper allowed to incorporate the zero-studies in the likelihood, and we found that the point estimates of the overall odds-ratio from these methods, were lower than the estimates reported in the literature Nissen and Wolski (2007). The use of ZIB model was to identify the percentage of excessive zeros, that is, the studies where the events could not occur, from the (Binomially) modeled zeros where the zero events occurred. Note that under ZIB, some zeros are observed with probability p_0 and some from Binomial model, making the probability of zero-event to be $p_0 + (1 - p_0)(1 - P_T)^{n_T}$ in the treatment arm. With the use of ZIB model, the Bayes estimates of the odds-ratio went slightly up than with the use of Binomial model, but still they were lower than the results from DerSimonian-Laird random effects model and the resulting estimates in Nissen and Wolski (2007). Note also that DP model being discrete with probability 1, has a clustering property, where the study effects, that are alike, fall in the same cluster. We also investigated the suitability of the DP prior over the conventional parametric normal prior on study effects. The LPML model selection indicated that DP prior is superior than the conventional parametric normal prior. Finally, as the results from ZIB model on the parameters p_0, q_0 and OR need to be interpreted together, a modified OR was introduced.

As a future direction of research, we would like to extend the approach discussed in this article for ordinal category data. For example, in some applications, the clinical trial end point could be a response variable in an ordinal scale with multiple categories such as Good/Moderate/Critical etc. This type of ordinal response data can be viewed as multivariate responses arising from continuous latent variables with cut-points. We assume that there is a continuous latent outcome behind these ordinal outcomes such that $X_i = (X_{i1}, \ldots, X_{im})' \sim \text{Normal}(\mu, \Sigma)$ where X's are the latent outcomes and m is the number of ordinal categories. Then the latent variables X_{ij}'s can be converted to the observed Y_{ij} using a cut-point vector λ. However the choice of cut-points and their priors need to be carefully selected as there are two arms and the counts on categories could be

Table 5 Modified odds ratios, standard deviations (in parentheses) and credible intervals under DP and normal priors

Model	Modified odds ratio	95% credible interval
Myocardial - DP Prior	1.448 (0.277)	(1.05, 2.11)
Myocardial - Normal Prior	1.446 (0.275)	(1.05, 2.11)
Cardiovascular - DP Prior	2.209 (0.830)	(1.15, 4.27)
Cardiovascular - Normal Prior	2.224 (0.837)	(1.16, 4.33)

Table 6 Simulation studies for the myocardial infarction data

Initial pair (p_0, q_0)	Mean of the posterior means of p_0 estimates	Standard error of p_0 estimates	Mean of the posterior means of q_0 estimates	Standard error of q_0 estimates
(0.1,0.1)	0.19888206	0.0907763	0.52069774	0.08299897
(0.2,0.2)	0.29206103	0.10530334	0.59059405	0.05131143
(0.3,0.3)	0.33836514	0.09115085	0.62706227	0.06062431
(0.4,0.4)	0.3587838	0.10839263	0.63652	0.06102938
(0.5,0.5)	0.4977337	0.1267577	0.7200964	0.04900527
(0.6,0.6)	0.6180761	0.1255134	0.7705962	0.06610487
(0.7,0.7)	0.6525938	0.1885614	0.8096615	0.07434429
(0.8,0.8)	0.796206	0.09240875	0.8696273	0.06536744
(0.9,0.9)	0.8958989	0.0501579	0.935196	0.03310256

sparse. In this case, one can consider an objective Bayes approach following the development in Bayarri et al. (2008). Yet another extension of the proposed model is where there are multinomial data with some particular cell(s) being observed excessively. This kind of data may arise from trials with patient reported outcomes.

Acknowledgments
The authors thank Editor-in-Chief and three anonymous reviewers whose comments helped to improve the manuscript. This article reflects the views of the authors and should not be attributed to FDA's views or policies.

Authors' contributions
All authors have contributed equally to the work and approved the final version of the paper.

Funding
Muthukumarana's research has been partially supported by a Discovery grant from the Natural Sciences and Engineering Research Council of Canada. Martell's research internship was funded by Mitacs Globalink program.

Competing interests
The authors declare that they have no competing interests.

Author details
[1]Department of Statistics, University of Manitoba, Machray Hall, Winnipeg, Canada. [2]ITAM, Mexico City, Mexico. [3]Office of Biostatistics, Center for Drug Evaluation and Research, Food and Drug Administration, 10903 New Hampshire Ave,Silver Spring, USA.

References

Albert, J.: Teaching Inference about Proportions Using Bayes and Discrete Models. J. Stat. Educ. **3** (1995). https://doi.org/10.1080/10691898.1995.11910494

Bayarri, M. J., Berger, J. O., Datta, G. S.: Objective Bayes testing of Poisson versus inflated poisson models. Inst. Math. Stat. **3**, 105–121 (2008)

Branscum, A. J., Hanson, T. E.: Bayesian nonparametric meta-analysis using Polya tree mixture models. Biometrics. **64**, 825–833 (2008)

Burr, D., Doss, H.: A Bayesian semiparametric model for random-effects meta-analysis. J. Am. Stat. Assoc. **100**, 242–251 (2005)

Carlin, J. B.: Meta-analysis for 2×2 tables: A bayesian approach. Stat. Med. **11**, 141–158 (1992)

Chang, B. H., Waternaux, C., Lipsitz, S.: Meta-analysis of binary data: which within study variance estimate to use? Stat. Med. **20**, 1947–1956 (2001)

DerSimonian, R., Laird, N.: Meta-analysis in clinical trials. Control. Clin. Trials. **7**, 177–188 (1986)

Ferguson, T. S.: Prior distributions on spaces of probability measures. Ann. Stat. **2**, 615–629 (1974)

Gamalo, M., Wu, R., Tiwari, R.: Bayesian approach to noninferiority trials for proportions. J. Biopharm. Stat. **21**, 902–919 (2011)

Geisser, S.: Predictive Inference: An Introduction. Chapman and Hall, London (1993)

Gelfand, A. E., Dey, D. K.: Bayesian Model Choice: Asymptotics and Exact Calculations. J. R. Stat. Soc. Ser. B. **56**, 501–514 (1994)

Gelfand, A. E., Dey, D. K., Chang, H.: Model determination using predictive distributions with implementation via sampling-based methods (with discussion). *Bayesian Statistics 4*(Bernardo, J. M., Berger, J. O., Dawid, A. P., Smith, A. F. M., eds.) Oxford University Press (1992)

Ghosh, S. K., Mukhopadhyay, P., Lu, J. C.: Bayesian analysis of zero-inflated regression models. J. Stat. Plan. Infer. **136**(4), 1360–1375 (2006)

Hall, D. B.: Zero-Inflated Poisson and Binomial Regression with Random Effects: A Case Study. Biometrics. **56**, 1030–1039 (2000)

Huang, L., Zheng, D., Zalkikar, J., Tiwari, R.: Zero-inflated Poisson model based likelihood ratio test for drug safety signal detection. Stat. Methods Med. Res. (2014). https://doi.org/10.1177/0962280214549590

Muthukumarana, S., Tiwari, R.: Meta-analysis using dirichlet process. Stat. Methods Med. Res. **25**(1), 352–365 (2016)

Neal, RM: Markov Chain Sampling Methods for Dirichlet Process Mixture Models. J. Comput. Graph. Stat. **9**(2), 249–265 (2000)

Nissen, S. E., Wolski, K.: Effect of rosiglitazone on the risk of myocardial infarction and death from cardiovascular causes. New Eng. J. Med. **356**, 2457–2471 (2007)

Smith, T. C., Spiegelhalter, D. J., Thomas, A.: Bayesian approaches to random-effects meta-analysis: a comparative study. Stat. Med. **14**, 2685–2699 (1995)

10

Joint distribution of k-tuple statistics in zero-one sequences of Markov-dependent trials

Anastasios N. Arapis[1], Frosso S. Makri[1*] and Zaharias M. Psillakis[2]

*Correspondence:
makri@math.upatras.gr
[1]Department of Mathematics, University of Patras, 26500 Patras, Greece
Full list of author information is available at the end of the article

Abstract
We consider a sequence of n, $n \geq 3$, zero (0) - one (1) Markov-dependent trials. We focus on k-tuples of 1s; i.e. runs of 1s of length at least equal to a fixed integer number k, $1 \leq k \leq n$. The statistics denoting the number of k-tuples of 1s, the number of 1s in them and the distance between the first and the last k-tuple of 1s in the sequence, are defined. The work provides, in a closed form, the exact conditional joint distribution of these statistics given that the number of k-tuples of 1s in the sequence is at least two. The case of independent and identical $0-1$ trials is also covered in the study. A numerical example illustrates further the theoretical results.

AMS Subject Classification: Primary 60E05, 62E15; Secondary 60J10, 60C05

Keywords: Exact Distributions, Runs, Binary trials, Markov chain

1 Introduction

Run counting statistics defined on a sequence of binary (zero (0) - one (1)) random variables (RVs), along with their exact and approximate distributions, have been extensively studied in the literature. Their popularity is due to the fact that such statistics appear as useful theoretical models in many research areas including statistics (e.g. hypothesis testing), engineering (e.g. system reliability and quality control), biology (e.g. population genetics and DNA sequence analysis), computer science (e.g. encoding/decoding/transmission of digital information) and financial engineering (e.g. insurance and risk analysis).

In such applications, a key point is the understanding how 1s and 0s are distributed and combined as elements of a $0-1$ sequence (finite or infinite, memoryless or not) and eventually forming runs of 1s or 0s which are enumerated according to certain counting schemes. Each scheme defines how runs of same symbols or strings (patterns) of both symbols are formed and consequently are enumerated. A counting scheme may depend on, among other considerations, whether overlapping counting is allowed or not as well as if the counting starts or not from scratch when a run/string of a certain size has been so far enumerated.

The counting scheme as well as the intrinsic uncertainty of a $0-1$ sequence are often suggested by the applications. Probabilistic models, in common use, for the internal structure of a $0-1$ sequence include the model of a sequence with elements independent of each other or a model for which it is assumed some kind of dependence among the elements of it. The methods used to derive exact/approximating, marginal/joint probability distributions include combinatorial analysis, generating functions, finite Markov chain imbedding technique, recursive schemes as well as normal, Poisson and large deviation approximations.

For extensive reviews of the recent literature on the distribution theory of runs and patterns we refer to Balakrishnan and Koutras (2002) and Fu and Lou (2003). Current works on the subject include, among others, those of Antzoulakos and Chadjiconstantinidis (2001); Eryilmaz (2006, 2015, 2016, 2017); Eryilmaz and Yalcin (2011); Johnson and Fu (2014); Koutras (2003); Koutras et al. (2016); Makri and Psillakis (2015); Makri et al. (2013) and Mytalas and Zazanis (2013, 2014).

In this article we derive expressions for a conditional distribution of a trivariate statistic. Its components denote the number of runs of 1s of length exceeding a fixed threshold number, the number of 1s in such runs of 1s and the length of the minimum sequence's segment in which these runs are concentrated. The study is developed on a sequence of two-state $(0-1)$ Markov-dependent trials. The runs are enumerated according to Mood's (1940) counting scheme.

More specifically, the manuscript is organized as follows. In Section 2 we present some preliminary material, including notation and definitions, necessary to develop our results which are obtained in Section 4. In Section 3 we give a motivation along with a statement of the aim of the work. A numerical example, showed in Section 5, clarifies the theoretical results of Section 4. A discussion on the results as well as a note on a future work are given in Section 6.

Throughout the article, for integers, n, m, $\binom{n}{m}$ denotes the extended binomial coefficient (see, Feller (1968), pp. 50, 63), $\lfloor x \rfloor$ stands for the greatest integer less than or equal to x and δ_{ij} denotes the Kronecker delta fuction of the integer arguments i and j. Further, for $\alpha > \beta$, we apply the conventions $\sum_{i=\alpha}^{\beta} y_i = 0$, $\prod_{i=\alpha}^{\beta} y_i = 1$, $\sum_{i=\alpha}^{\beta} \mathbf{Y}^{(i)} = \mathbf{O} \equiv \begin{pmatrix} 0 & 0 \\ 0 & 0 \end{pmatrix}$, $\prod_{i=\alpha}^{\beta} \mathbf{Y}^{(i)} = \mathbf{I} \equiv \begin{pmatrix} 1 & 0 \\ 0 & 1 \end{pmatrix}$, where y_i and $\mathbf{Y}^{(i)}$ are scalars and 2×2 matrices, respectively.

2 Preliminaries

2.1 Run counting statistics

Let $\{X_t\}_{t=1}^{n}$, $n \geq 1$, be the first n trials of a binary $(0-1)$ sequence of RVs, $X_t = x_t \in \{0, 1\}$. A run of 1s, is a (sub)sequence of $\{X_t\}_{t=1}^{n}$ consisting of consecutive 1s, the number of which is referred to as its length, preceded and succeeded by 0s or by nothing.

Given a fixed integer k, $1 \leq k \leq n$, a k-tuple of 1s is a run of 1s of length k or more. In the paper we will deal with the following statistics defined on a $0-1$ $\{X_t\}_{t=1}^{n}$. For details see, e.g. Makri et al. (2015) and the references therein.

(I) $G_{n,k}$ denoting the number of k-tuples of 1s, $1 \leq k \leq n$. In particular, $G_{n,1}$ denotes the number of 1-tuples of 1s, i.e. it represents the number $R_n \equiv G_{n,1}$ of all runs of 1s in the sequence. Using the convention $X_0 = X_{n+1} \equiv 0$, we can define $G_{n,k}$ as

$$G_{n,k} = \sum_{i=k}^{n} E_{n,i},\ 1 \leq k \leq n, \tag{1}$$

where

$$E_{n,i} = \sum_{j=i}^{n} J_j,\ J_j = (1 - X_{j-i})(1 - X_{j+1}) \prod_{r=j-i+1}^{j} X_r.$$

(II) $S_{n,k}$ denoting the number of 1s in the $G_{n,k}$ k-tuples of 1s; i.e. $S_{n,k}$ represents the sum of lengths of the $G_{n,k}$ k-tuples of 1s, $1 \le k \le n$. In particular $S_{n,1}$ represents the number of all 1s in the sequence; hence, the number of 0s, Y_n, in the sequence is $Y_n = n - S_{n,1}$. $S_{n,k}$ is formally defined as

$$S_{n,k} = \sum_{i=k}^{n} iE_{n,i},\ 1 \le k \le n. \tag{2}$$

Readily, $kG_{n,k} \le S_{n,k}$.

(III) L_n, $n \ge 1$, denoting the length of the longest run of 1s in the sequence. By setting

$$\Lambda_n = \{i : G_{n,i} > 0, 1 \le i \le n\},$$

we have that

$$L_n = \max\{k : k \in \Lambda_n\},\ \text{if}\ \Lambda_n \ne \emptyset;\ 0,\ \text{otherwise}. \tag{3}$$

Readily $L_n < k$ iff $G_{n,k} < 1$.

(IV) For $G_{n,k} \ge 1$, $1 \le k \le n$, $D_{n,k}$ denotes the distance (number of trials) between and including the first 1 of the first k-tuple of 1s and the last 1 of the last k-tuple of 1s in the sequence. If there is only one k-tuple of 1s in the sequence then $D_{n,k}$ denotes its length. That is, $D_{n,k}$ represents the size (length) of the minimum (sub)sequence of $\{X_t\}_{t=1}^{n}$ in which all $G_{n,k}$ k-tuple of 1s are concentrated. In particular, $D_{n,1}$ represents the length of the minimum segment of the sequence containing all R_n runs of 1s or in other words all $S_{n,1}$ 1s appearing in the sequence. For $G_{n,k} \ge 1, 1 \le k \le n$, $D_{n,k}$ can be formally defined as

$$D_{n,k} = U_{n,k}^{(2)} - U_{n,k}^{(1)} + 1, \tag{4}$$

where

$$U_{n,k}^{(1)} = \min\{j : I_j = 1, 1 \le j \le n - k + 1\},$$

$$U_{n,k}^{(2)} = \max\{j : I_{j-k+1} = 1, k \le j \le n\},$$

$$I_j = \prod_{r=j}^{j+k-1} X_r,\ 1 \le j \le n - k + 1.$$

Readily, $D_{n,k} = S_{n,k} = L_n$, if $G_{n,k} = 1$ and $D_{n,k} > S_{n,k} > L_n$, if $G_{n,k} > 1$.

(V) For $G_{n,k} \ge 1$, $1 \le k \le n$, set $\mathbf{V}_{n,k} = (D_{n,k}, G_{n,k}, S_{n,k})$. This is the RV we focus on in the article.

Example: By way of illustration consider the trials 1110001100010001010011101 111001001001001 numbered from 1 to 40. Then, $L_{40} = 4$ and $\mathbf{V}_{40,1} = (40, 11, 19)$, $\mathbf{V}_{40,2} = (28, 4, 12)$, $\mathbf{V}_{40,3} = (28, 3, 10)$, $\mathbf{V}_{40,4} = (4, 1, 4)$.

2.2 Internal structure's models

A general enough model for the internal structure of a $0 - 1$ $\{X_t\}_{t=1}^{n}$, $n \ge 2$, is that of the first n trials of a homogeneous $0-1$ Markov chain of first order (HMC1). On such a model we will develop our results. Accordingly, we next state the necessary notation/definitions.

Let $\{X_t\}_{t\geq 1}$ be a HMC1 with state space $\mathcal{A} = \{0,1\}$, one step transition probability matrix

$$\mathbf{P} = (p_{ij}) = \begin{pmatrix} p_{00} \; p_{01} \\ p_{10} \; p_{11} \end{pmatrix},$$

with

$$p_{ij} = P\left(X_t = j \mid X_{t-1} = i\right), \; i,j \in \mathcal{A}, \sum_{j\in\mathcal{A}} p_{ij} = 1, \; i \in \mathcal{A}, \; t \geq 2 \tag{5}$$

and probability distribution vector at time t

$$\mathbf{p}^{(t)} = \left(p_0^{(t)}, p_1^{(t)}\right),$$

with

$$p_i^{(t)} = P(X_t = i), \; i \in \mathcal{A}, \sum_{i\in\mathcal{A}} p_i^{(t)} = 1, \; t \geq 1. \tag{6}$$

Readily, because of the homogeneity of $\{X_t\}_{t\geq 1}$, it holds

$$\mathbf{p}^{(t)} = \mathbf{p}^{(t-1)}\mathbf{P} = \mathbf{p}^{(1)}\mathbf{P}^{t-1}, \; t \geq 2; \; \mathbf{p}^{(1)}, \; t = 1 \text{ and } \mathbf{P}^{t-1} = \left(p_{ij}^{(t-1)}\right), \; t \geq 2,$$

with

$$p_i^{(t)} = \mathbf{p}^{(t)}\mathbf{e}_i^{'}, \; i \in \mathcal{A}, \; t \geq 1,$$

$$p_{ij}^{(t-1)} = P(X_{t-1+m} = j \mid X_m = i) = \mathbf{e}_i\mathbf{P}^{t-1}\mathbf{e}_j^{'}, \; i,j \in \mathcal{A}, \; t \geq 2, \; m \geq 1, \tag{7}$$

where $\mathbf{e}_i^{'}$ is the transpose (i.e. the column vector) of the row vector $\mathbf{e}_i$, $i \in \mathcal{A}$, with $\mathbf{e}_0 = (1,0)$ and $\mathbf{e}_1 = (0,1)$.

In particular, for $p_{01} + p_{10} \neq 0$, i.e. $\mathbf{P} \neq \mathbf{I}$, it holds

$$\mathbf{P}^{t-1} = (p_{01} + p_{10})^{-1}\left\{\begin{pmatrix} p_{10} \; p_{01} \\ p_{10} \; p_{01} \end{pmatrix} + (1 - p_{01} - p_{10})^{t-1}\begin{pmatrix} p_{01} & -p_{01} \\ -p_{10} & p_{10} \end{pmatrix}\right\}, \; t \geq 2, \tag{8}$$

$$p_0^{(t)} = p_0^{(1)}(1 - p_{01} - p_{10})^{t-1} + p_{10}(p_{01} + p_{10})^{-1}\left[1 - (1 - p_{01} - p_{10})^{t-1}\right], \; t \geq 1. \tag{9}$$

The setup of a $0-1$ HMC1 $\{X_t\}_{t=1}^n$, $n \geq 2$, covers the case of a $0-1$ sequence of independent and identically distributed (IID) RVs, too. This is so, because a $0-1$ $\{X_t\}_{t=1}^n$, $n \geq 2$, IID sequence with

$$P(X_t = 1) = 1 - P(X_t = 0) = p_1, \; 1 \leq t \leq n, \tag{10}$$

is a particular HMC1 with

$$p_{ij} = 1 - p_1, \; j = 0; \; p_1, j = 1, \; i \in \mathcal{A}, \; p_{ij}^{(t-1)} = p_{ij}, \; i,j \in \mathcal{A}, \; t \geq 2,$$

$$p_1^{(t)} = p_1 = 1 - p_0^{(t)}, \; 1 \leq t \leq n. \tag{11}$$

2.3 A combinatorial result

In combinatorial analysis which will be used in Section 4, the following result, recalled from Makri et al. (2007), is useful. The coefficient

$$H_m(\alpha, r, k) = \sum_{j=0}^{\lfloor \frac{\alpha}{k+1} \rfloor} (-1)^j \binom{m}{j} \binom{\alpha - (k+1)j + r - 1}{\alpha - (k+1)j}, \tag{12}$$

represents the number of allocations of α indistinguishable balls into r distinguishable cells where each of the m, $0 \leq m \leq r$, specified cells is occupied by at most k balls. Equivalently, it gives the number of nonnegative integer solutions of the linear equation $x_1 + x_2 + \ldots + x_r = \alpha$ with the restrictions, for $m \geq 1$, $0 \leq x_{i_j} \leq k$, $1 \leq j \leq m$, for some specific m-combination $\{i_1, i_2, \ldots, i_m\}$ of $\{1, 2, \ldots, r\}$, and no restrictions on x_js, $1 \leq j \leq r$, for $m = 0$.

Moreover, $H_r(\alpha, r, k)$ is Riordan's (1964, p. 104) coefficient

$$C(\alpha, r, k) = \sum_{j=0}^{\lfloor \frac{\alpha}{k+1} \rfloor} (-1)^j \binom{r}{j} \binom{\alpha - (k+1)j + r - 1}{\alpha - (k+1)j}. \tag{13}$$

3 Motivation and aim of the work

In a study of a $0-1$ sequence $\{X_t\}_{t=1}^n$, $n \geq 3$, it is reasonable for one to be interested in the probabilistic behavior of RV $\mathbf{V}_{n,k} = (D_{n,k}, G_{n,k}, S_{n,k})$. This happens because jointly its components provide a more refined view of the internal clustering structure of the sequence than the information extracted by each one alone.

Interpreting a k-tuple of 1s as a cluster of consecutive 1s of size at least k, $D_{n,k}$ represents the size of the minimum segment of $\{X_t\}_{t=1}^n$ in which $G_{n,k}$ clusters of size at least k and at most L_n are concentrated. The overall density of $G_{n,k}$ clusters, with respect to the number of 1s in them, as well as of the minimum concentration segment is evaluated by $S_{n,k}$. Large values of $D_{n,k}$ suggest that these $G_{n,k}$ clusters spread over the interval between the left and the right side of the sequence whereas small values of $D_{n,k}$ indicate rather that the clusters are concentrated in a segment of the sequence of small size leaving the rest part(s) of the sequence empty of such clusters.

In addition to this information, a large value of $S_{n,k}$ paired with a small value of $G_{n,k}$ indicates the existence of clusters of 1s of a large size and therefore a trend whereas the same value of $S_{n,k}$ paired with a large value of $G_{n,k}$ indicates rather a distribution of clusters of small size in the (sub)sequence in which they are concentrated.

Therefore, based on the former interpretation, the motivation for the study as well as the usefulness of the statistic $\mathbf{V}_{n,k} = (D_{n,k}, G_{n,k}, S_{n,k})$ is apparent. In the sequel, we assume that $G_{n,k} \geq 2$ in order to have at least two k-tuples of 1s in the sequence and accordingly the distance $D_{n,k}$ is not a degenerate one. Moreover, this assumption is a common one in an application area of $D_{n,k}$; e.g., in detecting pattern (tandem or non-tandem direct) repeats in DNA sequences (Benson 1999).

For $1 \leq k \leq n$, set

$$\mathcal{M}_{n,k} = \{G_{n,k} \geq 2\}, \; \alpha_{n,k} = P\left(\mathcal{M}_{n,k}\right) \tag{14}$$

and for $n \geq 3$, $1 \leq k \leq \lfloor (n-1)/2 \rfloor$, define

$$\Omega_{n,k} = \{(d,m,s) : 2k+1 \le d \le n, 2k \le s \le d-1, 2 \le m \le \min\left(\lfloor s/k \rfloor, d-s+1\right)\} \tag{15}$$

and for $(d,m,s) \in \Omega_{n,k}$,

$$h_{n,k}(d,m,s) = P\left(\mathbf{V}_{n,k} = (d,m,s), \mathcal{M}_{n,k}\right),$$

$$v_{n,k}(d,m,s) = P\left(\mathbf{V}_{n,k} = (d,m,s) \mid \mathcal{M}_{n,k}\right) = h_{n,k}(d,m,s)/\alpha_{n,k}. \tag{16}$$

The paper provides exact closed form expressions for $\alpha_{n,k}$, $h_{n,k}(d,m,s)$ and eventually for $v_{n,k}(d,m,s)$ when $\mathbf{V}_{n,k}$ is defined on a $0-1$ HMC1/IID. The expressions are obtained via combinatorial analysis.

More specifically, closed formulae are established for the first time for $h_{n,k}(d,m,s)$, $1 \le k \le \lfloor (n-1)/2 \rfloor$, when $\mathbf{V}_{n,k}$ is defined on a $0-1$ HMC1 with given $\mathbf{P}$ and $\mathbf{p}^{(1)}$. Since, the general frame of HMC1 covers as a particular case IID sequences, the so implied expressions for $v_{n,k}(d,m,s)$ are alternative to those obtained for $v_{n,k}(d,m,s)$, $1 \le k \le \lfloor (n-1)/2 \rfloor$, by Makri et al. (2015) for IID sequences.

Moreover, for $n \ge 3$, $1 \le k \le \lfloor (n-1)/2 \rfloor$, $2k+1 \le d \le n$, let

$$f_{n,k}(d) = P\left(D_{n,k} = d \mid \mathcal{M}_{n,k}\right).$$

Therefore, since

$$f_{n,k}(d) = \sum_{s=2k}^{d-1} \sum_{m=2}^{\min(\lfloor s/k \rfloor, d-s+1)} v_{n,k}(d,m,s) = \alpha_{n,k}^{-1} \sum_{s=2k}^{d-1} \sum_{m=2}^{\min(\lfloor s/k \rfloor, d-s+1)} h_{n,k}(d,m,s), \tag{17}$$

hence, the work provides closed form expressions for determining $f_{n,k}(d)$ for HMC1 and IID $0-1$ $\{X_t\}_{t=1}^{n}$. These expressions are alternative to those derived, for IID sequences, by Makri et al. (2015) for $1 \le k \le \lfloor (n-1)/2 \rfloor$ as well as to those obtained, for HMC1, by Arapis et al. (2016) for $k=1$ and by Arapis et al. (2017) for $1 \le k \le \lfloor (n-1)/2 \rfloor$.

4 Results

In a $0-1$ sequence $\{X_t\}_{t=1}^{n}$, $n \ge 2$, for $0 \le y \le n$, $0 \le r \le \lfloor (n+1)/2 \rfloor$ and $(i,j) \in \{0,1\}^2$, define

$$B_n^{(i,j)}(y,r) = \{X_1 = i, X_n = j, Y_n = y, G_{n,1} = r\},$$

$$\pi_n^{(i,j)}(y,r) = P(B_n^{(i,j)}(y,r)).$$

Accordingly, for a HMC1 $\{X_t\}_{t=1}^{n}$, $n \ge 2$, with given $\mathbf{P}$ and $\mathbf{p}^{(1)}$, it holds

$$\pi_n^{(i,j)}(y,r) = \left(p_1^{(1)}\right)^i \left(1-p_1^{(1)}\right)^{1-i} p_{00}^{y-r-1+i+j} (1-p_{00})^{r-i} (1-p_{11})^{r-j} p_{11}^{n-y-r}, \tag{18}$$

for $2-(i+j) \le y \le n-(i+j)$, $1-\delta_{y,0}-\delta_{y,n}+\delta_{i+j,2} \le r \le \min\{n-y, y-1+i+j\}$ and $\pi_n^{(i,j)}(y,r) = 0$, otherwise.

Consequently, $\pi_n^{(i,j)}(y,r)$, for a $0-1$ IID sequence, reduces to

$$\pi_n^{(i,j)}(y,r) = \pi_n(y) = p_1^{n-y}(1-p_1)^y, \ 0 \le y \le n. \tag{19}$$

Theorem 1 *For* $n \geq 3$, $(d, m, s) \in \Omega_{n,1}$, $0 < p_1^{(1)} < 1$, *it holds*

$$h_{n,1}(d, m, s) = \binom{s-1}{m-1}\binom{d-s-1}{m-2}\pi_d^{(1,1)}(d-s, m)\varepsilon_n(d) \tag{20}$$

where $\varepsilon_n(d) = 1$, *if* $n = d$; $p_{00}^{n-d-2}\left\{p_{10}p_{00} + p_0^{(1)}(p_1^{(1)})^{-1}p_{01}\left[(n-d-1)p_{10} + p_{00}\right]\right\}$, *if* $n \geq d+1$.

Proof For $d = 3, \ldots, n-2, i = 2, 3, \ldots, n-d, s = 2, 3, \ldots, d-1, m = 2, 3, \ldots, \min\{s, d-s+1\}$ an element of the event $\Gamma_{i,d,m,s} = \{U_{n,1}^{(1)} = i, D_{n,1} = d, R_n = m, S_{n,1} = s\}$ is a $0-1$ sequence of length n with probability

$$p_0^{(1)}p_{00}^{i-2}p_{01}\left[\pi_d^{(1,1)}(d-s, m)\left(p_1^{(1)}\right)^{-1}\right]p_{10}p_{00}^{n-i-d}.$$

Fix i. Then the number of elements of the event $\Gamma_{i,d,m,s}$ is $\binom{s-1}{m-1}\binom{d-s-1}{m-2}$, since the number of allocations of s 1s in m runs of 1s is $\binom{s-1}{m-1}$ and the number of allocations of $d-s$ 0s in $m-1$ runs of 0s is $\binom{d-s-1}{m-2}$, so that

$$P\left(\Gamma_{i,d,m,s}\right) = \binom{s-1}{m-1}\binom{d-s-1}{m-2}p_0^{(1)}p_{01}\left[\pi_d^{(1,1)}(d-s, m)\left(p_1^{(1)}\right)^{-1}\right]p_{10}p_{00}^{n-d-2}.$$

We use similar reasoning for the rest cases. Then summing with respect to i we get the result. □

For a sequence $\{X_t\}_{t=1}^n$ of $0-1$ IID RVs, $h_{n,1}(d, m, s)$ reduces to the explicit formula given in the next Corollary.

Corollary 1 *For* $n \geq 3$, $(d, m, s) \in \Omega_{n,1}$, $0 < p_1 < 1$, *it is true that*

$$h_{n,1}(d, m, s) = (n-d+1)\binom{s-1}{m-1}\binom{d-s-1}{m-2}p_1^s(1-p_1)^{n-s}. \quad \diamond \tag{21}$$

In order to derive for HMC1, in the forthcoming Theorem 2, $h_{n,k}(d, m, s)$, $5 \leq 2k+1 \leq n$, we next recall, in Lemma 1, a result from (Makri et al.: On the concentration of runs of ones of length exceeding a threshold in a Markov chain, submitted).

Lemma 1 *For* $(i,j) \in \{0,1\}^2$, $n \geq 2$, *set* $\lambda_{n,k}^{(i,j)}(x) = P(G_{n,k} = x, X_1 = i, X_n = j)$, $x = 0, 1$. *Then, it holds that:*

(I) For $2 \leq k \leq n-2+i+j$,

$$\lambda_{n,k}^{(i,j)}(0) = \sum_{y=1}^{n-(i+j)}\sum_{r=i+j}^{y-1+i+j}\binom{y-1}{r-i-j}C(n-y-r, r, k-2)\pi_n^{(i,j)}(y, r),$$

$$\lambda_{n,k}^{(i,j)}(1) = \pi_n^{(i,j)}(0,1)\delta_{2,i+j} + \sum_{y=1}^{n-k}\sum_{r=1}^{y-1+i+j} r\binom{y-1}{r-i-j}H_{r-1}(n-y-r-k+1, r, k-2)\pi_n^{(i,j)}(y, r). \tag{22}$$

(II) For $k > n-2+i+j$,

$$\lambda_{n,k}^{(i,j)}(0) = \left(p_1^{(1)}\right)^i\left(1-p_1^{(1)}\right)^{1-i}p_{ij}^{(n-1)},$$

$$\lambda_{n,k}^{(i,j)}(1) = 0. \tag{23}$$

Theorem 2 *For $n \geq 5$, $2 \leq k \leq \lfloor (n-1)/2 \rfloor$, $(d, m, s) \in \Omega_{n,k}$, $0 < p_1^{(1)} < 1$, it holds*

$$h_{n,k}(d,m,s) = p_{11}^{2k-2}\left(p_1^{(1)}\right)^{-1} \sum_{i=1}^{n-d+1} \ell_{i-1,k}^{(\alpha)} \ell_{n-d-i+1,k}^{(\beta)} \sum_{r=m}^{m+\left\lfloor \frac{d-s-m+1}{2} \right\rfloor} \sum_{y=r-1}^{d-s-r+m} \gamma_{d,m,s}(y,r) \pi_{d-2k+2}^{(1,1)}(y,r), \tag{24}$$

where

$$\ell_{n,k}^{(\alpha)} = p_1^{(1)}, \textit{ for } n = 0; \quad p_0^{(n)} p_{01}, \textit{ for } 1 \leq n \leq k; \quad p_{01}\left[\lambda_{n,k}^{(0,0)}(0) + \lambda_{n,k}^{(1,0)}(0)\right], \textit{ for } n \geq k+1,$$

$$\ell_{n,k}^{(\beta)} = 1, \textit{ for } n = 0; \quad p_{10}, \textit{ for } 1 \leq n \leq k; \quad p_{10}(p_0^{(1)})^{-1}\left[\lambda_{n,k}^{(0,0)}(0) + \lambda_{n,k}^{(0,1)}(0)\right], \textit{ for } n \geq k+1 \tag{25}$$

and

$$\gamma_{d,m,s}(y,r) = \binom{y-1}{r-2}\binom{r-2}{m-2}\binom{s-mk+m-1}{m-1} C(d-y-s-r+m, r-m, k-2). \tag{26}$$

Proof For $1 \leq r_1 \leq r_2 \leq n$ let $Y_{r_1,r_2}, R_{r_1,r_2}, L_{r_1,r_2}, S_{r_1,r_2,k}, D_{r_1,r_2,k}, G_{r_1,r_2,k}$ be RVs defined on the subsequence $X_{r_1}, X_{r_1+1}, \ldots, X_{r_2}$ of $\{X_t\}_{t=1}^n$. For $m \geq 2$ define the event

$$\Delta_{r_1,r_1+d-1}(d,s,m,y,r) = \{D_{r_1,r_1+d-1,k} = d, G_{r_1,r_1+d-1,k} = m, S_{r_1,r_1+d-1,k} = s, Y_{r_1,r_1+d-1} = y, R_{r_1,r_1+d-1} = r\}.$$

An element of this event is a 0 - 1 sequence of length d, starting and ending with a 1, for which y_j's and z_j's, representing the lengths of the failure and success runs, respectively, satisfy the conditions:

(a) $y_1 + y_2 + \ldots + y_{r-1} = y, y_j \geq 1, 1 \leq j \leq r-1$.

(b) $z_1 + z_{i_1} + z_{i_2} + \ldots + z_{i_{m-2}} + z_r = s, z_j \geq k, j \in \{1, i_1, i_2, \ldots, i_{m-2}, r\}$, for some specific combination $\{1, i_1, i_2, \ldots, i_{m-2}, r\}$ of $\{1, 2, \ldots, r-1, r\}$ among the $\binom{r-2}{m-2}$ ones.

(c) $z_{i_{m-1}} + z_{i_m} + \ldots + z_{i_{r-2}} = d - y - s, 1 \leq z_{i_j} \leq k-1, m-1 \leq j \leq r-2$, for $\{i_{m-1}, \ldots, i_{r-2}\} \in \{1, 2, \ldots, r\} - \{1, i_1, i_2, \ldots, i_{m-2}, r\}$.

Fix $i_1, i_2, \ldots, i_{m-2}$. Then the number of such sequences, i.e. the number of solutions of the system (a)-(c), is

$$\binom{y-1}{r-2} C(d-y-s-r+m, r-m, k-2) \binom{s-mk+m-1}{m-1}$$

and each such sequence has probability

$$p_1^{(1)} p_{11}^{k-1} (p_1^{(1)})^{-1} \pi_{d-2k+2}^{(1,1)}(y,r) p_{11}^{k-1} = p_{11}^{2k-2} \pi_{d-2k+2}^{(1,1)}(y,r).$$

Hence,

$$P(\Delta_{r_1,r_1+d-1}(d,s,m,y,r)) = p_{11}^{2k-2} \pi_{d-2k+2}^{(1,1)}(y,r) \binom{r-2}{m-2}\binom{y-1}{r-2}\binom{s-mk+m-1}{m-1} \times C(d-y-s-r+m, r-m, k-2).$$

For $k+2 \leq i \leq n-k-d$, $m \geq 2$, we have that

$$
\begin{aligned}
&P\left(U_{n,k}^{(1)}=i, D_{n,k}=d, G_{n,k}=m, S_{n,k}=s, Y_{i,i+d-1}=y, R_{i,i+d-1}=r\right)\\
&= P\Big\{\left[(L_{1,i-1}<k, X_{i-1}=0)\cap[(X_1=0)\cup(X_1=1)]\right]\cap\Delta_{i,i+d-1}(d,s,m,y,r)\\
&\quad \cap\left[(L_{i+d,n}<k, X_{i+d}=0)\cap[(X_n=0)\cup(X_n=1)]\right]\Big\}\\
&= \left[\lambda_{i-1,k}^{(0,0)}(0)+\lambda_{i-1,k}^{(1,0)}(0)\right]p_{01}\\
&\quad \times\left(p_1^{(1)}\right)^{-1}P\left(\Delta_{i,i+d-1}(d,s,m,y,r)\right)p_{10}\left[\lambda_{n-i-d+1,k}^{(0,0)}(0)+\lambda_{n-i-d+1,k}^{(0,1)}(0)\right]/p_0^{(1)}\\
&= \left[\lambda_{i-1,k}^{(0,0)}(0)+\lambda_{i-1,k}^{(1,0)}(0)\right]p_{01}\left(p_1^{(1)}\right)^{-1}p_{11}^{2k-2}\pi_{d-2k+2}^{(1,1)}(y,r)\\
&\quad \times\binom{r-2}{m-2}\binom{y-1}{r-2}\binom{s-mk+m-1}{m-1}\\
&\quad \times C(d-y-s-r+m, r-m, k-2)p_{10}\left(p_0^{(1)}\right)^{-1}\left[\lambda_{n-i-d+1,k}^{(0,0)}(0)+\lambda_{n-i-d+1,k}^{(0,1)}(0)\right].
\end{aligned}
$$

By similar reasoning we get the remaining cases of i, i.e. $1 \leq i \leq k+1$ and $n-d+1-k \leq i \leq n-d+1$. Then summing with respect to i, y and r we get the result. □

Having found $h_{n,k}(d,m,s)$, we next proceed to obtain $v_{n,k}(d,m,s)$. In accomplishing it, the required probabilities $\alpha_{n,k}$ for HMC1 are recalled, in Lemma 2, from Arapis et al. (2016) for $k=1$, and they are computed via Lemma 1 for $2 \leq k \leq \lfloor (n-1)/2 \rfloor$.

Lemma 2 *For $n \geq k \geq 1$, the probability $\alpha_{n,k}$, for HMC1, is computed via the expressions:*
(I) For $k=1$,

$$
\alpha_{n,1} = 1-p_{00}^{n-3}\left\{p_{00}\left(1+(n-2)p_{01}\right)+\frac{(n-1)(n-2)}{2}p_0^{(1)}p_{01}^2\right\}, \quad \textit{if} \quad p_{00}=p_{11}
$$

and

$$
\begin{aligned}
\alpha_{n,1} &= 1-p_0^{(1)}p_{00}^{n-1}-p_{11}^{n-2}\left(p_1^{(1)}+p_0^{(1)}p_{01}\right)-p_{00}\left(p_0^{(1)}p_{01}+p_1^{(1)}p_{10}\right)\frac{p_{11}^{n-2}-p_{00}^{n-2}}{p_{11}-p_{00}}\\
&\quad -p_0^{(1)}p_{01}p_{10}\frac{p_{11}^{n-1}-p_{00}^{n-2}\left[p_{11}+(n-2)\left(p_{11}-p_{00}\right)\right]}{\left(p_{11}-p_{00}\right)^2}, \quad \textit{if} \quad p_{00}\neq p_{11}. \quad (27)
\end{aligned}
$$

(II) For $2 \leq k \leq n$,

$$
\alpha_{n,k} = 1-\sum_{(i,j)\in\{0,1\}^2}\left[\lambda_{n,k}^{(i,j)}(0)+\lambda_{n,k}^{(i,j)}(1)\right]. \quad (28)
$$

Theorem 3 *For $n \geq 3$, $1 \leq k \leq \lfloor (n-1)/2 \rfloor$, $(d,m,s) \in \Omega_{n,k}$, $0 < p_1^{(1)} < 1$, the PMF $v_{n,k}(d,m,s)$ for a HMC1, with given $\mathbf{P}$ and $\mathbf{p}^{(1)}$, is calculated by*

$$
v_{n,k}(d,m,s) = \alpha_{n,k}^{-1}h_{n,k}(d,m,s), \quad (29)
$$

where $\alpha_{n,k}$ and $h_{n,k}(d,m,s)$ are provided by Lemma 2 and Theorems 1 (for $k=1$) and 2 (for $2 \leq k \leq \lfloor (n-1)/2 \rfloor$), respectively.

Remark 1 *For IID sequences, in implementing Theorem 3, one has to take into consideration Eqs. (10) - (11), (19) and (21). Moreover, for speeding up calculations, one has to set $\pi_n(y)$ in front of the inner summation in (22).*

5 A numerical example

In this example we compute some indicative numerics concerning two model (i.e. HMC1 and IID) $0-1$ sequences $\{X_t\}_{t=1}^{n}$ which are considered in the paper. The common length of these was taken small, i.e. $n = 8$, so that the required computations can also be carried out by a hand/pocket calculator and thus it is possible to gain insight in the formulae developed in Section Results, and also because of space limitations. The sequences that have been used are as follows. Table 1: An IID sequence with $p_1 = 0.5$. Table 2: A HMC1 sequence with $p_{00} = p_{11} = 0.9, p_1^{(1)} = 0.5$.

Both tables depict for $k = 1, 2, 3$, $v_{8,k}(d, m, s)$, $(d, m, s) \in \Omega_{8,k}$ and $f_{8,k}(d)$, $2k+1 \leq d \leq 8$ illustrating the numeric values of the involved probabilities. $v_{8,k}(d, m, s)$ and $f_{8,k}(d)$ were computed via Eqs. (29) and (17), respectively.

6 Discussion and further study

In this article we have derived exact closed form expressions for PMF $v_{n,k}(d, m, s)$, $n \geq 3$, $1 \leq k \leq \lfloor (n-1)/2 \rfloor$, $(d, m, s) \in \Omega_{n,k}$, of the RV $\mathbf{V}_{n,k} \mid \mathcal{M}_{n,k}$ defined on a $0-1$ sequence of homogeneous Markov-dependent trials. The method used is a combinatorial one relied on results exploiting the internal structure of such a sequence.

As it is noticed in the Introduction the application domain of runs contains a diverse range of fields. Indicative potential ones are next discussed.

Table 1 $0-1$ IID sequence with $p_1 = 0.5$

s	m	$d=3$	$d=4$	$d=5$	$d=6$	$d=7$	$d=8$
				$v_{8,1}(d,m,s)$			
2	2	0.02739726	0.02283105	0.01826484	0.01369863	0.00913242	0.00456621
3	2		0.04566210	0.03652968	0.02739726	0.01826484	0.00913242
	3			0.01826484	0.02739726	0.02739726	0.01826484
4	2			0.05479452	0.04109589	0.02739726	0.01369863
	3				0.04109589	0.05479452	0.04109589
	4					0.00913242	0.01369863
5	2				0.05479452	0.03652968	0.01826484
	3					0.05479452	0.05479452
	4						0.01826484
6	2					0.04566210	0.02283105
	3						0.04566210
7	2						0.02739726
$f_{8,1}(d)$		0.02739726	0.06849315	0.12785388	0.20547945	0.28310503	0.28767123
				$v_{8,2}(d,m,s)$			
4	2			0.18518519	0.09259259	0.07407407	0.05555556
5	2				0.18518519	0.07407407	0.07407407
6	2					0.11111111	0.05555556
	3						0.01851852
7	2						0.07407407
$f_{8,2}(d)$				0.18518519	0.27777778	0.25925925	0.27777778
				$v_{8,3}(d,m,s)$			
6	2					0.40000000	0.20000000
7	2						0.40000000
$f_{8,3}(d)$						0.40000000	0.60000000

Table 2 $0-1$ HMC1 sequence with $p_{00}=p_{11}=0.9, p_1^{(1)}=0.5$

s	m	$d=3$	$d=4$	$d=5$	$d=6$	$d=7$	$d=8$
				$v_{8,1}(d,m,s)$			
2	2	0.00914441	0.00872875	0.00831310	0.00789744	0.00748179	0.03366804
3	2		0.01745750	0.01662619	0.01579488	0.01496357	0.06733609
	3			0.00010263	0.00019500	0.00027710	0.00166262
4	2			0.02493929	0.02369233	0.02244536	0.10100413
	3				0.00029250	0.00055421	0.00374089
	4					0.00000114	0.00001539
5	2				0.03158977	0.02992715	0.13467217
	3					0.00055421	0.00498786
	4						0.00002053
6	2					0.03740894	0.16834021
	3						0.00415655
7	2						0.20200826
$f_{8,1}(d)$		0.00914441	0.02618626	0.04998121	0.07946192	0.11361346	0.72161274
				$v_{8,2}(d,m,s)$			
4	2			0.02225160	0.02081956	0.01806565	0.08228685
5	2				0.04163913	0.03569068	0.16259088
6	2					0.05353602	0.24091210
	3						0.00099141
7	2						0.32121613
$f_{8,2}(d)$				0.02225160	0.06245869	0.10729236	0.80799735
				$v_{8,3}(d,m,s)$			
6	2					0.06896552	0.31034483
7	2						0.62068966
$f_{8,3}(d)$						0.06896552	0.93103448

Encoding, compression and transmission of digital information calls for the understanding the distributions of runs of 1s or 0s. Such a knowledge helps in analyzing, and also in comparing, several techniques used in communication networks. In such networks $0-1$ data ranging from a few kilobytes (e.g. e-mails) to many gigabytes of greedy multimedia applications (e.g. video on demand) are highly encoded, decoded and eventually proceeded under security. For details, see e.g., Sinha and Sinha (2009), Makri and Psillakis (2011a) and Tabatabaei and Zivic (2015).

An area where the study of runs of 1s and 0s has become increasingly useful is the field of bioinformatics or computational biology. For instance, molecular biologists design similarity tests between two DNA sequences where a 1 is interpreted as a match of the sequences at a given position and everything else as a 0. Moreover, the probabilistic analysis of such sequences according to the form, the length and the number of detected patterns as well as of the positions and the lengths of the segments of the sequence in which they are concentrated, probably suggests a functional reason for the internal structure of the examined sequence. The latter facts might be useful in suggesting a further investigation of the underline sequence(s) by biologists. See, e.g. Avery and Henderson (1999), Benson (1999) and Nuel et al. (2010).

Another active area where run statistics, in particular $G_{n,k}$ and $S_{n,k}$, have interesting statistical applications is that connected to hypothesis testing; e.g., in tests of randomness.

For a systematic study of such a topic, we refer among others, the works of Koutras and Alexandrou (1997) and Antzoulakos et al. (2003).

Accordingly, it is reasonable for one to use the exact expressions obtained for $v_{n,k}(d, m, s)$ in applications like the ones mentioned above. This is so, because this distribution, as a joint one, is more flexible than each one of its marginals which have been used in such applications. See, e.g. Lou (2003), Makri and Psillakis (2011b) and Arapis et al. (2016).

Moreover, in handling 0 - 1 sequences of a large length, with dependent or not elements, a Monte - Carlo simulation, based on Eqs. (1) - (4) would be a useful tool in obtaining approximate values for $v_{n,k}(d, m, s)$. In addition, the general approximating methods, suggested by Johnson and Fu (2014), might be helpful in deriving approximate values for $f_{n,k}(d)$.

Acknowledgements

The authors wish to thank the Editor for the thorough reading, and the anonymous reviewers for useful comments and suggestions which improved the article.

Authors' contributions

The authors, ANA, FSM and ZMP with the consultation of each other carried out this work and drafted the manuscript together. All authors read and approved the final manuscript.

Competing interests

The authors declare that they have no competing interests.

Author details

[1] Department of Mathematics, University of Patras, 26500 Patras, Greece. [2] Department of Physics, University of Patras, 26500 Patras, Greece.

References

Antzoulakos, DL, Bersimis, S, Koutras, MV: On the distribution of the total number of run lengths. Ann. Inst. Statist. Math. **55**, 865–884 (2003)

Antzoulakos, DL, Chadjiconstantinidis, S: Distributions of numbers of success runs of fixed length in Markov dependent trials. Ann. Inst. Statist. Math. **53**, 559–619 (2001)

Arapis, AN, Makri, FS, Psillakis, ZM: On the length and the position of the minimum sequence containing all runs of ones in a Markovian binary sequence. Statist. Probab. Lett. **116**, 45–54 (2016)

Arapis, AN, Makri, FS, Psillakis, ZM: Distribution of statistics describing concentration of runs in non homogeneous Markov-dependent trials. Commun. Statist. Theor. Meth. (2017). doi:10.1080/03610926.2017.1337144

Avery, PJ, Henderson, D: Fiting Markov chain models to discrete state series such as DNA sequences. Appl. Statist. **48** (Part 1), 53–61 (1999)

Balakrishnan, N, Koutras, MV: Runs and Scans with Applications. Wiley, New York (2002)

Benson, G: Tandem repeats finder: a program to analyze DNA sequences. Nucleic Acids Res. **27**, 573–580 (1999)

Eryilmaz, S: Some results associated with the longest run statistic in a sequence of Markov dependent trials. Appl. Math. Comput. **175**, 119–130 (2006)

Eryilmaz, S: Discrete time shock models involving runs. Statist. Probab. Lett. **107**, 93–100 (2015)

Eryilmaz, S: Generalized waiting time distributions associated with runs. Metrika. **79**, 357–368 (2016)

Eryilmaz, S: The concept of weak exchangeability and its applications. Metrika. **80**, 259–271 (2017)

Eryilmaz, S, Yalcin, F: Distribution of run statistics in partially exchangeable processes. Metrika. **73**, 293–304 (2011)

Feller, W: An Introduction to Probability Theory and Its Applications. 3rd Ed., Vol. I. Wiley, New York (1968)

Fu, JC, Lou, WYW: Distribution Theory of Runs and Patterns and Its Applications: A finite Markov chain imbedding approach. World Scientific, River Edge (2003)

Johnson, BC, Fu, JC: Approximating the distributions of runs and patterns. J. Stat. Distrib. Appl. **1:5**, 1–15 (2014)

Koutras, MV: Applications of Markov chains to the distribution of runs and patterns. In: Shanbhag, DN, Rao, CR (eds.) Handbook of Statistics, pp. 431–472. Elsevier, North-Holland, (2003)

Koutras, MV, Alexandrou, V: Non-parametric randomness tests based on success runs of fixed length. Statist. Probab. Lett. **32**, 393–404 (1997)

Koutras, VM, Koutras, MV, Yalcin, F: A simple compound scan statistic useful for modeling insurance and risk management problems. Insur. Math. Econ. **69**, 202–209 (2016)

Lou, WYW: The exact distribution of the *k*-tuple statistic for sequence homology. Statist. Probab. Lett. **61**, 51–59 (2003)

Makri, FS, Philippou, AN, Psillakis, ZM: Success run statistics defined on an urn model. Adv. Appl. Prob. **39**, 991–1019 (2007)
Makri, FS, Psillakis, ZM: On success runs of a fixed length in Bernoulli sequences: Exact and asymptotic results. Comput. Math. Appl. **61**, 761–772 (2011a)
Makri, FS, Psillakis, ZM: On runs of length exceeding a threshold: normal approximation. Stat. Papers. **52**, 531–551 (2011b)
Makri, FS, Psillakis, ZM: On ℓ-overlapping runs of ones of length k in sequences of independent binary random variables. Commun. Statist. Theor. Meth. **44**, 3865–3884 (2015)
Makri, FS, Psillakis, ZM, Arapis, AN: Counting runs of ones with overlapping parts in binary strings ordered linearly and circularly. Intern. J. Statist. Probab. **2**, 50–60 (2013)
Makri, FS, Psillakis, ZM, Arapis, AN: Length of the minimum sequence containing repeats of success runs. Statist. Probab. Lett. **96**, 28–37 (2015)
Mood, AM: The distribution theory of runs. Ann. Math. Statist. **11**, 367–392 (1940)
Mytalas, GC, Zazanis, MA: Central limit theorem approximations for the number of runs in Markov-dependent binary sequences. J. Statist. Plann. Infer. **143**, 321–333 (2013)
Mytalas, GC, Zazanis, MA: Central limit theorem approximations for the number of runs in Markov-dependent multi-type sequences. Commun. Statist. Theor. Meth. **43**, 1340–1350 (2014)
Nuel, G, Regad, L, Martin, J, Camproux, A-C: Exact distribution of a pattern in a set of random sequences generated by a Markov source: applications to biological data. Algorithm Mol. Biol. **5**, 1–18 (2010)
Riordan, AM: An Introduction to Combinatorial Analysis. Second Ed. John Wiley, New York (1964)
Sinha, K, Sinha, BP: On the distribution of runs of ones in binary trials. Comput. Math. Appl. **58**, 1816–1829 (2009)
Tabatabaei, SAH, Zivic, N: A review of approximate message authentication codes. In: Zivic, N (ed.) Robust Image Authentication in the Presence of Noise, pp. 106–127. Springer International Publishing AG, Cham (ZG), Switzerland, (2015)

11

The odd log-logistic logarithmic generated family of distributions with applications in different areas

Morad Alizadeh[1], S. M. T. K. MirMostafee[2], Edwin M. M. Ortega[3*], Thiago G. Ramires[3] and Gauss M. Cordeiro[4]

*Correspondence: edwin@usp.br
[3]Departamento de Ciências Exatas, Universidade de São Paulo, Piracicaba, Brazil
Full list of author information is available at the end of the article

Abstract

We introduce and study general mathematical properties of a new generator of continuous distributions with three extra parameters called the odd log-logistic logarithmic generated family of distributions. We present some special models and investigate the asymptotes and shapes. The new density function can be expressed as a linear combination of exponentiated densities based on the same baseline distribution. Explicit expressions for the ordinary and incomplete moments, quantile and generating functions, Shannon and Rényi entropies and order statistics, which hold for any baseline model, are determined. We discuss the estimation of the model parameters by maximum likelihood. Further, we introduce the new family in long-term survival models. We illustrate the potentiality of the proposed models by means of four applications to real data.

Keywords: Generated family, Maximum likelihood, Moment, Order statistic, Quantile function, Survival analysis

AMS Subject Classification: 97K50; 62N01; 62N02

1 Introduction

Statistical distributions are very useful in describing and predicting real world phenomena. Numerous extended distributions have been extensively used over the last decades for modeling data in several areas. Recent developments focus on defining new families that extend well-known distributions and at the same time provide greater flexibility in modeling data in practice. Hence, several classes to generate new distributions by adding one or more parameters have been proposed in the statistical literature. Some well-known generators are the Marshall-Olkin generated (MO-G) by Marshall and Olkin (1997), beta-G by Eugene et al. (2002), Kumaraswamy-G (Kw-G) by Cordeiro and de Castro (2011), Weibull-G by Bourguignon et al. (2014), exponentiated half-logistic-G by Cordeiro et al. (2014a), Lomax-G by Cordeiro et al. (2014b), among others.

Let $G(x;\boldsymbol{\xi})$ be a baseline cumulative distribution function (cdf) and $\boldsymbol{\xi}$ be the vector of associated parameters. Recently, Gleaton and Lynch (2004, 2006, 2010) defined the cdf of the *odd log-logistic* family with one extra shape parameter $\alpha > 0$ by

$$H(x) = \frac{G(x;\boldsymbol{\xi})^{\alpha}}{G(x;\boldsymbol{\xi})^{\alpha} + \bar{G}(x;\boldsymbol{\xi})^{\alpha}}, \tag{1}$$

where $\bar{G}(x;\xi) = 1 - G(x;\xi)$. More precisely, they showed the following facts:

- The set of generalized log-logistic (GLL) transformations form an Abelian group with the binary operation of composition;
- The transformation group partitions the set of all lifetime distributions into equivalence classes, so that any two distributions in an equivalence class are related through a GLL transformation;
- Either every distribution in an equivalence class has a moment generating function (mgf), or none does;
- Every distribution in an equivalence class has the same number of moments;
- Each equivalence class is linearly ordered according to the transformation parameter, with larger values of this parameter corresponding to smaller dispersion of the distribution about the common median class; and
- Within an equivalence class, the Kullback-Leibler information is an increasing function of the ratio of the transformation parameters.

In addition, Gleaton and Rahman (2010, 2014) obtained asymptotic results for the maximum likelihood estimates (MLEs) of the parameters of these two distributions. They proved that for distributions generated from either a two-parameter Weibull or a two-parameter inverse Gaussian distributions by a GLL transformation, the joint MLEs of the parameters are asymptotically normal and efficient, provided the GLL transformation parameter exceeds three.

We define the cdf of the *odd log-logistic logarithmic-G* (OLLL-G) family by

$$F(x) = \frac{\log\left[1 - \frac{\beta G(x,\xi)^{\alpha}}{G(x,\xi)^{\alpha} + \bar{G}(x,\xi)^{\alpha}}\right]}{\log(1-\beta)}, \tag{2}$$

where $G(x;\xi)$ is the baseline cdf depending on a parameter vector ξ and $\alpha > 0$ and $0 < \beta < 1$ are two additional shape parameters. It includes the odd log-logistic-G (OLL-G) family (Gleaton and Lynch 2004, 2006) and the logarithmic-G family. Some special models are given in Table 1.

This paper is organized as follows. In Section 2, we provide a physical interpretation of the OLLL-G family and define two special cases. In Section 3, two useful linear representations are derived. In Section 4, we obtain explicit expressions for the moments and generating function. In Section 5, general expressions for the Rényi and Shannon entropies and order statistics are presented. Estimation of the model parameters by maximum likelihood is investigated in Section 6. We also present the performance of the MLEs through a simulation study. In Section 7, the OLLL-G model is modified for possible presence of long-term survivors in the data. Four applications to real data illustrate the performance of the proposed models in Section 8. The paper is concluded in Section 9.

Table 1 Some special models

α	β	$G(x)$	Reduced distribution
-	$\uparrow 1$	$G(x)$	OLL-G family [Gleaton and Lynch (2004, 2006)]
1	-	$G(x)$	Logarithmic-G family baseline distribution
1	$\uparrow 1$	$G(x)$	$G(x)$

2 Motivation and special cases

The density function corresponding to (2) is given by

$$f(x;\alpha,\beta,\xi) = \frac{\alpha\beta g(x,\xi)G(x,\xi)^{\alpha-1}\bar{G}(x,\xi)^{\alpha-1}}{-\left[G(x,\xi)^{\alpha}+\bar{G}(x,\xi)^{\alpha}\right]\left[(1-\beta)G(x,\xi)^{\alpha}+\bar{G}(x,\xi)^{\alpha}\right]\log(1-\beta)}, \quad (3)$$

where $g(x;\xi)$ is the baseline pdf. Hereafter, a random variable X with density function (3) is denoted by $X \sim$ OLLL-G(α,β,ξ). Further, we can omit sometimes the dependence on the vector ξ of the parameters and write simply $G(x) = G(x;\xi)$.

A motivation of this family can be explained as follows. Suppose that a parallel system is made up of N components and the lifetimes of the components are independent and identically distributed (iid) random variables, denoted by $Z_1,\cdots,Z_N$, with common cdf (1). Then, the system fails as soon as the last component fails, namely the lifetime of the whole system is represented by $X = \max\{Z_1,\cdots,Z_N\}$. In many survival parallel systems, it is almost impossible to have a fixed number of components because some of them get lost or censored for various reasons. Therefore we may assume that N is a discrete random variable. Suppose that N has the logarithmic distribution with probability mass function given by

$$P(N=n) = \frac{-1}{\log(1-\beta)}\frac{\beta^n}{n}, \qquad n=1,2,\ldots, \quad 0<\beta<1.$$

Then, the cdf of the life length of the whole system, X, is obtained as

$$\begin{aligned} F(x) &= \sum_{n=1}^{\infty} P(X\le x|N=n)P(N=n) \\ &= \sum_{n=1}^{\infty}\left[\frac{G(x,\xi)^{\alpha}}{G(x,\xi)^{\alpha}+\bar{G}(x,\xi)^{\alpha}}\right]^n \frac{-1}{\log(1-\beta)}\frac{\beta^n}{n} = \frac{-\log\left[1-\frac{\beta G(x,\xi)^{\alpha}}{G(x,\xi)^{\alpha}+\bar{G}(x,\xi)^{\alpha}}\right]}{-\log(1-\beta)} \end{aligned}$$

which is identical to (2).

The hazard rate function (hrf) of X becomes

$$h(x;\alpha,\beta,\xi) = \frac{-\alpha\beta g(x,\xi)G(x,\xi)^{\alpha-1}\bar{G}(x,\xi)^{\alpha-1}\left[G(x,\xi)^{\alpha}+\bar{G}(x,\xi)^{\alpha}\right]^{-1}}{\left[(1-\beta)G(x,\xi)^{\alpha}+\bar{G}(x,\xi)^{\alpha}\right]\left\{\log\left[\frac{1-\beta}{1-\frac{\beta G(x,\xi)^{\alpha}}{G(x,\xi)^{\alpha}+\bar{G}(x,\xi)^{\alpha}})}\right]\right\}}. \quad (4)$$

The OLLL-G family of distributions is easily simulated by inverting (2) as follows: if U has a uniform $U(0,1)$ distribution, then

$$Q(U) = Q_G\left(\frac{\left[1-(1-\beta)^U\right]^{\frac{1}{\alpha}}}{\left[1-(1-\beta)^U\right]^{\frac{1}{\alpha}}+\left[(1-\beta)^U-(1-\beta)\right]^{\frac{1}{\alpha}}}\right), \quad (5)$$

has the density function (3), where $Q_G(u) = G^{-1}(u)$ is the quantile function (qf) of the baseline G.

Remark 1 *Although, we have stated that $\beta \in (0,1)$, Eq. (2) is still a cdf if $\beta < 0$. Hence, we can consider the OLLL-G family defined for any $\beta \in (-\infty,0)\cup(0,1)$.*

In Appendix 1, we present the asymptotes and shapes of the OLLL-G model.

2.1 Special OLLL-G distributions

The OLLL-G density function (3) allows for greater flexibility of its tails and can be widely applied in many areas of engineering and biology. It will be most tractable when $G(x; \boldsymbol{\xi})$ and $g(x; \boldsymbol{\xi})$ have simple analytic expressions. We now present and discuss some special cases of this family because it extends several widely-known distributions in the literature.

2.1.1 *Odd log-logistic logarithmic Weibull (OLLL-W) model*

Let $G(x; \boldsymbol{\xi}) = 1 - \mathrm{e}^{-(bx)^a}$ the Weibull cdf, with scale parameter $b > 0$ and shape parameter $a > 0$, where $\boldsymbol{\xi} = (a, b)$. The OLLL-W density function (for $x > 0$) is given by

$$f(x; a, b, \alpha, \beta) = \frac{a\,\alpha\,\beta\,b^a\,\mathrm{e}^{-\alpha(bx)^a}\left[1-\mathrm{e}^{-(bx)^a}\right]^{\alpha-1}}{-\left\{\mathrm{e}^{-\alpha(bx)^a}+\left[1-\mathrm{e}^{-(bx)^a}\right]^{\alpha}\right\}\left\{\mathrm{e}^{-\alpha(bx)^a}+(1-\beta)\left[1-\mathrm{e}^{-(bx)^a}\right]^{\alpha}\right\}\log(1-\beta)}.$$

Figures 1 and 2 display some shapes of the OLLL-W density and hazard functions, respectively. The plots in Fig. 1 indicate that this density function can be decreasing, unimodal and bimodal. Moreover, Fig. 2 reveals that the hrf of the OLLL-W model can be decreasing, increasing, increasing-decreasing-increasing, bathtub shaped and upside bathtub shaped.

2.1.2 *Odd log-logistic logarithmic normal (OLLLN) model*

Consider the normal model with location parameter $\mu \in \mathbb{R}$ and scale parameter $\sigma > 0$, whose pdf and cdf (for $x \in \mathbb{R}$) are given by

$$g(x; \boldsymbol{\xi}) = \frac{1}{\sigma}\,\phi\left(\frac{x-\mu}{\sigma}\right) \qquad \text{and} \qquad G(x; \boldsymbol{\xi}) = \Phi\left(\frac{x-\mu}{\sigma}\right),$$

respectively, where $\boldsymbol{\xi} = (\mu, \sigma)^T$. Inserting these expressions in (3), the OLLLN pdf is given by

$$f(x;\mu,\sigma,a,b) = \frac{\alpha\,\beta\,\phi\left(\frac{x-\mu}{\sigma}\right)\,\Phi\left(\frac{x-\mu}{\sigma}\right)^{\alpha-1}\left[1-\Phi\left(\frac{x-\mu}{\sigma}\right)\right]^{\alpha-1}\left[\log(1-\beta)\right]^{-1}}{-\sigma\left\{\Phi\left(\frac{x-\mu}{\sigma}\right)^{\alpha}+\left[1-\Phi\left(\frac{x-\mu}{\sigma}\right)\right]^{\alpha}\right\}\left\{(1-\beta)\Phi\left(\frac{x-\mu}{\sigma}\right)^{\alpha}+\left[1-\Phi\left(\frac{x-\mu}{\sigma}\right)\right]^{\alpha}\right\}},$$

where $x \in \mathbb{R}$, $\mu \in \mathbb{R}$ is a location parameter, $\sigma > 0$ is a scale parameter, α and β are the shape and scale parameters, and $\phi(\cdot)$ and $\Phi(\cdot)$ are the pdf and cdf of the standard normal

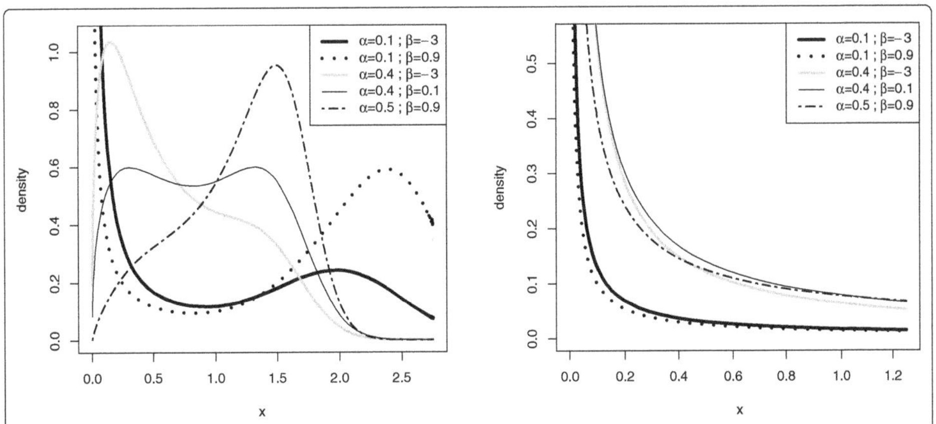

Fig. 1 Plots of the OLLL-W density for $a = 3.5$ and $b = 1$ (*left*) and for $a = 0.5$ and $b = 1$ (*right*)

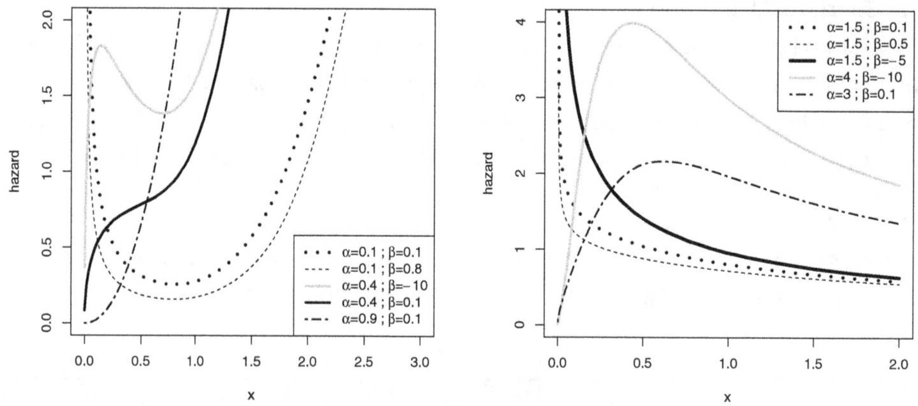

Fig. 2 Plots of the OLLL-W hrf for $a = 3.5$ and $b = 1$ (*left*) and for $a = 0.5$ and $b = 1$ (*right*)

distribution, respectively. For $\mu = 0$ and $\sigma = 1$, we obtain the OLLL-standard normal (OLLL-SN) distribution. Plots of the OLLL-SN density function for selected parameter values are displayed in Fig. 3. We note that this model is suitable for unimodal and bimodal data sets.

3 Linear representations

Let $A(u) = \frac{u^\alpha}{u^\alpha+(1-u)^\alpha}$ be the cdf of the odd log-logistic uniform (Gleaton and Lynch 2004, 2006, 2010) distribution. For $\beta \in (0, 1)$, we have $0 < \beta A(u) < 1$. Then, we can apply the power series reported in Appendix 2 by taking $u = G(x)^\alpha$, since they are always convergent in the interval $(0, 1)$, i.e., the power series are valid in the support of X. Henceforth, we consider that $0 < \beta < 1$, which is not a restrictive assumption since it is in agreement with the logarithm distribution defined for compounding the OLLL-G family.

Based on the power series $-\log(1-u) = \sum_{i=1}^{\infty} \frac{u^i}{i}$ (which converges for $|u| < 1$), the OLLL-G family cdf can be expanded as

$$F(x) = \frac{-1}{\log(1-\beta)} \sum_{i=1}^{\infty} \frac{\beta^i}{i} \left[\frac{G(x)^\alpha}{G(x)^\alpha + [1-G(x)]^\alpha}\right]^i.$$

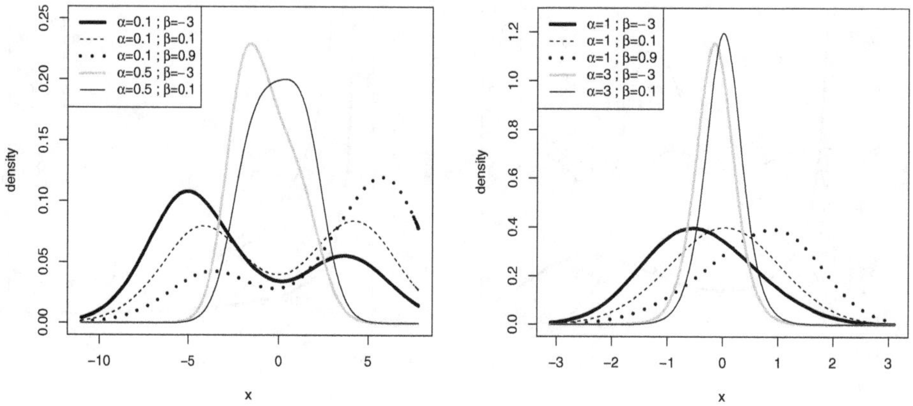

Fig. 3 Plots of the OLLL-SN density for selected values of α and β

By using Eq. (24) given in Appendix 2, we obtain

$$F(x) = \frac{-1}{\log(1-\beta)} \sum_{i=1}^{\infty} \frac{\beta^i}{i} \sum_{k=0}^{\infty} h_k^*(\alpha, i)\, G(x)^k,$$

where the coefficients $h_k^*(\alpha, i)$ can be determined from the recursive formula given after (24).

Further, we define the exponentiated-G ("Exp-G") distribution for an arbitrary parent distribution G, say $W \sim \text{Exp}^c G$, if W has cdf and pdf given by $H_c(x) = G(x)^c$ and $h_c(x) = c\, g(x)\, G(x)^{c-1}$, respectively. This transformed model is also called the Lehmann type I distribution, say $\text{Exp}^c(G)$.

Then, we can rewrite $F(x)$ as

$$F(x) = \sum_{k=0}^{\infty} d_k\, H_k(x), \tag{6}$$

where $d_k = \sum_{i=1}^{\infty} \frac{-\beta^i h_k^*(\alpha,i)}{i \log(1-\beta)}$ and $H_k(x)$ is the Exp-G cdf with power parameter k.

By differentiating (6), the pdf of X follows as

$$f(x) = \sum_{k=0}^{\infty} d_{k+1}\, h_{k+1}(x), \tag{7}$$

where $h_{k+1}(x) = (k+1)\, G(x)^k\, g(x)$ is the Exp-G density function with power parameter $(k+1)$.

Equation (7) reveals that the OLLL-G density function is a linear combination of Exp-G densities. Some structural properties of the new family such as the ordinary and incomplete moments and generating function can be determined from well-established properties of the Exp-G distribution. The properties of Exp-G distributions have been studied by many authors in recent years, see Mudholkar and Srivastava (1993) and Mudholkar et al. (1995) for exponentiated Weibull, Gupta and Kundu (1999) for exponentiated exponential and Nadarajah (2006) for exponentiated Gumbel, among others. The linear representations (6) and (7) are the main results of this section.

4 Moments and generating function

Let Y_k be a exp-G random variable with power parameter $k+1$, i.e., having density $h_{k+1}(x)$. The nth ordinary moment of $X \sim$ OLLL-G follows from (7) as

$$E(X^n) = \sum_{k=0}^{\infty} d_{k+1}\, E(Y_k^n) = \sum_{k=0}^{\infty} (k+1)\, d_{k+1}\, \tau(n,k), \tag{8}$$

where $\tau(n,k) = \int_{-\infty}^{\infty} x^n\, G(x)^k\, g(x) \mathrm{d}x = \int_0^1 Q_G(u)^n\, u^k \mathrm{d}u$. In fact, it is possible to exchange the infinite sum and the integral using the dominated convergence theorem for series.

Expressions for moments of several exp-G distributions are given by Nadarajah and Kotz (2006), which can be used to obtain $E(X^n)$. Cordeiro and Nadarajah (2011) determined $\tau(n,k)$ for some well-known distributions such as the normal, beta, gamma and Weibull distributions.

The variance, skewness and kurtosis measures are given by

$$\begin{aligned}
\text{Var}(X) &= E(X^2)-[E(X)]^2,\\
\text{Skewness}(X) &= \frac{E(X^3)-3E(X)E(X^2)+2[E(X)]^3}{[Var(X)]^{\frac{3}{2}}},\\
\text{Kurtosis}(X) &= \frac{E(X^4)-4E(X)E(X^3)+6E(X^2)[E(X)]^2-3[E(X)]^4}{[Var(X)]^2}.
\end{aligned}$$

Plots of the skewness and kurtosis of the OLLL-W distribution as functions of α and β for $a = 2$ and $b = 1$ are displayed in Fig. 4.

For empirical purposes, the shape of many distributions can be usefully described by what we call the incomplete moments. These types of moments play an important role for measuring inequality, for example, income quantiles and Lorenz and Bonferroni curves, which depend upon the incomplete moments of a distribution. The nth incomplete moment of X is calculated as

$$m_n(y) = \int_{-\infty}^{y} x^n f(x)\mathrm{d}x = \sum_{k=0}^{\infty}(k+1)\, d_{k+1} \int_0^{G(y)} Q_G(u)^n\, u^k \mathrm{d}u. \tag{9}$$

The last integral can be determined analytically or numerically for most baseline distributions. Equation (9) can be used to determine conditional moments, mean deviations and Bonferroni and Lorentz curves of X.

Let $M(t) = E(\mathrm{e}^{tX})$ be the mgf of X. We can obtain $M(t)$ from (7) as

$$M(t) = \sum_{k=0}^{\infty} d_{k+1}\, M_k(t) = \sum_{i=0}^{\infty}(k+1)\, d_{k+1}\, \rho(t,k), \tag{10}$$

where $M_k(t)$ is the mgf of Y_k and $\rho(t,k) = \int_{-\infty}^{\infty} \mathrm{e}^{tx}\, G(x)^k g(x)\mathrm{d}x = \int_0^1 \exp[\,t\, Q_G(u)]\, u^k \mathrm{d}u$.

We can determine the mgfs for several OLLL-G distributions directly from Eq. (10).

We present some mathematical properties of the odd log-logistic logarithmic exponential (OLLL-E) distribution in Appendix 3 to illustrate the applicability of the previous results.

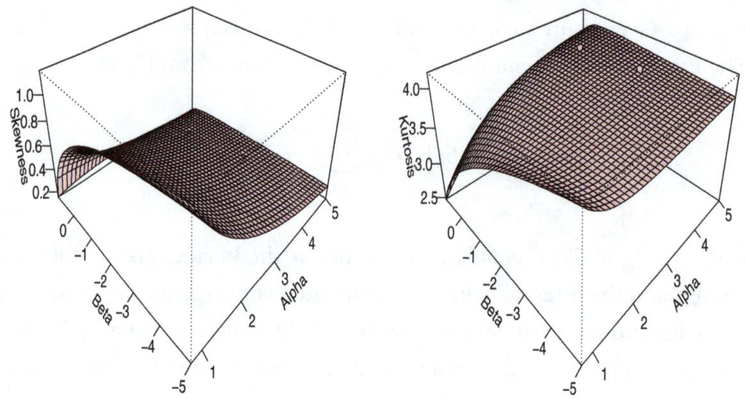

Fig. 4 Plots of the skewness (*left*) and kurtosis (*right*) of the OLLL-W distribution for $a = 2$ and $b = 1$

5 Other properties

We hardly need to emphasize the necessity and importance of entropies and order statistics in any statistical analysis especially in applied work.

5.1 Entropies

An entropy is a measure of variation or uncertainty of a random variable X. Two popular entropy measures are the Rényi and Shannon entropies (Shannon 1951; Rényi 1961). The Rényi entropy of a random variable with pdf $f(x)$ is defined by

$$I_R(\gamma) = \frac{1}{1-\gamma}\log\left(\int_0^{\infty} f^{\gamma}(x)dx\right),$$

for $\gamma > 0$ and $\gamma \neq 1$. The Shannon entropy of a random variable X is defined by $E\left\{-\log\left[f(X)\right]\right\}$. It is the special case of the Rényi entropy when $\gamma \uparrow 1$. Direct calculation gives

$$\begin{aligned}
\mathrm{E}\left\{-\log\left[f(X)\right]\right\} &= -\log\left[\frac{\alpha\beta}{-\log(1-\beta)}\right] - \mathrm{E}\left\{\log\left[g(X;\boldsymbol{\xi})\right]\right\} + (1-\alpha)\mathrm{E}\left\{\log\left[G(X;\boldsymbol{\xi})\right]\right\} \\
&+(1-\alpha)\mathrm{E}\left\{\log\left[\bar{G}(X;\boldsymbol{\xi})\right]\right\} + 2\mathrm{E}\left\{\log\left[G(X;\boldsymbol{\xi})^{\alpha} + \bar{G}(X;\boldsymbol{\xi})^{\alpha}\right]\right\} \\
&+\mathrm{E}\left\{\log\left[1 - \frac{\beta G(X,\boldsymbol{\xi})^{\alpha}}{G(X,\boldsymbol{\xi})^{\alpha} + \bar{G}(X,\boldsymbol{\xi})^{\alpha}}\right]\right\}.
\end{aligned}$$

First, we define

$$A(a_1,a_2,a_3,a_4;\alpha,\beta) = \int_0^1 \frac{u^{a_1}(1-u)^{a_2}}{[u^{\alpha}+(1-u)^{\alpha}]^{a_3}\left[1-\frac{\beta u^{\alpha}}{u^{\alpha}+(1-u)^{\alpha}}\right]^{a_4}}du.$$

By using the binomial expansion, we have

$$A(a_1,a_2,a_3,a_4;\alpha,\beta) = \sum_{i,j=0}^{\infty}(-1)^{i+j}\binom{-a_4}{i}\binom{a_2}{j}\int_0^1 \frac{\beta^i u^{a_1+\alpha i+j}}{[u^{\alpha}+(1-u)^{\alpha}]^{a_3+i}}du.$$

Second, we have the power series from Eq. (28) given in Appendix 2

$$\left[\frac{1}{u^{\alpha}+(1-u)^{\alpha}}\right]^{a_3+i} = \sum_{k=0}^{\infty} s_k^*(\alpha,a_3+i,0)\,u^k,$$

where $s_k^*(\alpha,a_3+i,0)$ is defined there. Then,

$$A(a_1,a_2,a_3,a_4;\alpha,\beta) = \sum_{i,j,k=0}^{\infty}\frac{(-1)^{i+j}\beta^i\binom{-a_4}{i}\binom{a_2}{j}s_k^*(\alpha,a_3+i,0)}{k+1+a_1+\alpha i+j}.$$

After some algebraic manipulations, we can write

$$\mathrm{E}\left\{\log\left[G(X)\right]\right\} = \frac{\alpha\beta}{-\log(1-\beta)}\frac{\partial}{\partial t}A(\alpha+t-1,\alpha-1,2,1;\alpha,\beta)\Big|_{t=0},$$

$$\mathrm{E}\left\{\log\left[\bar{G}(X)\right]\right\} = \frac{\alpha\beta}{-\log(1-\beta)}\frac{\partial}{\partial t}A(\alpha-1,\alpha+t-1,2,1;\alpha,\beta)\Big|_{t=0},$$

$$\mathrm{E}\left\{\log\left\{G(X;\boldsymbol{\xi})^{\alpha}+\bar{G}(X;\boldsymbol{\xi})^{\alpha}\right\}\right\} = \frac{\alpha\beta}{-\log(1-\beta)}\frac{\partial}{\partial t}A(\alpha-1,\alpha-1,2-t,1;\alpha,\beta)\Big|_{t=0},$$

$$\mathrm{E}\left\{\log\left[1-\frac{\beta G(X,\boldsymbol{\xi})^{\alpha}}{G(X,\boldsymbol{\xi})^{\alpha}+\bar{G}(X,\boldsymbol{\xi})^{\alpha}}\right]\right\} = \frac{\alpha\beta}{-\log(1-\beta)}\frac{\partial}{\partial t}A(\alpha-1,\alpha-1,2,1-t;\alpha,\beta)\Big|_{t=0}.$$

Then, the Shannon entropy of X reduces to

$$\begin{aligned}\mathrm{E}\{-\log[f(X)]\} &= -\log\left[\frac{\alpha\beta}{-\log(1-\beta)}\right] - \mathrm{E}\{\log[g(X;\boldsymbol{\xi})]\}\\ &-\frac{\alpha(1-\alpha)\beta}{\log(1-\beta)}\frac{\partial}{\partial t}A(\alpha+t-1,\alpha-1,2,1;\alpha,\beta)\Big|_{t=0}\\ &-\frac{\alpha(1-\alpha)\beta}{\log(1-\beta)}\frac{\partial}{\partial t}A(\alpha-1,\alpha+t-1,2,1;\alpha,\beta)\Big|_{t=0}\\ &-\frac{2\alpha\beta}{\log(1-\beta)}\frac{\partial}{\partial t}A(\alpha-1,\alpha-1,2-t,1;\alpha,\beta)\Big|_{t=0}\\ &-\frac{\alpha\beta}{\log(1-\beta)}\frac{\partial}{\partial t}A(\alpha-1,\alpha-1,2,1-t;\alpha,\beta)\Big|_{t=0}.\end{aligned}$$

For the Rényi entropy, after some algebraic developments, we obtain

$$I_R(\gamma) = \tfrac{\gamma}{1-\gamma}\log\left[\tfrac{\alpha\beta}{-\log(1-\beta)}\right] + \tfrac{1}{1-\gamma}\log\left[B(\alpha,\beta,\gamma)\right],$$

where

$$B(\alpha,\beta,\gamma) = \int_0^1 \frac{g^{\gamma-1}\left(G^{-1}(u)\right)u^{\gamma(\alpha-1)}(1-u)^{\gamma(\alpha-1)}}{[u^\alpha+(1-u)^\alpha]^{2\gamma}\left[1-\frac{\beta\,u^\alpha}{u^\alpha+(1-u)^\alpha}\right]^\gamma}du.$$

By using the generalized binomial expansion, we have

$$B(\alpha,\beta,\gamma) = \sum_{i,j=0}^{\infty}(-1)^{i+j}\beta^i\binom{-\gamma}{i}\binom{\gamma(\alpha-1)}{j}\int_0^1\frac{g^{\gamma-1}\left(G^{-1}(u)\right)u^{\gamma(\alpha-1)+\alpha\,i+j}}{[u^\alpha+(1-u)^\alpha]^{2\gamma+i}}du.$$

Further, we can write from Eq. (28)

$$\frac{u^{-\gamma}}{[u^\alpha+(1-u)^\alpha]^{2\gamma+i}} = \sum_{k=0}^{\infty}s_k^*(\alpha,2\gamma+i,\gamma)\,u^k,$$

where $s_k^*(\alpha,2\gamma+i,\gamma)$ is defined there. Finally,

$$B(\alpha,\beta,\gamma) = \sum_{i,j,k=0}^{\infty}\frac{(-1)^{i+j}\beta^i}{k+1+\alpha\,(\gamma+i)+j}\binom{-\gamma}{i}\binom{\gamma(\alpha-1)}{j}s_k^*(\alpha,2\gamma+i,\gamma)\mathrm{E}_{Y_k}\{g^{\gamma-1}\left(G^{-1}(Y)\right)\},$$

where $Y_k \sim \mathrm{Beta}(k+1+\alpha\,(\gamma+i)+j,1)$. Figure 5 displays plots of the Rényi entropy versus γ for $a=3.5$, $b=1$ and selected values of α and β.

5.2 Order statistics

Order statistics make their appearance in many areas of statistical theory and practice. Suppose that $X_1,\cdots,X_n$ is a random sample from the OLLL-G distribution. Let $X_{i:n}$ denote the ith order statistic. From Eqs. (6) and (7), the pdf of $X_{i:n}$ can be written as

$$f_{i:n}(x) = K\sum_{j=0}^{n-i}(-1)^j\binom{n-i}{j}\left[\sum_{r=0}^{\infty}d_{r+1}\,(r+1)\,G(x)^r g(x)\right]\left[\sum_{k=0}^{\infty}d_k\,G(x)^k\right]^{j+i-1},$$

where $K = n!/[\,(i-1)!\;(n-i)!\,]$. By using a result of Gradshteyn and Ryzhik (2000, Section 0.314) for a power series raised to a positive integer number, we obtain

$$\left[\sum_{k=0}^{\infty}d_k\,G(x)^k\right]^{j+i-1} = \sum_{k=0}^{\infty}e_{j+i-1,k}\,G(x)^k,$$

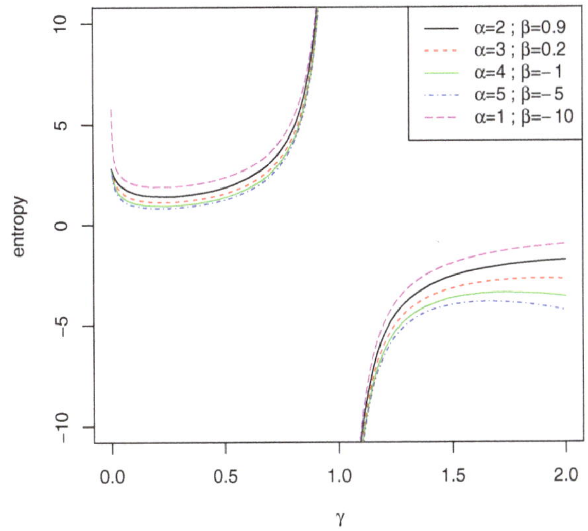

Fig. 5 Plots of Rényi entropy of the OLLL-W distribution versus γ for $a = 3.5$, $b = 1$ and selected values of α and β

where $e_{j+i-1,0} = d_0^{j+i-1}$ and, for $k \geq 1$,

$$e_{j+i-1,k} = (k\,d_0)^{-1} \sum_{q=1}^{k} [\,q(j+i) - k]\; d_q\, e_{j+i-1,k-q}.$$

Setting $d_r^* = (r+1)d_{r+1}$ and using a result of Gradshteyn and Ryzhik (2000, Section 0.316) for multiplying two power series, we have

$$\begin{aligned} f_{i:n}(x) &= K \sum_{j=0}^{n-i} (-1)^j \binom{n-i}{j} g(x) \left[\sum_{r=0}^{\infty} d_r^*\; G(x)^r\right] \left[\sum_{k=0}^{\infty} e_{j+i-1,k}\; G(x)^k\right] \\ &= K \sum_{j=0}^{n-i} (-1)^j \binom{n-i}{j} g(x) \sum_{k=0}^{\infty} e_k^*\, G(x)^k, \end{aligned}$$

where $e_k^* = \sum_{q=0}^{k} e_{j+i-1,q}\, d_{k-q}^*$. Hence, we can write

$$f_{i:n}(x) = \sum_{k=0}^{\infty} s_k\, h_{k+1}(x), \tag{11}$$

where (for $k \geq 0$)

$$s_k = \frac{k}{k+1} \sum_{j=0}^{n-i} (-1)^j \binom{n-i}{j} e_k^*.$$

Equation (11) is the main result of this section. It reveals that the pdf of the OLLL-G order statistics is a linear combination of Exp-G densities. So, several mathematical quantities of the OLLL-G order statistics such as ordinary, incomplete and factorial moments, mgf, mean deviations, among others, can be obtained from those quantities of the Exp-G distribution.

6 Estimation

In this section, we determine the MLEs of the model parameters of the new family from complete samples. Let $x_1, \cdots, x_n$ be the observed values from the OLLL-G distribution with parameters α, β and $\boldsymbol{\xi}$. Let $\boldsymbol{\theta} = (\alpha, \beta, \boldsymbol{\xi})^\top$ be the parameter vector. The total log-likelihood function for $\boldsymbol{\theta}$ is given by

$$\ell_n = \ell_n(\boldsymbol{\theta}) = n \log\left[\frac{\alpha\beta}{-\log(1-\beta)}\right] + \sum_{i=1}^{n} \log[g(x_i; \boldsymbol{\xi})] + (\alpha - 1)\sum_{i=1}^{n} \log\left[G(x_i; \boldsymbol{\xi})\bar{G}(x_i; \boldsymbol{\xi})\right]$$
$$-\sum_{i=1}^{n} \log\left[G(x_i; \boldsymbol{\xi})^\alpha + \bar{G}(x_i; \boldsymbol{\xi})^\alpha\right] - \sum_{i=1}^{n} \log\left[(1-\beta)G(x_i; \boldsymbol{\xi})^\alpha + \bar{G}(x_i; \boldsymbol{\xi})^\alpha\right]. \qquad (12)$$

The log-likelihood function can be maximized either directly or by solving the nonlinear likelihood equations obtained by differentiating (12). We use the goodness.fit function in R (R Development Core Team 2013) and NLMixed procedure in SAS to obtain the MLEs. The components of the score function $U_n(\theta) = (\partial\ell_n/\partial\alpha, \partial\ell_n/\partial\beta, \partial\ell_n/\partial\boldsymbol{\xi})^\top$ are

$$\frac{\partial\ell_n}{\partial\alpha} = \frac{n}{\alpha} + \sum_{i=1}^{n} \log\left[G(x_i; \boldsymbol{\xi})\bar{G}(x_i; \boldsymbol{\xi})\right]$$
$$-\sum_{i=1}^{n} \frac{G(x_i; \boldsymbol{\xi})^\alpha \log\left[G(x_i; \boldsymbol{\xi})\right] + \bar{G}(x_i; \boldsymbol{\xi})^\alpha \log\left[\bar{G}(x_i; \boldsymbol{\xi})\right]}{G(x_i; \boldsymbol{\xi})^\alpha + \bar{G}(x_i; \boldsymbol{\xi})^\alpha}$$
$$-\sum_{i=1}^{n} \frac{(1-\beta)G(x_i; \boldsymbol{\xi})^\alpha \log\left[G(x_i; \boldsymbol{\xi})\right] + \bar{G}(x_i; \boldsymbol{\xi})^\alpha \log\left[\bar{G}(x_i; \boldsymbol{\xi})\right]}{(1-\beta)G(x_i; \boldsymbol{\xi})^\alpha + \bar{G}(x_i; \boldsymbol{\xi})^\alpha},$$

$$\frac{\partial\ell_n}{\partial\beta} = \frac{n}{\beta} + \frac{n}{(1-\beta)\log(1-\beta)} + \sum_{i=1}^{n} \frac{G(x_i; \boldsymbol{\xi})^\alpha}{(1-\beta)G(x_i; \boldsymbol{\xi})^\alpha + \bar{G}(x_i; \boldsymbol{\xi})^\alpha}$$

and

$$\frac{\partial\ell_n}{\partial\boldsymbol{\xi}} = \sum_{i=1}^{n} \frac{g^{(\xi)}(x_i, \boldsymbol{\xi})}{g(x_i, \boldsymbol{\xi})} + (\alpha - 1)\sum_{i=1}^{n} \frac{G^{(\xi)}(x_i, \boldsymbol{\xi})}{G(x_i, \boldsymbol{\xi})} + (1-\alpha)\sum_{i=1}^{n} \frac{G^{(\xi)}(x_i, \boldsymbol{\xi})}{\bar{G}(x_i, \boldsymbol{\xi})}$$
$$-\alpha\sum_{i=1}^{n} \frac{G^{(\xi)}(x_i, \boldsymbol{\xi})\left[G(x_i; \boldsymbol{\xi})^{\alpha-1} - \bar{G}(x_i; \boldsymbol{\xi})^{\alpha-1}\right]}{G(x_i; \boldsymbol{\xi})^\alpha + \bar{G}(x_i; \boldsymbol{\xi})^\alpha}$$
$$-\alpha\sum_{i=1}^{n} \frac{G^{(\xi)}(x_i, \boldsymbol{\xi})\left[(1-\beta)G(x_i; \boldsymbol{\xi})^{\alpha-1} - \bar{G}(x_i; \boldsymbol{\xi})^{\alpha-1}\right]}{(1-\beta)G(x_i; \boldsymbol{\xi})^\alpha + \bar{G}(x_i; \boldsymbol{\xi})^\alpha},$$

where $h^{(\xi)}(\cdot)$ means the derivative of the function h with respect to ξ.

6.1 Simulation study

We simulate the OLLL-SN distribution (with $\mu = 0, \sigma = 1, \alpha = 0.2, 0.5, \beta = -0.5, 0.7$) from Eq. (5) by using a random variable U having a uniform distribution in $(0, 1)$. We simulate n= 50, 150 and 300 variates and, for each replication, we evaluate the MLEs $\hat{\mu}$, $\hat{\sigma}$, $\hat{\alpha}$ and $\hat{\beta}$. We repeat this process 2000 times and determine the average estimates (AEs), biases and means squared errors (MSEs). The results are reported in Table 2.

The figures in Table 2 indicate that the MSEs and the AEs of the estimates of μ, σ, α and β decay toward zero when the sample size increases, as expected under first-order asymptotic theory. As n increases, the AEs of the parameters tend to be closer to the true parameter values. This fact supports that the asymptotic normal distribution provides an adequate approximation to the finite sample distribution of the MLEs.

Table 2 The AEs, biases and MSEs of the OLLLN distribution for μ=0, σ=1, $\alpha = 0.2, 0.5$ and $\beta = -0.5, 0.7$ varying n

		$\alpha = 0.2$ $\beta = -0.5$				$\alpha = 0.5$ $\beta = -0.5$		
n	Parameter	AE	Bias	MSE	Parameter	AE	Bias	MSE
50	μ	-0.0499	-0.0499	0.3033	μ	-0.0602	-0.0602	0.2529
	σ	0.9025	-0.0975	0.2581	σ	1.1766	0.1766	42.5479
	α	0.1926	-0.0074	0.0358	α	0.6543	0.1543	20.1315
	β	-0.7939	-0.2939	1.6520	β	-1.0228	-0.5228	3.4332
150	μ	-0.0518	-0.0518	0.2303	μ	-0.0003	-0.0003	0.1117
	σ	0.8911	-0.1089	0.1107	σ	0.9765	-0.0235	0.1005
	α	0.1771	-0.0229	0.0137	α	0.4934	-0.0066	0.0487
	β	-0.5812	-0.0812	0.4428	β	-0.9119	-0.4119	2.1433
300	μ	-0.0395	-0.0395	0.1105	μ	0.0037	0.0037	0.0551
	σ	0.8787	-0.1213	0.0772	σ	0.9877	-0.0123	0.0452
	α	0.1737	-0.0263	0.0099	α	0.4973	-0.0027	0.0211
	β	-0.5532	-0.0532	0.1873	β	-0.7216	-0.2216	0.9713
		$\alpha = 0.2$ $\beta = 0.7$				$\alpha = 0.5$ $\beta = 0.7$		
n	Parameter	AE	Bias	MSE	Parameter	AE	Bias	MSE
50	μ	0.0436	0.0436	0.3052	μ	0.0792	0.0792	0.2650
	σ	0.9008	-0.0992	0.1778	σ	1.1175	0.1175	83.1835
	α	0.1933	-0.0067	0.0251	α	0.6144	0.1144	42.5964
	β	0.5943	-0.1057	0.1579	β	0.4014	-0.2986	0.8603
150	μ	-0.0064	-0.0064	0.2759	μ	0.0152	0.0152	0.1170
	σ	0.8930	-0.1070	0.1148	σ	0.9899	-0.0101	0.0938
	α	0.1864	-0.0136	0.0151	α	0.5029	0.0029	0.0474
	β	0.6668	-0.0332	0.0500	β	0.6043	-0.0957	0.1170
300	μ	0.0624	0.0624	0.1436	μ	-0.0038	-0.0038	0.0616
	σ	0.8393	-0.1607	0.0897	σ	0.9916	-0.0084	0.0460
	α	0.1713	-0.0287	0.0117	α	0.5012	0.0012	0.0227
	β	0.6797	-0.0203	0.0310	β	0.6603	-0.0397	0.0356

7 The OLLL-G family with long-term survival

Models for survival analysis typically consider that every subject in the population under study is susceptible to the event of interest and will eventually experience such event if follow-up is sufficiently long. However, there are situations when a fraction of individuals are not expected to experience the event of interest, that is, those individuals are cured or not susceptible. Cure rate models for survival data have been used to model time-to-event data for various types of cancers, including breast cancer, non-Hodgkins lymphoma, leukemia, prostate cancer and melanoma. These models have become very popular due to significant progress in treatment therapies leading to enhanced cure rates.

Models to accommodate a cured fraction have been widely developed. Perhaps the most popular type of cure rate models are the mixture models (MMs) developed by Boag (1949), Berkson and Gage (1952) and Farewell (1982). Some proposals have been made recently in the literature by more long term survival to model lifetimes with covariates. For example, Ortega et al. (2012) considered the problem of assessing local influence in the negative binomial beta Weibull regression model to predict the cure of prostate cancer, Hashimoto et al. (2013) derived curvature quantities under various perturbation schemes in Neyman type A beta-Weibull model for long-term survivors, Fachini et al. (2014) adapted local influence methods to a bivariate regression model with cure fraction and, recently, Ortega et al. (2015) used local influence methods to the power series beta-Weibull regression model for predicting breast carcinoma. The MMs allow simultaneously estimating whether the event of interest will occur, which is called incidence, and when it will occur, given that it can occur, which is called latency. Let N_i (for $i = 1,\ldots,n$) be the indicator denoting that the ith individual is susceptible ($N_i = 1$) or non-susceptible ($N_i = 0$), i.e., the population is divided into two sub-populations so that an individual either is cured with probability $0 < p < 1$, or has a proper survival function $S(x)$ with probability $(1 - p)$. In this work, we do not consider the regression structure, although future research using covariates on the probability p may be investigated. The MM can be represented by

$$S_{pop}(x_i) = p + (1-p)S(x_i|N_i = 1), \tag{13}$$

where $S_{pop}(x_i)$ is the unconditional survival function of x_i for the entire population, $S(x_i|N_i = 1) = 1 - F(x_i|N_i = 1)$ is the survival function for the susceptible individuals and $p = P(N_i = 0)$ is the probability of cure of an individual. The pdf corresponding to (13) is given by

$$f_{pop}(x_i) = -\frac{d\,S_{pop}(x_i)}{dt} = (1-p)f(x_i|N_i = 1), \tag{14}$$

where $f(x_i|N_i = 1)$ is the baseline pdf (see Section 2.1) for the susceptible individuals. Equations (13) and (14) are improper functions, since $S_{pop}(x)$ is not a proper survival function. We can omit sometimes the dependence on the indicator N_i and write simply $S(x_i|N_i = 1) = S(x), f(x_i|N_i = 1) = f(x)$, etc.

Inserting (3) in (14) and (2) in (13), the pdf and survival function of the OLLL-G cure rate family are given, respectively, by

$$f_{pop}(x;\alpha,\beta,\xi,p) = \frac{(1-p)\,\alpha\beta g(x,\xi)G(x,\xi)^{\alpha-1}\bar{G}(x,\xi)^{\alpha-1}}{-\left[G(x,\xi)^{\alpha} + \bar{G}(x,\xi)^{\alpha}\right]\left[(1-\beta)G(x,\xi)^{\alpha} + \bar{G}(x,\xi)^{\alpha}\right]\log(1-\beta)} \tag{15}$$

and

$$S_{pop}(x_i) = p + \frac{(1-p)}{\log(1-\beta)}\left[\log(1-\beta) - \log\left(1 - \frac{\beta G(x,\xi)^{\alpha}}{G(x,\xi)^{\alpha} + \bar{G}(x,\xi)^{\alpha}}\right)\right]. \tag{16}$$

A random variable having density (15) is denoted by $X \sim \text{OLLGcr}(\alpha,\beta,\xi,p)$. The hrf of the OLLGcr model is given by $h_{pop}(t) = f_{pop}(t)/S_{pop}(t)$.

7.1 Estimation

We consider the situation when the time-to-event is not completely observed and is subject to right censoring. Let c_i denote the censoring time. We observe $x_i = \min\{x_i, c_i\}$ and $\delta_i = I(x_i \leq c_i)$, where $\delta_i = 1$ if x_i is a time-to-event and $\delta_i = 0$ if x_i is right censored (for $i = 1,\ldots,n$). From n pairs of times and censoring indicators $(x_1,\delta_1),\cdots,(x_n,\delta_n)$, the log-likelihood function under non-informative censoring is given by $\ell_n(\boldsymbol{\theta}) = \sum_{i\in F}\log f_{pop}(x_i;\boldsymbol{\theta}) + \sum_{i\in C}\log S_{pop}(x_i;\boldsymbol{\theta})$, where $\boldsymbol{\theta} = (\alpha,\beta,\xi,p)^T$ denotes the parameter vector and F and C denote the uncensored and censored sets of observations, respectively. Replacing $f_{pop}(x_i;\boldsymbol{\theta})$ and $S_{pop}(x;\boldsymbol{\theta})$ by (15) and (16), respectively, the log-likelihood reduces to

$$\begin{aligned}\ell_n(\boldsymbol{\theta}) &= r\log\left[\frac{(1-p)\,\alpha\beta}{-\log(1-\beta)}\right]+\sum_{i\in F}\log\left[g(x_i;\boldsymbol{\xi})\right]+(\alpha-1)\sum_{i\in F}\log\left[G(x_i;\boldsymbol{\xi})\bar{G}(x_i;\boldsymbol{\xi})\right]\\ &-\sum_{i\in F}\log\left[G(x_i;\boldsymbol{\xi})^{\alpha}+\bar{G}(x_i;\boldsymbol{\xi})^{\alpha}\right]-\sum_{i\in F}\log\left[(1-\beta)G(x_i;\boldsymbol{\xi})^{\alpha}+\bar{G}(x_i;\boldsymbol{\xi})^{\alpha}\right]\\ &+\sum_{i\in C}\log\left\{p+\frac{(1-p)}{\log(1-\beta)}\left[\log(1-\beta)-\log\left(1-\frac{\beta G(x,\boldsymbol{\xi})^{\alpha}}{G(x,\boldsymbol{\xi})^{\alpha}+\bar{G}(x,\boldsymbol{\xi})^{\alpha}}\right)\right]\right\},\end{aligned} \quad (17)$$

where r is the number of failures (uncensored observations). We can obtain the MLE $\widehat{\boldsymbol{\theta}}$ of $\boldsymbol{\theta}$ by maximizing the log-likelihood (17) either directly in R using the `optim` function, in SAS using the `NLMixed` procedure and in other statistical software or by solving the nonlinear likelihood equations obtained by differentiating (17).

8 Applications

In this section, we provide four applications to real data. In the first three applications, we present some results by fitting special models defined in Section 2.1. In the fourth application, we present an application using the long-term survival model defined in Section 7.

For the first three applications, the goodness-of-fit statistics including the Cramér-von Mises (W^*) and Anderson-Darling (A^*) test statistics are used to compare the fitted models; see Chen and Balakrishnan (1995) for more details. The smaller the values of A^* and W^*, the better the fit to the data. We also consider the Kolmogrov-Smirnov (K-S) statistic (and its corresponding p-value) and minus the maximized log-likelihood ($\hat{\ell}_n$) for the sake of comparison. For the fourth application (censored data), we adopt the AIC and BIC statistics to compare the fitted models since the A^* and W^* statistics are not suitable for censored data.

For the next three applications, we consider the OLLLN distribution and, for the purpose of comparison, we fit the following models to the data sets described below:

- The normal distribution.
- The exponentiated normal (EN) distribution.
- The logarithmic normal (LN) distribution, the special case of the OLLLN distribution when $\alpha = 1$.
- The beta normal (BN) distribution (Eugene et al. 2002) with density

$$f_{BN}(x)=\frac{1}{\sigma B(\alpha,\beta)}\left[\Phi\left(\frac{x-\mu}{\sigma}\right)\right]^{\alpha-1}\left[1-\Phi\left(\frac{x-\mu}{\sigma}\right)\right]^{\beta-1}\phi\left(\frac{x-\mu}{\sigma}\right).$$

- The gamma normal (GN) distribution (Alzaatreh et al. 2014) with density

$$f_{GN}(x)=\frac{\beta^{\alpha}}{\sigma\Gamma(\alpha)}\left[-\log\left\{1-\Phi\left(\frac{x-\mu}{\sigma}\right)\right\}\right]^{\alpha-1}\left[1-\Phi\left(\frac{x-\mu}{\sigma}\right)\right]^{\beta-1}\phi\left(\frac{x-\mu}{\sigma}\right).$$

- The Kumaraswamy normal (KN) distribution (Cordeiro and de Castro 2011) with density

$$f_{KN}(x)=\frac{\alpha\beta}{\sigma}\left\{\Phi\left[\left(\frac{x-\mu}{\sigma}\right)\right]\right\}^{\alpha-1}\left\{1-\left[\Phi\left(\frac{x-\mu}{\sigma}\right)\right]^{\alpha}\right\}^{\beta-1}\phi\left(\frac{x-\mu}{\sigma}\right).$$

- The odd log-logistic normal (OLL-N) distribution (special case of OLLLN distribution when $\beta \to 1$) with density (Braga et al. 2016)

$$f_{OLL-N}(x) = \frac{\alpha\,\phi\left(\frac{x-\mu}{\sigma}\right)\left[\Phi\left(\frac{x-\mu}{\sigma}\right)\right]^{\alpha-1}\left[1-\Phi\left(\frac{x-\mu}{\sigma}\right)\right]^{\alpha-1}}{\sigma\{[1-\Phi\left(\frac{x-\mu}{\sigma}\right)]^{\alpha}+[\Phi\left(\frac{x-\mu}{\sigma}\right)]^{\alpha}\}^{2}},$$

where $x \in \mathbb{R}$, $\mu \in \mathbb{R}$, $\alpha > 0$, $\beta > 0$ and $\sigma > 0$.

8.1 Application 1

First, we consider the data set representing the failure times of a particular windshield device. These data were also studied by Blischke and Murthy (2000) and Murthy et al. (2004). The data, referred as D1, are: 0.040, 1.866, 2.385, 3.443, 0.301, 1.876, 2.481, 3.467, 0.309, 1.899, 2.610, 3.478, 0.557, 1.911, 2.625, 3.578, 0.943, 1.912, 2.632, 3.595, 1.070, 1.914, 2.646, 3.699, 1.124, 1.981, 2.661, 3.779, 1.248, 2.010, 2.688, 3.924, 1.281, 2.038, 2.823, 4.035, 1.281, 2.085, 2.890, 4.121, 1.303, 2.089, 2.902, 4.167, 1.432, 2.097, 2.934, 4.240, 1.480, 2.135, 2.962, 4.255, 1.505, 2.154, 2.964, 4.278, 1.506, 2.190, 3.000, 4.305, 1.568, 2.194, 3.103, 4.376, 1.615, 2.223, 3.114, 4.449, 1.619, 2.224, 3.117, 4.485, 1.652, 2.229, 3.166, 4.570, 1.652, 2.300, 3.344, 4.602, 1.757, 2.324, 3.376, 4.663.

The MLEs of the parameters and the standard errors (SEs) in parentheses and the goodness-of-fit statistics for D1 are listed in Table 3. We note that the OLLLN model outperforms all the fitted competitive models under these statistics.

The histogram of the data D1 and fitted densities are displayed in Fig. 6. We note that the fitted OLLLN distribution best captures the empirical histogram. Based on the equations given in Section 4, we give some measures based on the moments of the OLLLN distribution. The expected value and the variance of the failure times of windshield devices are: $E(X) = 2.58$ and $Var(X) = 1.24$., respectively. Also, the skewness

Table 3 The MLEs of the parameters and SEs in parentheses and the goodness-of-fit statistics for D1

Model	μ	σ	α	β	$\hat{\ell}_n$	W^*	A^*	K-S	p-value
OLLLN	2.887	0.532	0.379	-1.639	126.178	0.035	0.331	0.053	0.963
	(0.212)	(0.180)	(0.192)	(1.947)					
Normal	2.557	1.112			128.119	0.092	0.608	0.092	0.445
	(0.121)	(0.086)							
EN	1.832	1.337	1.939		128.064	0.075	0.522	0.085	0.557
	(2.366)	(0.708)	(3.878)						
BN	0.808	2.443	7.113	2.469	128.085	0.074	0.520	0.084	0.562
	(7.151)	(8.162)	(48.582)	(14.619)					
GN	2.805	0.541	0.290	0.197	127.757	0.058	0.438	0.075	0.7105
	(1.059)	(0.265)	(0.382)	(0.216)					
KN	1.654	0.748	0.920	0.320	127.848	0.063	0.469	0.079	0.642
	(1.067)	(0.539)	(1.017)	(0.524)					
LN	3.172	1.079		-7.091	127.570	0.048	0.391	0.066	0.828
	(0.562)	(0.094)		(16.134)					
OLL-N	2.626	0.603	0.452		127.062	0.076	0.523	0.095	0.408
	(0.126)	(0.218)	(0.232)						

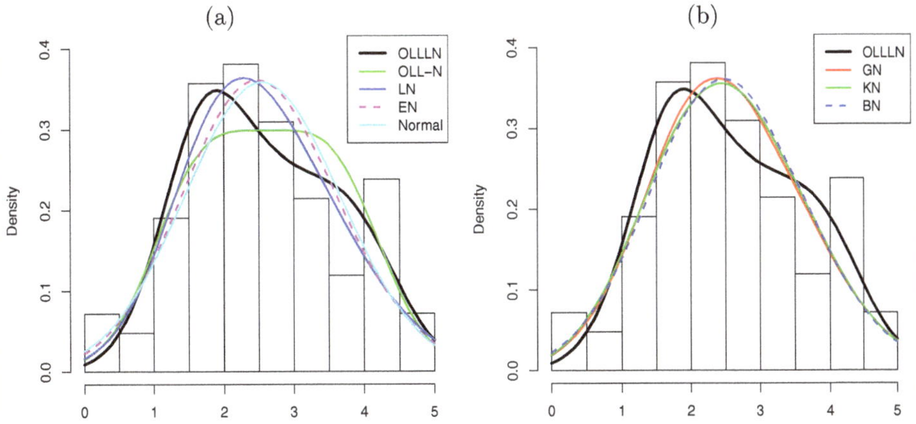

Fig. 6 For D1 data, the histogram and the fitted densities for: **a** OLLLN model and sub-models and **b** competitive models

and kurtosis measures are given by $Skewness(X) = 0.23$ and $Kurtosis(X) = 2.42$, thus indicating that the tail on the right side is longer and then it is a platykurtic distribution.

8.2 Application 2

The second data set D2 consists of lifetimes of 43 blood cancer patients (in days) from one of the Health Hospitals in Saudi Arabia (Abouammoh and Abdulghani 1994). These data are: 115, 181, 255, 418, 441, 461, 516, 739, 743, 789, 807, 865, 924, 983, 1025, 1062, 1063, 1165, 1191, 1222, 1222, 1251, 1277, 1290, 1357, 1369, 1408, 1455, 1478, 1519, 1578, 1578, 1599, 1603, 1605, 1696, 1735, 1799, 1815 ,1852, 1899, 1925, 1965.

The MLEs of the parameters and SEs (in parentheses) and the goodness-of-fit statistics for D2 are listed in Table 4. We conclude that the KN model has the smallest $\hat{\ell}_n$ and the OLLLN has the third smallest $\hat{\ell}_n$, the BN model has the smallest W^* and the OLLLN and KN models have the second smallest W^*'s. The OLLLN and BN models have the smallest A^*'s. The OLLLN and GN models have the smallest K-S statistics. The GN model has the largest p-value and the OLLLN and KN models have the second largest p-values. Therefore, we can conclude that the OLLLN model has either the best fit or is very close to the best fit with respect to the current criterions. The histogram of D2 and the fitted densities are displayed in Fig. 7.

8.3 Application 3

The third data set D3 includes the lower discharge of at least seven consecutive days and return period (time) of ten years of the Cuiabá River, Cuiabá, Mato Grosso, Brazil. These data have also been studied by Cordeiro et al. (2012). The MLEs of the parameters and SEs (in parentheses) and the goodness-of-fit statistics for D3 are listed in Table 5. We note that the OLLLN model outperforms all other fitted models.

The fitted densities for the models listed in Table 5 are displayed in Fig. 8. We verify that the fitted OLLLN distribution best captures the histogram of these data.

In summary, we conclude that the OLLLN distribution outperforms all the fitted competitive models under the selected criterion for D1, D2 and D3. For all three data sets, we verify that the fitted OLLLN distribution best captures the three histograms, especially for the third data set, which indicates the outstanding performance of this distribution.

Table 4 The MLEs of the parameters and SEs in parentheses and the goodness-of-fit statistics for D2

Model	μ	σ	α	β	$\hat{\ell}_n$	W^*	A^*	K-S	*p*-value
OLLLN	930.893	215.288	0.323	0.854	325.684	0.018	0.157	0.057	0.997
	(151.780)	(93.152)	(0.224)	(0.202)					
Normal	1191.628	500.709			328.303	0.068	0.489	0.083	0.901
	(76.357)	(53.993)							
EN	2083.910	52.5006	0.005		325.847	0.020	0.174	0.076	0.9485
	(81.898)	(14.640)	(0.003)						
BN	2153.003	173.89	0.0496	4.139	325.680	0.017	0.157	0.059	0.996
	(313.548)	(512.195)	(0.268)	(6.681)					
GN	2138.76	251.154	0.101	2.607	326.237	0.020	0.183	0.057	0.998
	(0.036)	(0.036)	(0.016)	(1.319)					
KN	1825.282	71.445	0.0097	0.529	325.4645	0.018	0.162	0.058	0.997
	(288.596)	(65.481)	(0.019)	(0.319)					
LN	748.777	472.416		0.974	326.628	0.026	0.221	0.064	0.991
	(275.414)	(71.311)		(0.070)					
OLL-N	1131.115	223.172	0.338		327.292	0.058	0.415	0.125	0.473
	(75.566)	(99.640)	(0.231)						

8.4 Application 4: OLLL-G long-term survival models

These data consist of $n = 493$ lifetimes (t_i in months) of patients diagnosed with breast cancer. The steps to construct these data can be found in Gendoo et al. (2015). In many applications there is qualitative information about the hazard shape, which can help for selecting a particular polyhazard model. In this context, a device called the total time on test (TTT) plot is useful. The TTT plot is obtained by plotting $\mathrm{G}(r/n) = \left[\left(\sum_{i=1}^{r} T_{i:n}\right) + (n-r)T_{r:n}\right] / \left(\sum_{i=1}^{n} T_{i:n}\right)$, where $r = 1, \ldots, n$, and $T_{i:n}$ (for $i = 1, \ldots, n$) are the order statistics of the sample, against r/n. It is a straight diagonal for constant hazards leading to an exponential model. It is convex for decreasing hazards and concave for increasing hazards leading to a single-Weibull model. It is first convex and then concave if the hazard is bathtub-shaped leading to a bi-Weibull model. It is first concave and

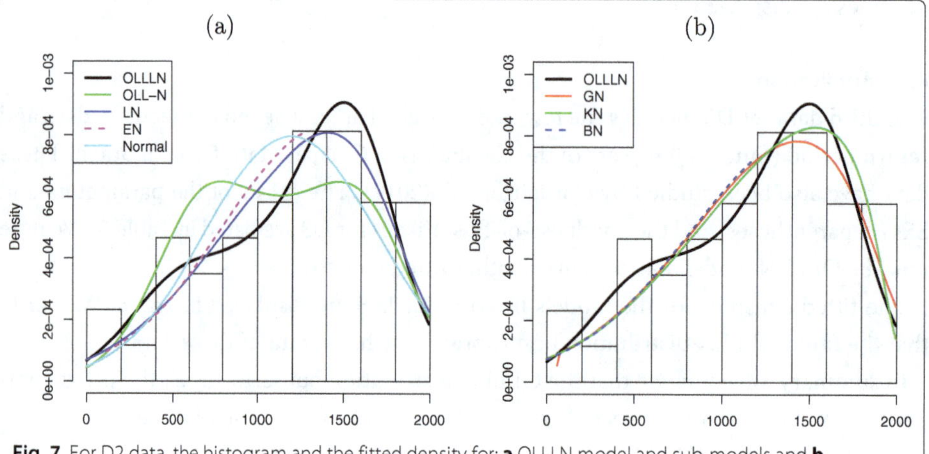

Fig. 7 For D2 data, the histogram and the fitted density for: **a** OLLLN model and sub-models and **b** competitive models

Table 5 The MLEs of the parameters and SEs in parentheses and the goodness-of-fit statistics for D3

Model	μ	σ	α	β	$\hat{\ell}_n$	W^*	A^*	K-S	p-value
OLLLN	102.050	7.879	0.100	0.521	188.193	0.024	0.206	0.076	0.9705
	(5.611)	(5.275)	(0.119)	(0.381)					
Normal	110.214	37.873			192.021	0.109	0.671	0.144	0.378
	(6.144)	(4.344)							
EN	169.145	13.261	0.062		191.360	0.074	0.491	0.113	0.674
	(26.432)	(13.067)	(0.142)						
BN	172.325	18.326	0.109	1.785	191.273	0.068	0.465	0.105	0.759
	(22.216)	(14.485)	(0.160)	(2.635)					
GN	172.393	17.696	0.102	1.340	191.263	0.068	0.465	0.106	0.752
	(20.263)	(15.038)	(0.159)	(2.564)					
KN	129.188	13.163	0.082	0.361	191.071	0.065	0.450	0.111	0.695
	(30.610)	(5.980)	(0.104)	(0.297)					
LN	62.065	33.056		0.997	190.819	0.059	0.417	0.109	0.717
	(23.996)	(6.044)		(0.011)					
OLL-N	105.546	7.830	0.098		188.655	0.034	0.271	0.110	0.7025
	(4.163)	(4.921)	(0.110)						

then convex if the hazard is bimodal-shaped leading to a log-logistic model. For multimodal hazards, the TTT plot contains several concave and convex regions. The TTT plot in Fig. 10a indicates an increasing-decreasing-increasing hrf. So, the OLLLW distribution would be a good option to model these data.

Next, we compare the results by fitting the OLLLWcr model and some of its sub-models such as: the odd log-logistic Weibull cure rate (OLL-Wcr) model (OLLLNcr distribution when $\beta \to 1$), the logarithmic Weibull cure rate (LWcr) model (OLLLNcr distribution when $\alpha = 1$) and Weibull cure rate model (OLLLNcr distribution when $\alpha = 1$ and $\beta \to 1$). Table 6 provides the MLEs (and the corresponding SEs in parentheses) of the model parameters and the values of the AIC and BIC statistics. The results indicate that the OLLLWcr model has the lowest values of these statistics among those values of the

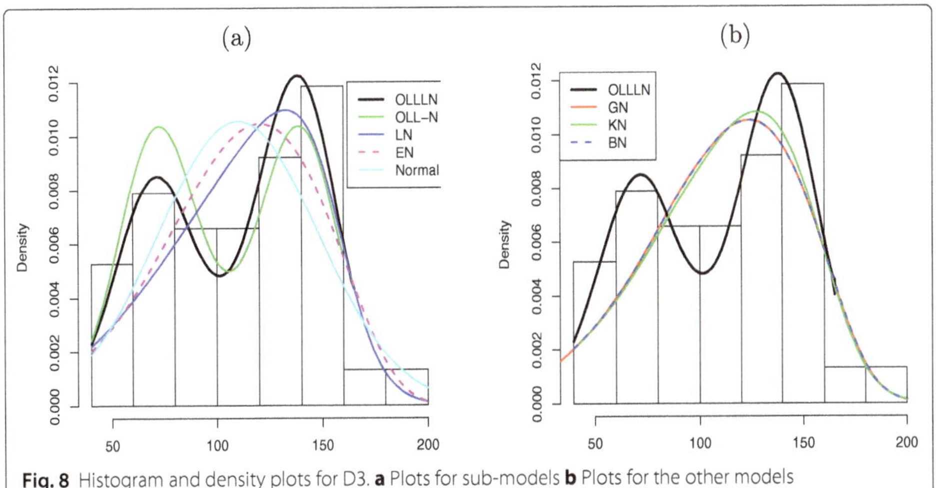

Fig. 8 Histogram and density plots for D3. **a** Plots for sub-models **b** Plots for the other models

Table 6 MLEs, SEs in parentheses and goodness-of-fit statistics for breast cancer data

Model	μ	σ	α	β	p	AIC	BIC
OLLLWcr	5.933	0.006	0.317	-9.131	0.575	1878.15	1899.15
	(0.079)	(0.001)	(0.084)	(4.319)	(0.061)		
OLL-Wcr	0.778	0.036	1.788	-	0.569	1886.37	1903.17
	(0.085)	(0.009)	(0.426)	-	(0.059)		
LWcr	1.581	0.007	-	-1.334	0.597	1885.27	1902.07
	(0.188)	(0.001)	-	(1.712)	(0.044)		
Weibull cure rate	1.027	0.040	-	-	0.550	1900.03	1912.63
	(0.082)	(0.008)	-	-	(0.064)		

fitted models, and therefore it could be chosen as the best model. On the other hand, the proportion of cured individuals obtained by the Kaplan-Meier estimator is 0.577. Thus, we can conclude based on the figures in Table 6 that the OLLLNcr model gives a more accurate estimate for the proportion of cured individuals.

The adequacy of the fitted models can also be noted in Fig. 9, which presents the empirical and estimated survival functions. Based on these plots, we can conclude that the OLLLWcr model provides a good fit for the breast cancer data. In additional, the empirical scaled TTT transform can be used to identify the shape of the hazard function. The fitted hazard function for the OLLLWcr model is displayed in Fig. 10b, which we observe bimodal shapes, thus indicating a good fit.

9 Conclusions

We study some mathematical properties of the odd log-logistic logarithmic-G family of distributions with two extra shape parameters $\alpha > 0$ and $\beta \in (0, 1)$. We provide some special models, a very useful linear representation for the density function in terms of exponentiated densities, explicit expressions for the moments, generating function, entropies and order statistics. The model parameters are estimated by the method of maximum likelihood. We perform a simulation study to verify the adequacy of the estimators. We also introduce a long-term survival model based on the new family. The importance

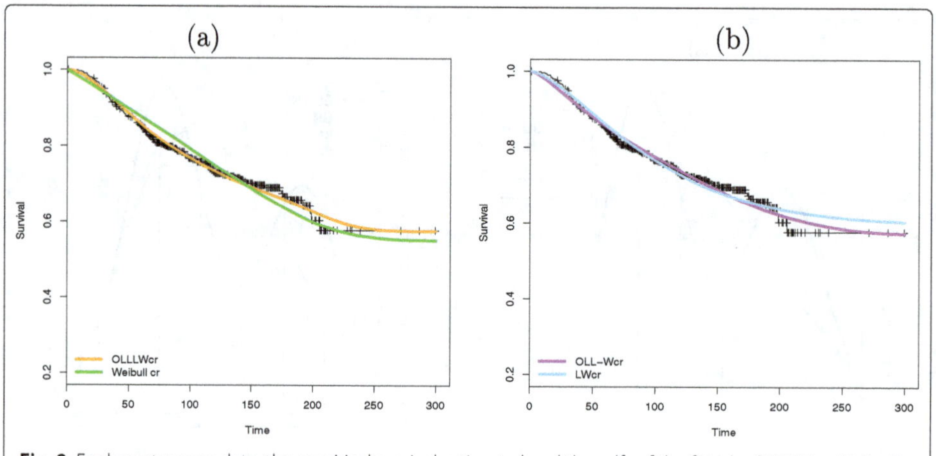

Fig. 9 For breast cancer data, the empirical survival estimated and the cdfs of the fitted **a** OLLLWcr, Weibull cure rate, **b** OLL-Wcr and LWcr models

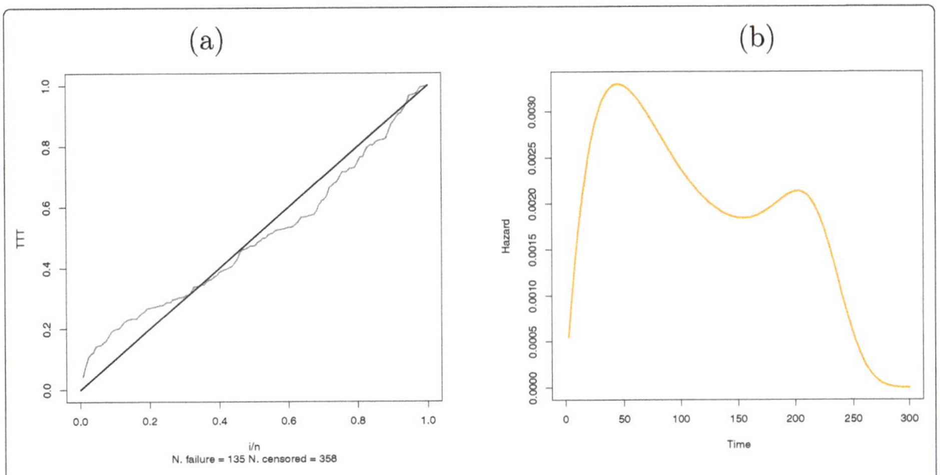

Fig. 10 For breast cancer data, the **a** TTT plot and **b** estimated hazard function for the OLLLWcr model

of the proposed models is illustrated by means of four real life data sets. The new models provide consistently better fits than other competitive models for the current data.

Appendix 1: Asymptotes and shapes

Corollary 1 *Let* $a = \inf\{x|F(x) > 0\}$. *The asymptotics of Eqs.* (2), (3) *and* (4) *when* $x \to a$ *are given by*

$$F(x) \sim \frac{\beta G(x)^{\alpha}}{-\log(1-\beta)} \quad \text{as } x \to a,$$
$$f(x) \sim \frac{\alpha\beta\, g(x)\, G(x)^{\alpha-1}}{-\log(1-\beta)} \quad \text{as } x \to a,$$
$$h(x) \sim \frac{\alpha\beta\, g(x)\, G(x)^{\alpha-1}}{-\log(1-\beta)} \quad \text{as } x \to a.$$

Corollary 2 *The asymptotics of Eqs.* (2), (3) *and* (4) *when* $x \to \infty$ *are given by*

$$1 - F(x) \sim 1 - \frac{\log\left[1 - \beta\, G(x)^{\alpha}\right]}{\log(1-\beta)} \quad as \ x \to \infty,$$
$$f(x) \sim \frac{\alpha\, \beta\, g(x)\, G(x)^{-1}}{-(1-\beta)\log(1-\beta)} \quad as \ x \to \infty,$$
$$h(x) \sim \frac{\alpha\, \beta\, g(x)\, G(x)^{-1}}{-(1-\beta)\log\left[\frac{1-\beta}{1-\beta\, G(x)^{\alpha}}\right]} \quad as \ x \to \infty.$$

The shapes of the density and hazard rate functions can be described analytically. The critical points of the OLLL-G density function are the roots of the equation:

$$\frac{g'(x)}{g(x)} + (\alpha-1)\frac{g(x)}{G(x)} + (1-\alpha)\frac{g(x)}{\bar{G}(x)} - \alpha\, g(x)\, \frac{G(x)^{\alpha-1} - \bar{G}(x)^{\alpha-1}}{G(x)^{\alpha} + \bar{G}(x)^{\alpha}}$$
$$-\alpha\, g(x)\, \frac{(1-\beta)G(x)^{\alpha-1} - \bar{G}(x)^{\alpha-1}}{(1-\beta)G(x)^{\alpha} + \bar{G}(x)^{\alpha}} = 0. \qquad (18)$$

There may be more than one root to (18). Let $\lambda(x) = \frac{d^2 \log[f(x)]}{dx^2}$. We have

$$\begin{aligned}\lambda(x) = {} & \frac{g''(x)g(x) - g'(x)^2}{g(x)^2} + (\alpha-1)\frac{g'(x)G(x) - g(x)^2}{G(x)^2} + (1-\alpha)\frac{g'(x)\bar{G}(x) + g(x)^2}{\bar{G}(x)^2} \\ & -\alpha\, g'(x)\,\frac{G(x)^{\alpha-1} - \bar{G}(x)^{\alpha-1}}{G(x)^{\alpha} + \bar{G}(x)^{\alpha}} - \alpha(\alpha-1)\, g(x)^2\,\frac{G(x)^{\alpha-2} + \bar{G}(x)^{\alpha-2}}{G(x)^{\alpha} + \bar{G}(x)^{\alpha}} \\ & + \left\{\alpha\, g(x)\,\frac{G(x)^{\alpha-1} - \bar{G}(x)^{\alpha-1}}{G(x)^{\alpha} + \bar{G}(x)^{\alpha}}\right\}^2 - \alpha\, g'(x)\,\frac{(1-\beta)G(x)^{\alpha-1} - \bar{G}(x)^{\alpha-1}}{(1-\beta)G(x)^{\alpha} + \bar{G}(x)^{\alpha}} \\ & -\alpha(\alpha-1)\, g(x)^2\,\frac{(1-\beta)G(x)^{\alpha-2} + \bar{G}(x)^{\alpha-2}}{(1-\beta)G(x)^{\alpha} + \bar{G}(x)^{\alpha}} \\ & + \left\{\alpha\, g(x)\,\frac{(1-\beta)G(x)^{\alpha-1} - \bar{G}(x)^{\alpha-1}}{(1-\beta)G(x)^{\alpha} + \bar{G}(x)^{\alpha}}\right\}^2.\end{aligned}$$

If $x = x_0$ is a root of (18) then it corresponds to a local maximum (minimum) if $\lambda(x_0) < 0$ ($\lambda(x_0) > 0$). It refers to a point of inflexion if $\lambda(x_0) = 0$.

The critical point of $h(x)$ are obtained from the equation

$$\begin{aligned}& \frac{g'(x)}{g(x)} + (\alpha-1)\frac{g(x)}{G(x)} + (1-\alpha)\frac{g(x)}{\bar{G}(x)} - \alpha\, g(x)\,\frac{G(x)^{\alpha-1} - \bar{G}(x)^{\alpha-1}}{G(x)^{\alpha} + \bar{G}(x)^{\alpha}} - \\ & \alpha\, g(x)\,\frac{(1-\beta)G(x)^{\alpha-1} - \bar{G}(x)^{\alpha-1}}{(1-\beta)G(x)^{\alpha} + \bar{G}(x)^{\alpha}} - \\ & \frac{\alpha\beta g(x)G(x)^{\alpha-1}\bar{G}(x)^{\alpha-1}}{\left[G(x)^{\alpha} + \bar{G}(x)^{\alpha}\right]\left[(1-\beta)G(x)^{\alpha} + \bar{G}(x)^{\alpha}\right]\left\{\log\left[\frac{1-\beta}{1-\frac{\beta\, G(x)^{\alpha}}{G(x)^{\alpha}+\bar{G}(x)^{\alpha}})}\right]\right\}} = 0. \qquad (19)\end{aligned}$$

There may be more than one root to (19).

Appendix 2: Useful power series

The power series derived in this appendix are required for the proofs of the linear representations in Section 3. All power series given below are convergent for $|u| \leq 1$. In Sections 3 and 5.1, they can be applied for the support of X since the quantity $\beta G(x,\xi)^{\alpha}/[\,G(x,\xi)^{\alpha} + \bar{G}(x,\xi)^{\alpha}]$ does belong to the interval $(0, 1)$ when $\beta \in (0, 1)$.

First, for $a > 0$, we have the generalized binomial expansion

$$(1-u)^{a} = \sum_{j=0}^{\infty} (-1)^{j} \binom{a}{j} u^{j}, \qquad (20)$$

which holds for $|u| \leq 1$.

Second, we obtain an expression for $\left[\frac{u^{\alpha}}{u^{\alpha}+(1-u)^{\alpha}}\right]^{m}$, where $\alpha > 0$ is a real number, m is a natural number and $|u| \leq 1$. We can write

$$u^{\alpha} = \sum_{k=0}^{\infty} a_k(\alpha)\, u^{k}, \qquad (21)$$

where

$$a_k(\alpha) = \sum_{j=k}^{\infty} (-1)^{k+j} \binom{\alpha}{j} \binom{j}{k}.$$

The power series (20) and (21) and the others derived from them converge everywhere. For any real $\alpha > 0$, the power series follows from (20) and (21)

$$u^{\alpha} + (1-u)^{\alpha} = \sum_{k=0}^{\infty} b_k(\alpha)\, u^k, \tag{22}$$

where $b_k(\alpha) = a_k(\alpha) + (-1)^k \binom{\alpha}{k}$. Combining (21) and (22), we have (see Gradshteyn and Ryzhik 2000, Section 0.313)

$$\frac{u^{\alpha}}{u^{\alpha} + (1-u)^{\alpha}} = \frac{\sum_{k=0}^{\infty} a_k(\alpha)\, u^k}{\sum_{k=0}^{\infty} b_k(\alpha)\, u^k} = \sum_{k=0}^{\infty} c_k(\alpha)\, u^k, \tag{23}$$

where $c_0(\alpha) = a_0(\alpha)/b_0(\alpha)$ and the $c_k(\alpha)$'s (for $k \geq 1$) are determined from the recurrence equation

$$c_k(\alpha) = \frac{1}{b_0(\alpha)} \left[a_k(\alpha) - \frac{1}{b_0(\alpha)} \sum_{r=1}^{k} b_r\, c_{k-r}(\alpha) \right].$$

Third, based on (23) and using a result of Gradshteyn and Ryzhik (2000 Section 0.314) for a power series raised to a positive integer number, we obtain

$$\left[\frac{u^{\alpha}}{u^{\alpha} + (1-u)^{\alpha}} \right]^m = \left(\sum_{k=0}^{\infty} c_k(\alpha) u^k \right)^m = \sum_{k=0}^{\infty} h_k^*(\alpha, m)\, u^k, \tag{24}$$

where $h_0^*(\alpha, m) = c_0(\alpha)^m$ and, for $k \geq 1$,

$$h_k^*(\alpha, m) = \frac{1}{k c_0(\alpha)} \sum_{q=1}^{k} [\,(m+1)q - k]\; c_q(\alpha)\, h_{k-q}^*(\alpha, m).$$

Fourth, we obtain a power series for $\frac{u^{-\gamma}}{[u^{\alpha} + (1-u)^{\alpha}]^w}$, which is applied in Section 5.1, where w and α are positive real numbers, $\gamma > 0$ and $0 < u < 1$. We have

$$\frac{u^{-\gamma}}{[u^{\alpha} + (1-u)^{\alpha}]^w} = u^{-\alpha w - \gamma} \left[\frac{u^{\alpha}}{u^{\alpha} + (1-u)^{\alpha}} \right]^w. \tag{25}$$

The second term in (25) can be expanded as

$$\left[\frac{u^{\alpha}}{u^{\alpha} + (1-u)^{\alpha}} \right]^w = \sum_{i=0}^{\infty} a_i(w) \left[\frac{u^{\alpha}}{u^{\alpha} + (1-u)^{\alpha}} \right]^i,$$

where $a_i(w) = \sum_{j=i}^{\infty} (-1)^{i+j} \binom{w}{j} \binom{j}{i}$. Now, from (24), we have

$$\left[\frac{u^{\alpha}}{u^{\alpha} + (1-u)^{\alpha}} \right]^w = \sum_{k=0}^{\infty} s_{1,k}^*(\alpha, w)\, u^k, \tag{26}$$

where

$$s_{1,k}^*(\alpha, w) = \sum_{i=0}^{\infty} a_i(w) h_k^*(\alpha, i).$$

The first term $u^{-\alpha w - \gamma}$ in (25) can be expanded as

$$u^{-\alpha w - \gamma} = \sum_{k=0}^{\infty} s_{2,k}^*(\alpha, w, \gamma) u^k, \tag{27}$$

where $s_{2,k}^{*}(\alpha,w,\gamma) = \sum_{i=k}^{\infty}(-1)^{i+k}\binom{-\alpha w-\gamma}{i}\binom{i}{k}$. Finally, from (25), (26), (27) and using a result of Gradshteyn and Ryzhik (2000 Section 0.316) for multiplication of two power series, we obtain

$$\frac{u^{-\gamma}}{[u^{\alpha}+(1-u)^{\alpha}]^{w}} = \left(\sum_{k=0}^{\infty} s_{1,k}^{*}(\alpha,w)\,u^{k}\right)\left(\sum_{k=0}^{\infty} s_{2,k}^{*}(\alpha,w,\gamma)\,u^{k}\right) = \sum_{k=0}^{\infty} s_{k}^{*}(\alpha,w,\gamma)\,u^{k}, \quad (28)$$

where $s_{k}^{*}(\alpha,w,\gamma) = \sum_{j=0}^{k} s_{1,j}^{*}(\alpha,w)\,s_{2,k-j}^{*}(\alpha,w,\gamma)$. Equation (28) is the main result to obtain the Rényi entropy in Section 5.1.

Appendix 3: Properties for a special model

The OLLL-E distribution is defined by inserting $G(x) = 1-\mathrm{e}^{-\lambda x}$ and $g(x) = \lambda\,\mathrm{e}^{-\lambda x}$ in Eq. (3), where $x > 0$ and $\lambda > 0$. Let X be the random variable representing this distribution. We derive some statistical measures of X from the asymptotics in Appendix 1 and the general results in Sections 4 and 5.2

Corollary 3 *The asymptotics of Eqs.* (2), (3) *and* (4) *for the OLLL-E distribution when* $x \to 0$ *are given by*

$$F(x) \sim \frac{\beta(\lambda x)^{\alpha}}{-\log(1-\beta)} \quad \text{as } x \to 0,$$
$$f(x) \sim \frac{\alpha\beta\,\lambda^{\alpha}\,x^{\alpha-1}}{-\log(1-\beta)} \quad \text{as } x \to 0,$$
$$h(x) \sim \frac{\alpha\beta\,\lambda^{\alpha}\,x^{\alpha-1}}{-\log(1-\beta)} \quad \text{as } x \to 0.$$

Corollary 4 *The asymptotics of Eqs.* (2), (3) *and* (4) *for the OLLL-E distribution when* $x \to \infty$ *are given by*

$$1-F(x) \sim 1-\frac{\log\left[1-\beta\,(1-\mathrm{e}^{-\lambda x})^{\alpha}\right]}{\log(1-\beta)} \quad \textit{as } x \to \infty,$$
$$f(x) \sim \frac{\alpha\,\beta\,\lambda\mathrm{e}^{-\lambda x}\,(1-\mathrm{e}^{-\lambda x})^{-1}}{-(1-\beta)\log(1-\beta)} \quad \textit{as } x \to \infty,$$
$$h(x) \sim \frac{\alpha\,\beta\,\lambda\mathrm{e}^{-\lambda x}\,(1-\mathrm{e}^{-\lambda x})^{-1}}{-(1-\beta)\log\left[\frac{1-\beta}{1-\beta\,G(x)^{\alpha}}\right]} \quad \textit{as } x \to \infty.$$

These equations can provide the effects of the parameters on the tails of the OLLL-E distribution.

We provide some statistical measures of X. Its nth ordinary moment follows from (8) as

$$E(X^{n}) = \lambda\,n!\sum_{k=0}^{\infty}\sum_{m=0}^{k}\frac{(-1)^{m}\,(k+1)\,d_{k+1}}{[\,(m+1)\lambda]^{n+1}}\binom{k}{m},$$

where d_k is defined in Eq. (6).

Further, the nth incomplete moment of the OLLL-E distribution is obtained from (9) as

$$m_{n}(y) = \lambda\sum_{k=0}^{\infty}\sum_{m=0}^{k}\frac{(-1)^{m}\,(k+1)\,d_{k+1}\,\gamma(n+1,y)}{[\,(m+1)\lambda]^{n+1}}\binom{k}{m},$$

where $\gamma(a,z) = \int_{0}^{z} t^{a-1}\,\mathrm{e}^{-t}\,dt$ denotes the incomplete gamma function.

The mgf of X comes immediately from (10)

$$M(t) = \lambda \sum_{k=0}^{\infty} \sum_{m=0}^{k} \frac{(-1)^m (k+1) d_{k+1}}{(m+1)\lambda - t} \binom{k}{m}.$$

Finally, the rth ordinary moment of the ith OLLL-E order statistic reduces to

$$E\left(X_{i:n}^r\right) = \lambda\, r! \sum_{k=0}^{\infty} \sum_{m=0}^{k} \frac{(-1)^m (k+1) s_k}{[(m+1)\lambda]^{r+1}} \binom{k}{m},$$

where s_k is given by (11).

Acknowledgements
The authors are very grateful to the editor Felix Famoye for helpful comments and suggestions.

Authors' contributions
The five authors jointly participated of the study and elaborated the research. Gauss Cordeiro corrected the manuscript throughout. All authors read and approved the final manuscript.

Competing interests
The authors declare that they have no competing interests.

Author details
[1]Department of Statistics, Persian Gulf university, Bushehr, Iran. [2]Department of Statistics, University of Mazandaran, Babolsar, Iran. [3]Departamento de Ciências Exatas, Universidade de São Paulo, Piracicaba, Brazil. [4]Departamento de Estatística, Universidade Federal de Pernambuco, Recife, Brazil.

References
Abouammoh, AM, Abdulghani, SA: On partial orderings and testing of new better than renewal used classes. Reliability Eng. Syst. Safety. **43**, 37–41 (1994)
Alzaatreh, A, Famoye, F, Lee, C: The gamma-normal distribution: Properties and applications. Comput. Stat. Data Anal. **69**, 67–80 (2014)
Berkson, J, Gage, RP: Survival curve for cancer patients following treatment. J. Am. Stat. Assoc. **47**, 501–515 (1952)
Blischke, WR, Murthy, DNP: Reliability: Modeling, Prediction and Optimization. 1st ed. Wiley, New York (2000)
Boag, JW: Maximum likelihood estimates of the proportion of patients cured by cancer therapy. J. R. Stat. Soc. Series B. **11**, 15–53 (1949)
Bourguignon, M, Silva, RB, Cordeiro, GM: The Weibull-G family of probability distributions. J. Data Sci. **12**, 53–68 (2014)
Braga, AS, Cordeiro, GM, Ortega, GMM, da Cruz, JN: The odd log-logistic normal distribution: Theory and applications in analysis of experiments. J. Stat. Theory Prac (2016). doi:10.1080/15598608.2016.1141127
Chen, G, Balakrishnan, N: A general purpose approximate goodness-of-fit test. J. Q. Technol. **27**, 154–161 (1995)
Cordeiro, GM, Alizadeh, M, Ortega, EMM: The exponentiated half-logistic family of distributions: Properties and applications. J Probab Stat. **1**, 1–21 (2014a)
Cordeiro, GM, de Castro, M: A new family of generalized distributions. J. Stat. Comput. Simul. **81**, 883–893 (2011)
Cordeiro, GM, Nadarajah, S: Closed-form expressions for moments of a class of beta generalized distributions. Braz. J. Prob. Stat. **25**, 14–33 (2011)
Cordeiro, GM, Nadarajah, S, Ortega, EMM: The Kumaraswamy Gumbel distribution. Stat. Methods Appl. **21**, 139–168 (2012)
Cordeiro, GM, Ortega, EMM, Bozidar, PV, Pescim, RR: The Lomax generator of distributions: Properties, minification process and regression model. Appl Math Comput. **247**, 465-486 (2014b)
Eugene, N, Lee, C, Famoye, F: Beta-normal distribution and its applications. Commun. Statistics-Theory Methods. **31**, 497–512 (2002)
Fachini, JB, Ortega, EMM, Cordeiro, GM: A bivariate regression model with cure fraction. J. Stat. Comput. Simul. **84**, 1580–1595 (2014)
Farewell, VT: The use of mixture models for the analysis of survival data with long-term survivors. Biometrics. **38**, 1041–1046 (1982)
Gendoo, DMA, Ratanasirigulchai, N, Schröder, M, Pare, L, Parker, JS, Prat, A, N Haibe-Kains, B: genefu: a package for breast cancer gene expression analysis (2015). Retrieved 2016-03-30, from https://bioc.ism.ac.jp/packages/devel/bioc/vignettes/genefu/inst/doc/genefu.pdf, https://goo.gl/jngJMY
Gleaton, JU, Lynch, JD: On the distribution of the breaking strain of a bundle of brittle elastic fibers. Adv. Appl. Probab. **36**, 98–115 (2004)
Gleaton, JU, Lynch, JD: Properties of generalized log-logistic families of lifetime distributions. J. Probab. Stat. Sci. **4**, 51–64 (2006)
Gleaton, JU, Lynch, JD: Extended generalized log-logistic families of lifetime distributions with an application. J. Probab. Stat. Sci. **8**, 1–17 (2010)

Gleaton, JU, Rahman, MM: Asymptotic properties of MLE's for distributions generated from a 2-parameter Weibull distribution by a generalized log-logistic transformation. J. Probab. Stat. Sci. **8**, 199–214 (2010)
Gleaton, JU, Rahman, MM: Asymptotic properties of MLE's for distributions generated from a 2-parameter inverse Gaussian distribution by a generalized log-logistic transformation. J. Probab. Stat. Sci. **12**, 85–99 (2014)
Gradshteyn, IS, Ryzhik, IM: Table of Integrals, Series, and Products. 7th ed. Academic Press, San Diego (2000)
Gupta, RD, Kundu, D: Generalized exponential distributions. Aust. New Zealand J. Stat. **42**, 173–188 (1999)
Hashimoto, EM, Cordeiro, GM, Ortega, EMM: The new Neyman type A beta Weibull model with long-term survivors. Comput. Stat. **28**, 933–954 (2013)
Marshall, AN, Olkin, I: A new method for adding a parameter to a family of distributions with applications to the exponential and Weibull families. Biometrika. **84**, 641–652 (1997)
Mudholkar, GS, Srivastava, DK: Exponentiated Weibull family for analyzing bathtub failure-rate data. IEEE Trans. Reliab. **42**, 299–302 (1993)
Mudholkar, GS, Srivastava, DK, Freimer, M: The exponentiated Weibull family: A reanalysis of the bus-motor-failure data. Technometrics. **37**, 436–445 (1995)
Murthy, DNP, Xie, M, Jiang, R: Weibull Models. 1st ed. Wiley, Hoboken (2004)
Nadarajah, S: The exponentiated Gumbel distribution with climate application. Environmetrics. **17**, 13–23 (2006)
Nadarajah, S, Kotz, S: The exponentiated type distributions. Acta Applicandae Mathematicae. **92**, 97–111 (2006)
Ortega, EMM, Cordeiro, GM, Kattan, MW: The negative binomial-beta Weibull regression model to predict the cure of prostate cancer. J. Appl. Stat. **39**, 1191–1210 (2012)
Ortega, EMM, Cordeiro, GM, Campelo, AK, Kattan, MW, Cancho, VG: A power series beta-Weibull regression model for predicting breast carcinoma. Stat. Med. **34**, 1366–1388 (2015)
Rényi, A: On measures of entropy and information. Proc. Fourth Berkeley Symp. Math. Stat. Probab. **1**, 547–561 (1961)
R Development Core Team: R: A Language and Environment for Statistical Computing (2013)
Shannon, CE: Prediction and entropy of printed English. Bell Syst. Tech. J. **30**, 50–64 (1951)

12

Simple robust parameter estimation for the Birnbaum-Saunders distribution

Min Wang[1]*, Chanseok Park[2] and Xiaoqian Sun[3]

*Correspondence: minwang@mtu.edu
[1] Department of Mathematical Sciences, Michigan Technological University, Houghton, MI, USA
Full list of author information is available at the end of the article

Abstract

We study the problem of robust estimation for the two-parameter Birnbaum-Saunders distribution. It is well known that the maximum likelihood estimator (MLE) is efficient when the underlying model is true but at the same time it is quite sensitive to data contamination that is often encountered in practice. In this paper, we propose several estimators which have simple closed forms and are also robust to data contamination. We study the breakdown points and asymptotic properties of the proposed estimators. These estimators are then applied to both simulated and real datasets. Numerical results show that the proposed estimators are attractive alternative to the MLE in that they are quite robust to data contamination and also highly efficient when the underlying model is true.

Keywords: Hodges-Lehmann estimator, Rousseeuw-Croux estimator, Robustness, Breakdown point, Data contamination

Mathematics subject classification: 62F10, 62F12, 62F35

1 Introduction

The two-parameter Birnbaum-Saunders (for short, BS) distribution was originally proposed by Birnbaum and Saunders (1969a) as a failure time distribution to describe the total time until the damage caused by the development and growth of a dominant crack grows to a critical level that would cause fracture of failure. The random variable T is said to follow the BS distribution with parameters α and β if its cumulative distribution function (cdf) is given by

$$F(t \mid \alpha, \beta) = \Phi\left[\frac{1}{\alpha}\left(\sqrt{\frac{t}{\beta}} - \sqrt{\frac{\beta}{t}}\right)\right], \quad 0 < t < \infty, \quad \alpha, \beta > 0, \tag{1}$$

where $\Phi(\cdot)$ is the cdf of the standard normal distribution, and the parameters α and β are the shape and scale parameters, respectively. This distribution has been widely applied to a variety of quality and reliability engineering problems. For example, Bhattacharyya and Fries (1982) developed fatigue failure models and also discussed the intrinsic relation between the inverse Gaussian distribution and the BS distribution. Using the inverse Gaussian approximation to the BS distribution, Park and Padgett (2006) developed various new cumulative damage models and degradation models. Lio and Park (2008) developed a bootstrap control chart based on the BS distribution.

Estimation of the BS parameters is of great interest to researchers and has recently received much attention in the literature. Birnbaum and Saunders (1969b) firstly discussed the MLEs of α and β. Engelhardt et al. (1981) derived the asymptotic joint distribution of the MLEs and showed that the MLEs are asymptotically independent. Achcar (1993) studied the Bayesian estimators based on Jeffreys' prior and the reference prior and adopted Laplace's approximations to the posterior marginal distribution of the parameters of interest. Ng et al. (2003) proposed the method-of-moment estimator (MME) of the two parameters. All the above mentioned approaches, except for the MME, have no explicit closed-form solutions, so some numerical approaches are often required to obtain these estimators.

The quality of data is extremely important for parameter estimation and complete data without any contamination is always preferred to achieve a high accuracy on parameter estimation. Unfortunately, in the engineering sciences, there is no guarantee that the collected data exactly follow the assumed model. In other words, reliability engineers often face data with deviations from the assumed model especially when studying failure data. Note that the commonly used estimators such as the MLEs are very sensitive to data contamination and model departure that are often encountered in many practical situations. Small deviations may induce a large impact on these estimators and a single outlier can even make them break down. This phenomenon motivates many authors to study robust estimation for various distributions; see, for example, Agostinelli et al. (2014), Boudt et al. (2011), Lawson et al. (1997), among others. Of particular note is that many researchers mainly focus on robust estimation for the Weibull distribution and that robust estimation for the BS distribution is quite scant, even though the BS distribution is prevalent in the engineering sciences as an effective model of studying fatigue data.

It deserves mentioning that Dupuis and Mills (1998) developed the robust estimation procedures based on an optimal bias-robust estimator (OBRE) of the BS distribution, whereas the OBRE is not in explicit form and it may suffer from the problem of convergence. These observations give rise to the need for some alternative estimators, which should be easy to calculate for practitioners and are also quite robust against a certain amount of data contamination. Recently, Wang et al. (2013) developed a new method for robustly estimating the parameters of the BS distribution, whereas their method is somewhat complicated for practitioners.

In this paper, we propose several alternative estimators for the BS distribution and study the asymptotic properties of the proposed estimators and their breakdown points. These estimators are explicit closed-form expressions of the sample observations and are thus easy to calculate without involving any computational complexity. Several of the proposed estimators have a very high breakdown point. Here, the breakdown point, a criterion often used to measure robustness of an estimator, is defined as the proportion of incorrect observations (i.e., arbitrarily small or large observations), the estimators of α and β can handle before giving estimated values arbitrary close to zero or infinity. As an illustration, the median has a breakdown point of 50 %, whereas the mean has a breakdown point of 0 %. The results of numerical studies show that the proposed estimators are attractive alternatives to the MLE in that they are quite robust to data contamination and also highly efficient when the underlying model is true.

The remainder of this paper is organized as follows. In Section 2, we firstly present a summary of the BS distribution and the MLEs, and then provide several alternative estimators in explicit closed-form expressions. We also investigate the asymptotic properties of these estimators along with their breakdown points. Section 3 contains an extensive Monte Carlo simulation to compare the behavior of the proposed estimators and that of the MLEs. In Section 4, we illustrate the practical application of all the estimators under consideration through using a real dataset. Finally, some concluding remarks and discussions are given in Section 5.

2 Parameter estimation

If a random variable T follows the BS distribution with the cdf in (1), then T has the probability density function (pdf) given by

$$f(t \mid \alpha, \beta) = \frac{1}{2\sqrt{2}\alpha\beta}\left[\left(\frac{\beta}{t}\right)^{1/2} + \left(\frac{\beta}{t}\right)^{3/2}\right]\exp\left\{-\frac{1}{2\alpha^2}\left(\frac{t}{\beta}+\frac{\beta}{t}-2\right)\right\}.$$

It is well known that this distribution has many attractive properties. (i) The scale parameter β is the median of the distribution, that is, $F(\beta) = \Phi(0) = 0.5$; (ii) it is positively skewed, with degree of skewness decreasing with α; (iii) for any constant $\kappa > 0$, it follows that $\kappa T \sim \text{BS}(\alpha, \kappa\beta)$; (iv) its reciprocal property (Saunders 1974) holds, that is, $T^{-1} \sim \text{BS}(\alpha, \beta^{-1})$; and (v) if we make the transformation $T = \beta\left[1 + 2X^2 + 2X(1+X^2)^{1/2}\right]$, then X is normally distributed with mean zero and variance $\alpha^2/4$.

We firstly provide a brief review to the MLEs of the BS distribution. Let $\mathcal{T} = \{t_1, t_2, \cdots, t_n\}$ be a random sample of size n from the BS distribution. The sample arithmetic and harmonic means are given by

$$s = \frac{1}{n}\sum_{i=1}^{n} t_i \quad \text{and} \quad r = \left[\frac{1}{n}\sum_{i=1}^{n} t_i^{-1}\right]^{-1},$$

respectively. Let $h(\cdot)$ be the harmonic mean function given by

$$h(x) = \left[\frac{1}{n}\sum_{i=1}^{n}(x+t_i)^{-1}\right]^{-1}.$$

Then the MLE of β, denoted by $\hat{\beta}^{\text{MLE}}$, can be obtained as the unique positive root of the following equation

$$\beta^2 - \beta[2r + h(\beta)] + r[s + h(\beta)] = 0. \qquad (2)$$

Once $\hat{\beta}^{\text{MLE}}$ is obtained, the MLE of α, denoted by $\hat{\alpha}^{\text{MLE}}$, is given by

$$\hat{\alpha}^{\text{MLE}} = \frac{s}{\hat{\beta}^{\text{MLE}}} + \frac{\hat{\beta}^{\text{MLE}}}{r} - 2.$$

Because the uniqueness of the solution for Eq. (2) is guaranteed in the interval (r, s), a one-dimensional root search is adopted in this paper to compute the MLEs instead of their methods; see, for example, Lio and Park (2008).

Although the MLEs enjoy several optimal properties, they have no explicit expressions and are highly sensitive to model departure that often occurs in practice. These observations motivate us to develop alternative estimators. Specifically, we propose several alternative estimators in the following two subsections. One is dedicated to developing two estimators of β; the other is to develop two estimators of α. The proposed estimators are explicit closed-form expressions in terms of the sample observations and can thus be easily computed in practical situations. Furthermore, it can be shown that some of them are asymptotically normally distributed and have the highest breakdown point of 50 %.

2.1 The estimators of the parameter β

Since the parameter β is the median of the BS distribution which satisfies $F(\beta) = \Phi(0) = 0.5$, the median is a natural choice as an estimator of β. We denote the sample median estimator of β by

$$\hat{\beta}^{\mathrm{M}} = \mathrm{median}\{t_1, t_2, \cdots, t_n\}.$$

The above median estimator is quite robust and has the highest breakdown point of 50 %. From the asymptotic normality of the median estimator, it can be shown that

$$\sqrt{n}\left(\hat{\beta}^{\mathrm{M}} - \beta\right) \xrightarrow{D} N\left(0, \frac{\pi(\alpha\beta)^2}{2}\right).$$

Rieck and Nedelman (1991) investigated the relationship between the BS distribution and the sinh-normal distribution and showed that if a random variable $T \sim \mathrm{BS}(\alpha, \beta)$, then $Y = \log(T)$ has the sinh-normal distribution with the cdf and pdf given by

$$F(y \mid \mu, \alpha) = \Phi\left[\frac{2}{\alpha}\sinh\left(\frac{y-\mu}{2}\right)\right] \tag{3}$$

and

$$f(y \mid \mu, \alpha) = \frac{1}{\alpha\sqrt{2\pi}}\cosh\left(\frac{y-\mu}{2}\right)\exp\left[-\frac{2}{\alpha^2}\sinh\left(\frac{y-\mu}{2}\right)\right], \tag{4}$$

respectively. This distribution has a shape parameter α and a location parameter $\mu = \log(\beta)$, which is just the natural logarithm of the scale parameter of the BS distribution. One appealing advantage of using the above relationship is that we transform the *asymmetric* BS distribution to the *symmetric* sinh-normal distribution. It is well known that in the location framework, the Hodges and Lehmann (1963) estimator (for short, HL) of (Hodges and Lehmann 1963) has greater efficiency than does the sample median for symmetric distributions. Thus, it appears to be more beneficial to adopt the HL estimator for the parameter μ under the sinh-normal distribution, instead of the sample median of the original BS distribution. The HL estimator is defined as the median of $n(n-1)/2$ pairwise averages of observations and can be written as

$$\tilde{\mu} = \mathrm{median}_{i<j}\left\{\frac{y_i + y_j}{2}\right\}, \tag{5}$$

which can easily be calculated using the *hl.loc*($\cdot$) function in the package of *ICSNP* in *R* language; see R Development Core Team (2011). Also, this estimator has 29 % breakdown point, which means that it remains consistent even if about 29 % percent of the data have been contaminated. In addition, as stated by Rousseeuw and Croux (1993), '*the Hodges-Lehmann estimator might be viewed as a "smooth" version of the median*'. It can be shown that this estimator is asymptotically normally distributed, namely,

$$\sqrt{n}(\tilde{\mu} - \mu) \xrightarrow{D} N\left(0, \frac{1}{12\int_0^\infty f(x)^2\,dx}\right),$$

where $f(x)$ is the pdf of the sinh-normal distribution in (4). Also, it is immediate from the above that the estimator $\tilde{\mu}$ converges in probability to $\mu = \log(\beta)$, that is, $\tilde{\mu} \xrightarrow{p} \mu$.

Based on the HL estimator in (5), we obtain the back-transformed estimator of the BS parameter β, denoted by $\hat{\beta}^{\mathrm{HL}}$, which is simply given by

$$\hat{\beta}^{\mathrm{HL}} = \exp(\tilde{\mu}) = \exp\left[\mathrm{median}_{i<j}\left(\frac{y_i + y_j}{2}\right)\right]. \tag{6}$$

This estimator also has a breakdown point of 29 %. Note that the estimator $\hat{\beta}^{\mathrm{HL}}$ approximately follows the log-normal distribution and that the estimator $\hat{\beta}^{\mathrm{HL}}$ is biased, but consistent for β.

In what follows, we will adopt the above two estimators of β to develop the estimator of α due to their simplicity and computational ease as well as high breakdown points.

2.2 The estimators for the parameter α

Here, we proposed two robust estimators for the shape parameter α.

Let $\{t_1, t_2, \cdots, t_n\}$ be a random sample from the BS distribution with the cdf in (1). Note that $\Phi^{-1}(0.75) - \Phi^{-1}(0.25) = 1.34898$ is the distance between the two quartiles of the standard normal distribution, the interquartile range (IQR). The sample IQR is equal to the difference between the third sample quartile (Q_3) and the first sample quartile (Q_1). To have a normal consistency, we consider the following estimator of α given by

$$\hat{\alpha}_j^{\mathrm{IQR}} = \frac{\mathrm{IQR}(y_j)}{1.34898}, \quad j = 1, 2,$$

The estimator $\hat{\alpha}_j^{\mathrm{IQR}}$ is in simple closed-form with a higher breakdown point of 25 %, indicating that it remains consistent even if about 25 % percent of the data have been contaminated. Furthermore, by using the Bahadur's representation theorem (Ghosh 1971), we have

$$\sqrt{n}\left(\hat{\alpha}_j^{\mathrm{IQR}} - \alpha\right) \xrightarrow{D} N\left(0,\ c\alpha^2\right), \quad j = 1, 2,$$

where c is a constant approximately equal to $2.48/1.349^2 \approx 1.363$. The quartile estimator $\hat{\alpha}_j^{\mathrm{IQR}}$ converges in probability to the parameter α, that is, $\hat{\alpha}_j^{\mathrm{IQR}} \xrightarrow{p} \alpha$ for $j = 1, 2$.

We observe from Eq. (3) that by letting $Y = \log(T)$, we change the estimation problem of the *shape* and scale parameters to that of the *location* and scale parameters. We advocate the use of (reference) estimator (shortly RC) for the estimator of α. The RC estimator is given by

$$\hat{\alpha}_j^{\mathrm{RC}} = b\ \left\{|\log t_i - \log t_j|;\ i < j\right\}_{(k)},$$

where b is a constant factor equal to $1/(\sqrt{2}\Phi^{-1}(5/8)) \approx 2.2219$ to achieve consistency for standard deviation of α in the case of the normal distribution above, and $k = \binom{h}{2} \approx \binom{n}{2}/4$ with $h = [n/2] + 1$ being roughly half the number of observations. The RC estimator is an explicit expression and can be easily calculated using the $Q_n(\cdot)$ function in the package of *robustbase* in *R* language. In addition, it has the highest breakdown point of 50 % and a very high normal efficiency about 82 % under the normal distribution. An analogous scale estimator, called the Shamos estimator proposed by Shamos (1976) and studied by Bickel and Lehmann (1976), can also be adopted for α. However, the Shamos estimator has only a 29 % breakdown point and 86 % efficiency which is slightly higher than the RC estimator. Readers interested in full details of the Shamos estimator should refer to these papers.

3 Monte Carlo simulations

To illustrate the proposed method, we consider Monte Carlo simulation studies with one without contamination and the other involving different kinds of contaminations.

3.1 Numerical results without contamination

We carry out Monte Carlo simulations to compare the performances of the estimators under consideration. We take the sample size $n = 10, 50, 100$, and the shape parameter $\alpha = 0.5, 1.0, 2.0$. Since β is the scale parameter, its value was kept fixed at $\beta = 1.0$, without loss of any generality.

Inverting the cdf of the BS distribution $F(T \mid \alpha, \beta) = p$ given by (3), we obtain the following inverse cdf of the form

$$T = F^{-1}(p) = \frac{1}{4}\left(\alpha\sqrt{\beta}\Phi^{-1}(p) + \sqrt{\alpha^2\beta\left(\Phi^{-1}(p)\right)^2 + 4\beta}\right)^2.$$

Hence, random numbers following the BS distribution can be generated by using a direct inverse method, that is, $T = F^{-1}(U)$ with $U \sim \text{Uniform}(0,\ 1)$. Replicate each case 10,000 times. In each simulation, we compute the average biases of each estimator under consideration with its corresponding square root of the mean square error (RMSE) given by

$$\text{RMSE}_\alpha = \sqrt{\frac{1}{M}\sum_{i=1}^{M}\left(\hat{\alpha}_i - \alpha_T\right)^2} \quad \text{and} \quad \text{RMSE}_\beta = \sqrt{\frac{1}{M}\sum_{i=1}^{M}\left(\hat{\beta}_i - \beta_T\right)^2}.$$

The results based on $\hat{\beta}^{\text{M}}$ and $\hat{\beta}^{\text{HL}}$ are presented in Tables 1 and 2, respectively. Several conclusions from the numerical results can be drawn as follows.

(i) The average bias and RMSE of all the estimators significantly decrease as n increases. As expected, for large sample sizes, the performance of the proposed estimators and that of the MLEs are very close in terms of the average bias and RMSE.

(ii) The RC estimator $\hat{\alpha}_j^{\text{RC}}$ for $j = 1, 2$ outperforms others in terms of the average bias for all the considered cases. Note also that estimation of α using the HL estimator $\hat{\beta}^{\text{HL}}$ is superior than the one using the median estimator $\hat{\beta}^{\text{M}}$ in terms of the RMSE.

(iii) For the cases without contamination, the estimator $\hat{\beta}^{\text{MLE}}$ performs the best as expected. We observe that the performance of the HL estimator $\hat{\beta}^{\text{HL}}$ is better than that of the median estimator $\hat{\beta}^{\text{M}}$. All the estimators of β become closer together as n increases.

3.2 Numerical results with contamination

In practice, there is no guarantee that the collected data exactly follow the BS distribution, especially when considering fatigue data. Specifically, data contamination may occur due to several reasons such as the measurement errors, departure from the true model, etc. Accordingly, it becomes necessary to investigate the performance of the estimators under a scenario in which data are contaminated with outliers. In a similar way as done by (Dupuis and Mills 1998), we compare the performances of the proposed estimators and that of the MLEs based on the following four possible models.

Table 1 Average bias and RMSE (in parentheses) of estimates for α and β

		Estimator of α			Estimator of β	
n	α	$\hat{\alpha}^{MLE}$	$\hat{\alpha}_1^{IQR}$	$\hat{\alpha}_1^{RC}$	$\hat{\beta}^{MLE}$	$\hat{\beta}^{M}$
10	0.5	−0.039	−0.065	−0.000	−0.012	−0.022
		(0.117)	(0.172)	(0.150)	(0.156)	(0.192)
	1.0	−0.087	−0.136	−0.017	−0.040	−0.080
		(0.235)	(0.343)	(0.298)	(0.299)	(0.412)
	2.0	−0.186	−0.273	−0.088	−0.105	−0.331
		(0.481)	(0.726)	(0.597)	(0.534)	(1.065)
50	0.5	−0.007	−0.015	−0.000	−0.002	−0.003
		(0.051)	(0.081)	(0.059)	(0.068)	(0.086)
	1.0	−0.017	−0.029	−0.003	−0.008	−0.017
		(0.101)	(0.163)	(0.116)	(0.128)	(0.179)
	2.0	−0.035	−0.060	−0.021	−0.018	−0.060
		(0.206)	(0.334)	(0.238)	(0.195)	(0.379)
100	0.5	−0.004	−0.007	−0.001	−0.001	−0.002
		(0.036)	(0.058)	(0.041)	(0.048)	(0.063)
	1.0	−0.008	−0.014	−0.001	−0.004	−0.007
		(0.070)	(0.116)	(0.080)	(0.088)	(0.124)
	2.0	−0.019	−0.028	−0.011	−0.010	−0.033
		(0.144)	(0.234)	(0.163)	(0.135)	(0.260)

Model 1 A model with no contamination.

Model 2 A model with 5 % of severe contamination; the upper 5 % of order statistics are multiplied by 5.

Model 3 A model with 5 % of severe contamination; the lower 5 % of order statistics are multiplied by $1/5$.

Table 2 Average bias and RMSE (in parentheses) of estimates for α and β

		Estimator of α			Estimator of β	
n	α	$\hat{\alpha}^{MLE}$	$\hat{\alpha}_2^{IQR}$	$\hat{\alpha}_2^{RC}$	$\hat{\beta}^{MLE}$	$\hat{\beta}^{HL}$
10	0.5	−0.039	−0.066	−0.000	−0.012	−0.014
		(0.117)	(0.171)	(0.149)	(0.156)	(0.164)
	1.0	−0.087	−0.141	−0.017	−0.040	−0.048
		(0.235)	(0.340)	(0.297)	(0.299)	(0.332)
	2.0	−0.186	−0.314	−0.078	−0.105	−0.165
		(0.481)	(0.700)	(0.595)	(0.534)	(0.714)
50	0.5	−0.007	−0.015	−0.000	−0.002	−0.002
		(0.051)	(0.081)	(0.059)	(0.068)	(0.071)
	1.0	−0.017	−0.031	−0.003	−0.008	−0.010
		(0.101)	(0.162)	(0.116)	(0.128)	(0.139)
	2.0	−0.035	−0.071	−0.018	−0.018	−0.029
		(0.206)	(0.330)	(0.238)	(0.195)	(0.256)
100	0.5	−0.004	−0.004	−0.000	−0.001	−0.001
		(0.036)	(0.058)	(0.041)	(0.048)	(0.051)
	1.0	−0.008	−0.015	−0.001	−0.004	−0.004
		(0.070)	(0.115)	(0.080)	(0.088)	(0.096)
	2.0	−0.019	−0.034	−0.009	−0.010	−0.016
		(0.144)	(0.233)	(0.163)	(0.135)	(0.176)

Model 4 A model with 5 % of more extreme contamination from a point mass distribution at 50.

The reference distribution is the BS distribution with the parameters $\alpha = 0.5$ and $\beta = 1$. We generate 10,000 samples of size $n = 100$ according to the above four scenarios and then calculate the average bias and RMSE of each estimator. The results are shown in Table 3. Some conclusions can be drawn as follows.

(i) For Model 1, that is, there is no contamination, the RC estimator $\hat{\alpha}_j^{RC}$ for $j = 1, 2$ performs the best for estimating α in terms of the average bias; the MLE $\hat{\beta}$ is the best one for estimating β, but the HL estimator $\hat{\beta}^{HL}$ behaves much similarly.

(ii) For Models 2, 3, and 4, that is, contamination presents in the dataset, we observe that contamination induces a large influence on the average bias and RMSE of the non-robust estimators including the MLEs, especially in the presence of extreme outliers such as the fourth scenario, whereas it has a smaller impact on the proposed estimators.

(iii) For the scale parameter estimation, the HL estimator $\hat{\beta}^{HL}$ outperforms the median estimators in terms of the RMSE, whereas both are quite robust against data contamination.

Which of the two robust estimators, the HL estimator or the median estimator, is preferable for the parameter β in the analysis of real lifetime data? Numerical results show that for all the cases considered in this paper, the HL estimator $\hat{\beta}^{HL}$ outperforms the median estimator $\hat{\beta}^{M}$ in term of the RMSE. Additionally, the estimator of α developed based on $\hat{\beta}^{HL}$ also slightly outperforms the one using $\hat{\beta}^{M}$ in most cases. We thus have a preference to recommend the HL estimator for β. It should be mentioned that other simulation results with respect to several other values of the parameter α and different sample sizes have also been conducted, and the conclusions are quite similar and are thus not provided here for brevity.

Table 3 Average bias and RMSE (in parentheses) of estimates for α and β with $n = 100$ under the four models

Estimator of α					Estimator of β		
$\hat{\alpha}^{MLE}$	$\hat{\alpha}_1^{IQR}$	$\hat{\alpha}_1^{RC}$	$\hat{\alpha}_2^{IQR}$	$\hat{\alpha}_2^{RC}$	$\hat{\beta}^{MLE}$	$\tilde{\beta}^{M}$	$\hat{\beta}^{HL}$
Model 1							
−0.0038	−0.0068	−0.0001	−0.0040	−0.0001	0.0013	0.0024	0.0014
(0.0353)	(0.0583)	(0.0405)	(0.0445)	(0.0405)	(0.0483)	(0.0625)	(0.0504)
Model 2							
0.3424	−0.0068	0.0075	0.9484	0.0075	0.2512	0.0024	0.0040
(0.3452)	(0.9570)	(0.3742)	(0.0583)	(0.0426)	(0.2593)	(0.0625)	(0.0511)
Model 3							
0.3423	−0.0068	0.0075	0.0139	0.0075	−0.1983	0.0024	−0.0011
(0.3451)	(0.0583)	(0.0426)	(0.0448)	(0.0426)	(0.2026)	(0.0625)	(0.0507)
Model 4							
0.8931	0.0253	0.0523	2.4594	0.0523	1.0944	0.0364	0.0501
(0.8936)	(0.0659)	(0.0693)	(2.4607)	(0.0693)	(1.0957)	(0.0751)	(0.0736)

Table 4 Fatigue lifetime data by Birnbaum and Saunders (1969b)

70	90	96	97	99	100	103	104	104	105	107	108	108	108	109	109	112
112	113	114	114	114	116	119	120	120	120	121	121	123	124	124	124	124
124	128	128	129	129	130	130	130	131	131	131	131	131	132	132	132	133
134	134	134	134	134	136	136	137	138	138	138	139	139	141	141	142	142
142	142	142	142	144	144	145	146	148	148	149	151	151	152	155	156	157
157	157	157	158	159	162	163	163	164	166	166	168	170	174	196	212	

4 An illustrative example

We illustrate the practical application of the proposed estimators using a real data example. The dataset from Birnbaum and Saunders (1969b) is the fatigue lifetime of $6061 - T6$ aluminum coupons cut parallel to the direction of rolling and oscillated at 18 cycles per second. The dataset consists of 101 observations with maximum stress per cycle 31,000 psi and is presented in Table 4.

The parameter estimates of α and β by all the methods under consideration are presented in Table 5. As mentioned in Section 3, we have a preference over the HL estimator $\hat{\beta}^{\text{HL}}$ for β, and thus we just analyze the results based on this estimator for simplicity. It can be seen from Table 5 that in the case of no contamination, most of the proposed estimators are in good agreement with the MLEs and that the estimators $\left(\hat{\alpha}_2^{\text{RC}}, \hat{\beta}^{\text{HL}}\right)$ are slightly different.

To evaluate robustness of the proposed methods, we follow the same scenario by Dupuis and Mills (1998) and assume that the 51st observation t_{51} was misrecorded as 633, instead of 133. It is desirable that the estimated shape and scale parameters should be very similar under the two scenarios, because we already know that the observation t_{51} is a recording error. However, it has been observed from Table 5 that the MLEs are heavily distorted by this single outlier and resulted in $\hat{\alpha}^{\text{MLE}} = 0.2415$ and $\hat{\beta}^{\text{MLE}} = 134.7689$, far from the MLEs with $t_{51} = 133$, whereas the proposed robust estimators $\left(\hat{\alpha}_2^{\text{RC}}, \hat{\beta}^{\text{HL}}\right)$ and $\left(\hat{\alpha}_2^{\text{IQR}}, \hat{\beta}^{\text{HL}}\right)$ still provided more reasonable results, which are quite close to the estimated values with $t_{51} = 133$.

In Fig. 1, we plot the estimated parameters with the same data but where we replace the 51st observation t_{51} by a range of values between 1 and 700. We observe that changing the value of t_{51} induces a large impact on the behavior of the MLEs, whereas it has little influence on the proposed robust estimators. As expected, the proposed estimators $\left(\hat{\alpha}_2^{\text{IQR}}, \hat{\beta}^{\text{HL}}\right)$ and $\left(\hat{\alpha}_2^{\text{RC}}, \hat{\beta}^{\text{HL}}\right)$ have some built-in protection against a certain amount of deviation due to data contamination or the measurement errors. In conclusion, the performance of all the proposed estimators is quite satisfactory.

Table 5 Comparison between the developed estimators and the MLE through fatigue lifetime data by Birnbaum and Saunders (1969b)

	Table 4 data		Misrecorded data	
Method	α	β	α	β
$(\hat{\alpha}^{\text{MLE}}, \hat{\beta}^{\text{MLE}})$	0.1704	131.8188	0.2415	134.7689
$(\hat{\alpha}_1^{\text{IQR}}, \hat{\beta}^{\text{M}})$	0.1454	133.0000	0.1555	134.0000
$(\hat{\alpha}_1^{\text{RC}}, \hat{\beta}^{\text{M}})$	0.1601	133.0000	0.1677	134.0000
$(\hat{\alpha}_2^{\text{IQR}}, \hat{\beta}^{\text{HL}})$	0.1454	132.6047	0.1555	132.8834
$(\hat{\alpha}_2^{\text{RC}}, \hat{\beta}^{\text{HL}})$	0.1601	132.6047	0.1677	132.8834

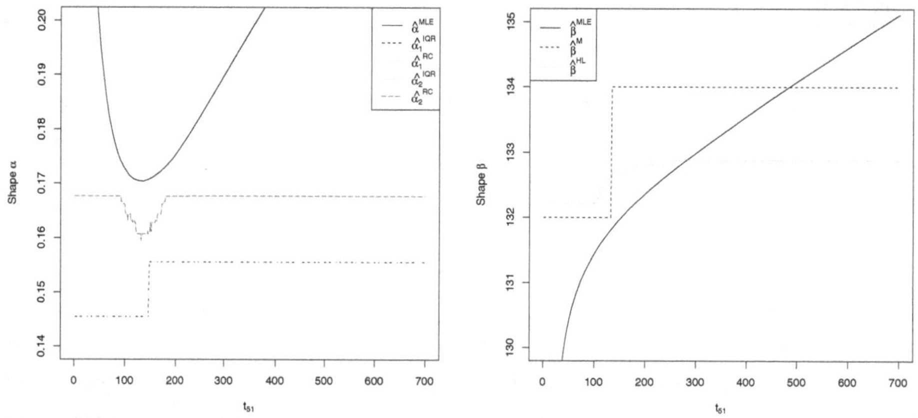

Fig. 1 Estimates for the BS parameters for the fatigue life data in (Birnbaum and Saunders 1969b), where the 51st observation is replaced by $t_{51} = 1, 2, \cdots, 700$

5 Concluding remarks

In this paper, we have developed the two families of the estimators for the BS distribution, which are quite robust to data contamination. Unlike the MLEs, these estimators have simple closed-form expressions with higher breakdown points. For estimation of β, we have a preference for the use of the HL estimator $\hat{\beta}^{\text{HL}}$, because numerical results show that it remains more accurate than the median estimator $\hat{\beta}^{\text{M}}$. Of all the considered estimators for α using the estimator $\hat{\beta}^{\text{HL}}$, we recommend the RC estimator $\hat{\alpha}_2^{\text{RC}}$, since it has a good trade-off between efficiency and robustness. It deserves to be mentioned that other proposed estimators of α are also attractive alternatives to the MLE in that they are highly efficient when the underlying model is true.

In summary, we have a preference for the RC and HL estimators $\left(\hat{\alpha}_2^{\text{RC}},\ \hat{\beta}^{\text{HL}}\right)$ for estimating $(\alpha,\ \beta)$, because it has been shown to be simple, very effective, and quite robust against model departure that often occurs in many practical situations. Note that censored data occur commonly in the field data from reliability tests, so a possible extension of the proposed estimators for the censored data will be investigated in the future.

Competing interests
The authors declare that they have no competing interests.

Authors' contributions
MW initiated and carried out the study. MW drafted the manuscript. XS and CP participated in the discussion and proofread the manuscript. All authors read and approved the final manuscript.

Acknowledgements
The authors thank the Editor Carl Lee and the two anonymous reviewers for their comments which have improved the appearance of this paper.

Author details
[1]Department of Mathematical Sciences, Michigan Technological University, Houghton, MI, USA. [2]Department of Industrial Engineering, Pusan National University, Geumjeong-gu, Busan, Republic of Korea. [3]Department of Mathematical Sciences, Clemson University, Clemson, SC, USA.

References
Achcar, JA: Inferences for the Birnbaum-Saunders fatigue life model using Bayesian methods. Comput. Statist. Data Anal. **15**, 367–380 (1993)

Agostinelli, C, Marazzi, A, Yohai, VJ: Robust estimators of the generalized loggamma distribution. Technometrics. **56**, 92–101 (2014)

Bhattacharyya, G, Fries, A: Fatigue failure models — birnbaum-saunders vs. inverse Gaussian. Reliability IEEE Trans. **31**, 439–441 (1982)

Bickel, PJ, Lehmann, EL: Descriptive statistics for non-parametric models III: Dispersion. Ann. Statist. **4**, 1139–1158 (1976)
Birnbaum, ZW, Saunders, SC: A new family of life distributions. J. Appl. Probability. **6**, 319–327 (1969a)
Birnbaum, ZW, Saunders, SC: Estimation for a family of life distributions with applications to fatigue. J. Appl. Probability. **6**, 328–347 (1969b)
Boudt, K, Caliskan, D, Croux, C: Robust explicit estimators of Weibull parameters. Metrika. **73**, 187–209 (2011)
Dupuis, D, Mills, J: Robust estimation of the Birnbaum-Saunders distribution. IEEE Trans. Reliab. **47**, 88–95 (1998)
Engelhardt, M, Bain, LJ, Wright, FT: Inferences on the parameters of the Birnbaum-Saunders fatigue life distribution based on maximum likelihood estimation. Technometrics. **23**, 251–256 (1981)
Ghosh, JK: A new proof of the Bahadur representation of quantiles and an application. Ann. Math. Statist. **42**, 1957–1961 (1971)
Hodges, JL Jr, Lehmann, EL: Estimates of location based on rank tests. Ann. Math. Statist. **34**, 598–611 (1963)
Lawson, C, Keats, J, Montgomery, D: Comparison of robust and least-squares regression in computer-generated probability plots. Reliability IEEE Trans. **46**, 108–115 (1997)
Lio, YL, Park, C: A bootstrap control chart for Birnbaum-Saunders percentiles. Qual. Reliability Eng Int. **24**, 585–600 (2008)
Ng, HKT, Kundu D, Balakrishnan, N: Modified moment estimation for the two-parameter Birnbaum-Saunders distribution. Comput. Statist. Data Anal. **43**, 283–298 (2003)
Park, C, Padgett, W: Stochastic degradation models with several accelerating variables. Reliability, IEEE Trans. **55**, 379–390 (2006)
R Development Core Team, R: A Language and Environment for Statistical Computing. R Foundation for Statistical Computing, Vienna, Austria (2011). ISBN 3-900051-07-0
Rieck, JR, Nedelman, JR: A log-linear model for the Birnbaum-Saunders distribution. Technometrics. **33**, 51–60 (1991)
Rousseeuw, PJ, Croux, C: Alternatives to the median absolute deviation. J. Amer. Statist. Assoc. **88**, 1273–1283 (1993)
Saunders, SC: A family of random variables closed under reciprocation. J. Amer. Statist. Assoc. **69**, 533–539 (1974)
Shamos, MI: Geometry and statistics: problems at the interface. In Algorithms and complexity (Proc. Sympos., Carnegie-Mellon Univ., Pittsburgh, Pa., 1976). Academic Press, New York (1976)
Wang, M, Zhao, J, Sun, X, Park, C: Robust explicit estimation of the two-parameter Birnbaum-Saunders distribution. J Appl. Stat. **40**, 2259–2274 (2013)

Permissions

All chapters in this book were first published in JSDA, by Springer; hereby published with permission under the Creative Commons Attribution License or equivalent. Every chapter published in this book has been scrutinized by our experts. Their significance has been extensively debated. The topics covered herein carry significant findings which will fuel the growth of the discipline. They may even be implemented as practical applications or may be referred to as a beginning point for another development.

The contributors of this book come from diverse backgrounds, making this book a truly international effort. This book will bring forth new frontiers with its revolutionizing research information and detailed analysis of the nascent developments around the world.

We would like to thank all the contributing authors for lending their expertise to make the book truly unique. They have played a crucial role in the development of this book. Without their invaluable contributions this book wouldn't have been possible. They have made vital efforts to compile up to date information on the varied aspects of this subject to make this book a valuable addition to the collection of many professionals and students.

This book was conceptualized with the vision of imparting up-to-date information and advanced data in this field. To ensure the same, a matchless editorial board was set up. Every individual on the board went through rigorous rounds of assessment to prove their worth. After which they invested a large part of their time researching and compiling the most relevant data for our readers.

The editorial board has been involved in producing this book since its inception. They have spent rigorous hours researching and exploring the diverse topics which have resulted in the successful publishing of this book. They have passed on their knowledge of decades through this book. To expedite this challenging task, the publisher supported the team at every step. A small team of assistant editors was also appointed to further simplify the editing procedure and attain best results for the readers.

Apart from the editorial board, the designing team has also invested a significant amount of their time in understanding the subject and creating the most relevant covers. They scrutinized every image to scout for the most suitable representation of the subject and create an appropriate cover for the book.

The publishing team has been an ardent support to the editorial, designing and production team. Their endless efforts to recruit the best for this project, has resulted in the accomplishment of this book. They are a veteran in the field of academics and their pool of knowledge is as vast as their experience in printing. Their expertise and guidance has proved useful at every step. Their uncompromising quality standards have made this book an exceptional effort. Their encouragement from time to time has been an inspiration for everyone.

The publisher and the editorial board hope that this book will prove to be a valuable piece of knowledge for researchers, students, practitioners and scholars across the globe.

List of Contributors

Wolf-Dieter Richter and Kay Schicker
Institute of Mathematics, University of Rostock, Ulmenstr. 69, Haus 3, Rostock, 18057, Germany

Klaus Müller and Wolf-Dieter Richter
University of Rostock, Institute of Mathematics, Ulmenstraße 69, Rostock, Germany

Wolf-Dieter Richter
University of Rostock, Institute of Mathematics, Ulmenstraße 69, Haus 3, 18057 Rostock, Germany

Frantisek Duris
Geneton s.r.o., Galvaniho 7, 82104 Bratislava, Slovakia
Slovak Centre of Scientific and Technical Information, Lamacska cesta 7315/8A, 81104 Bratislava, Slovakia

Juraj Gazdarica, Iveta Gazdaricova and Lucia Strieskova
Comenius University, Faculty of Natural Sciences, Ilkovicova 3278/6, 84104 Bratislava, Slovakia

Jaroslav Budis
Comenius University Faculty of Mathematics, Physics and Informatics, Mlynska dolina, 84248 Bratislava, Slovakia

Jan Turna
Comenius University, Science Park, Ilkovicova 8, 84104 Bratislava, Slovakia

Tomas Szemes
Geneton s.r.o., Galvaniho 7, 82104 Bratislava, Slovakia
Comenius University, Faculty of Natural Sciences, Ilkovicova 3278/6, 84104 Bratislava, Slovakia
Comenius University, Science Park, Ilkovicova 8, 84104 Bratislava, Slovakia

Maria A. Veretennikova
Department of Statistics and Data Analysis, Faculty of Economic Science, National Research University, Higher School of Economics, Shabolovka 28/11, 9, 19049 Moscow, Russia

Alla Sikorskii
Department of Psychiatry and Department of Statistics and Probability, Michigan State University, 909 Wilson Road, 48824, East Lansing MI, USA

Michael J. Boivin
Department of Psychiatry and Department of Neurology and Ophtalmology, Michigan State University, 909 Wilson Road, 48824 East Lansing MI, USA

Sher B. Chhetri and Hongwei Long
Department of Mathematical Sciences, Florida Atlantic University, Bacon Raton, FL, USA

Alfred A. Akinsete
Department of Mathematics, Marshall University, Huntington, WV, USA

Gokarna Aryal
Department of Mathematics, Statistics and Computer Science, Purdue University Northwest, Hammond, IN, USA

Mei Ling Huang
Department of Mathematics & Statistics, Brock University, St. Catharines, Ontario L2S 3A1, Canada

Christine Nguyen
Apotex Inc., Toronto, Ontario M9L 1T9, Canada

Saman Muthukumarana
Department of Statistics, University of Manitoba, Machray Hall, Winnipeg, Canada

David Martell
ITAM, Mexico City, Mexico

Ram Tiwari
Office of Biostatistics, Center for Drug Evaluation and Research, Food and Drug Administration, 10903 New Hampshire Ave, Silver Spring, USA

Anastasios N. Arapis and Frosso S. Makri
Department of Mathematics, University of Patras, 26500 Patras, Greece

Zaharias M. Psillakis
Department of Physics, University of Patras, 26500 Patras, Greece

Morad Alizadeh
Department of Statistics, Persian Gulf university, Bushehr, Iran

S. M. T. K. MirMostafee
Department of Statistics, University of Mazandaran, Babolsar, Iran

Edwin M. M. Ortega and Thiago G. Ramires
Departamento de Ciências Exatas, Universidade de São Paulo, Piracicaba, Brazil

Gauss M. Cordeiro
Departamento de Estatística, Universidade Federal de Pernambuco, Recife, Brazil

Min Wang
Department of Mathematical Sciences, Michigan Technological University, Houghton, MI, USA

Chanseok Park
Department of Industrial Engineering, Pusan National University, Geumjeong-gu, Busan, Republic of Korea

Xiaoqian Sun
Department of Mathematical Sciences, Clemson University, Clemson, SC, USA

Index

www.ingramcontent.com/pod-product-compliance
Lightning Source LLC
LaVergne TN
LVHW080915210926
848485LV00004B/654